PRACTICAL MICROSTRIP CIRCUIT DESIGN

ELLIS HORWOOD SERIES IN ELECTRICAL AND ELECTRONIC ENGINEERING

Series Editor: PETER BRANDON,
Emeritus Professor of Electrical and Electronic Engineering, University of Cambridge

P.R. Adby	APPLIED CIRCUIT THEORY: Matrix and Computer Methods
J. Beynon	PRINCIPLES OF ELECTRONICS: A User-Friendly Approach
M. J. Buckingham	NOISE IN ELECTRONIC DEVICES AND SYSTEMS
A.G. Butkovskiy & L.M. Pustylnikov	THE MOBILE CONTROL OF DISTRIBUTED SYSTEMS
S.J. Cahill	DIGITAL AND MICROPROCESSOR ENGINEERING
R. Chatterjee	ADVANCED MICROWAVE ENGINEERING: Special Advanced Topics
R. Chatterjee	ELEMENTS OF MICROWAVE ENGINEERING
R.H. Clarke	DIFFRACTION THEORY AND ANTENNAS
P.G. Ducksbury	PARALLEL ARRAY PROCESSING
J.-F. Eloy	POWER LASERS
D.A. Fraser	INTRODUCTION TO MICROCOMPUTER ENGINEERING
J. Jordan, P. Bishop & B. Kiani	CORRELATION-BASED MEASUREMENT SYSTEMS
F. Kouril & K. Vrba	THEORY OF NON-LINEAR AND PARAMETRIC CIRCUITS
S.S. Lawson & A. Mirzai	WAVE DIGITAL FILTERS
P.G. McLaren	ELEMENTARY ELECTRIC POWER AND MACHINES
J.L. Min & J.J. Schrage	DESIGNING ANALOG AND DIGITAL CONTROL SYSTEMS
P. Naish & P. Bishop	DESIGNING ASICS
J. R. Oswald	DIACRITICAL ANALYSIS OF SYSTEMS: A Treatise on Information Theory
M. Ramamoorty	COMPUTER-AIDED DESIGN OF ELECTRICAL EQUIPMENT
J. Richardson & G. Reader	ANALOGUE ELECTRONICS CIRCUIT ANALYSIS
J. Richardson & G. Reader	DIGITAL ELECTRONICS CIRCUIT ANALYSIS
J. Richardson & G. Reader	ELECTRICAL CIRCUIT ANALYSIS
J. Seidler	PRINCIPLES OF COMPUTER COMMUNICATION NETWORK DESIGN
P. Sinha	MICROPROCESSORS FOR ENGINEERS: Interfacing for Real Time Applications
E. Thornton	ELECTRICAL INTERFERENCE AND PROTECTION
L.A. Trinogga, K.Z. Guo & I.C. Hunter	PRACTICAL MICROSTRIP CIRCUIT DESIGN
C. Vannes	LASERS AND INDUSTRIES OF TRANSFORMATION
R.E. Webb	ELECTRONICS FOR SCIENTISTS
J. E. Whitehouse	PRINCIPLES OF NETWORK ANALYSIS
Wen Xun Zhang	ENGINEERING ELECTROMAGNETISM: Functional Methods
A.M. Zikic	DIGITAL CONTROL

ELECTRONIC AND COMMUNICATION ENGINEERING

R.L. Brewster	TELECOMMUNICATIONS TECHNOLOGY
R.L. Brewster	COMMUNICATION SYSTEMS AND COMPUTER NETWORKS
J.N. Slater	CABLE TELEVISION TECHNOLOGY
J.N. Slater & L.A. Trinogga	SATELLITE BROADCASTING SYSTEMS: Planning and Design
J.G. Wade	SIGNAL CODING AND PROCESSING: An Introduction Based on Video Systems

PRACTICAL MICROSTRIP CIRCUIT DESIGN

L. A. TRINOGGA
Leeds Polytechnic, Faculty of Information and Engineering Systems
GUO KAIZHOU
Academia Sinica, Institute of Electronics, Beijing
I. C. HUNTER
Department of Electrical Engineering, University of Bradford

ELLIS HORWOOD
NEW YORK LONDON TORONTO SYDNEY TOKYO SINGAPORE

First published in 1991 by
ELLIS HORWOOD LIMITED
Market Cross House, Cooper Street,
Chichester, West Sussex, PO19 1EB, England

A division of
Simon & Schuster International Group
A Paramount Communications Company

Typeset in Times by Ellis Horwood Limited
Printed and bound in Great Britain
by Hartnolls, Bodmin, Cornwall

British Library Cataloguing-in-Publication Data

Trinogga, L. A.
Practical microstrip circuit design. —
(Ellis Horwood series in electrical and electronic engineering)
I. Title. II. Guo, K. Z. III. Hunter, L. C. IV. Series
ISBN 0-13-580077-3
ISBN 0-13-720160-5 pbk

Library of Congress Cataloging-in-Publication Data

Trinogga, L. A. (Lothar Alfred), 1940–
Practical microstrip circuit design / L. A. Trinogga, K. Z. Guo, I. C. Hunter.
p. cm. — (Ellis Horwood series in electrical and electronic engineering)
Includes bibliographical references and index.
ISBN 0-13-580077-3
ISBN 0-13-720160-5 pbk
1. Microwave integrated circuits. 2. Strip transmission lines. I. Guo, K. Z. II. Hunter, I. C. III. Title. IV. Series.
TK7876.T68 1991
621.381'32-dc20 91-30870
CIP

Table of contents

Foreword

In 1986 Professor Guo and Dr Trinogga met at the first Asia Pacific Microwave Conference in India. Because of their common subject area, Professor Guo accepted an invitation to undertake research at Leeds Polytechnic. During this joint research phase the idea of writing a book was conceived.

It has been my pleasure to observe the progress of the research and the development of the book into the present form. It is commendable that the original team of authors expanded into three in order to encourage younger members of staff to engage in scholarly activity and to widen the scope of the book.

This book gives a good coverage of microstrip circuit design, based to a large extent on the authors' own expertise. What distinguishes it from many other books is its practical bias and the inclusion of proprietary measured results. This makes it most useful for undergraduate students and young researchers as well as for professional engineers.

The long educational and practical experience of the authors manifests itself in a logically structured book and a lucid style. Furthermore, the authors have drawn on their own contributions which are not available in book form elsewhere.

The text of the book is supplemented with many diagrams to assist understanding. Most chapters include solved sample problems and key references. This will prove useful to those readers wishing to widen their in-depth knowledge. I hope that the very nature of the book will be found useful by the readership it has set out to address.

July 1991

Professor CHAI Zhenming
Director, Institute of Electronics
Academia Sinica, Beijing

Preface

Many good books have been written for educational purposes on the subject of microwave circuits and computer-aided design. The question thus begs itself why another book is required and why indeed more books will inevitably be written. Two answers to these questions immediately spring to mind. The first, predictably, is that technology changes constantly and hence the need for up-to-date publications. The second one, at least in my view, is more fundamental. There is no one source of information which is absolute and all-encompassing. Different books and hence different authors have different strengths within the same subject area and different approaches to it. Thus, a more comprehensive and balanced knowledge can be obtained only by drawing on different books, benefiting from their various strengths.

In the microwave and computer-aided design area (CAD) the majority of books deal mainly with the theoretical side of the subject, although lately a few books have been published which include some computer programs. It is a trend which I hope to promote with this book. In addition I have set out to explain the important aspect of the practical realization of the microwave circuit designed and have included typical measured results. It is felt that this more rounded presentation will result in the better understanding of the subject.

Although some of the information presented here is the spin-off from research, no effort has been made to increase the correlation or accuracy between calculated and measured results. This is left to the available specialized research literature. The purpose of the book is to provide a well-rounded body of information on the subject matter.

In order that the book be relatively self-contained, relevant standard theory has been included. References are given for further reading and acknowledgements are freely made where due. Whenever possible, the notation and symbols most frequently used in the microwave community have been employed. I very much hope that this book will provide a good practical introduction to microstrip circuit design.

I should like to acknowledge the support provided under the Royal Society British Telecom China Fellowship which allowed for cooperation with Chinese colleagues, especially on the subject of dielectric resonators. I should also like to record my appreciation to some of the students on our course in Communication

Engineering, as well as the research students Mr Jaspal Singh Nakhwal and Mr Mohammad Nasser-Moghadasi for making valuable improvements to some of the computer programs.

Leeds
May, 1991

L. A. Trinogga

Principal symbols

a	=	radius of DR
B	=	magnetic flux density [Vs/m] or [G]
B	=	coupling factor
c	=	velocity of light [$3*10^{10}$ cm/s]
d	=	distance microstrip to DR
d_1	=	thickness of substrate
D	=	electric flux density [As/m]
E	=	electric field [V/m]
ε_a	=	dielectric constant of air
H	=	magnetic field [A/m]
H	=	height of DR
i	=	subscript counter 1,2,3, etc.
J	=	current density [A/m^2]
J_0, J_1	=	Bessel function of the first kind
k	=	ω/c, constant giving Eigenvalues
k_r	=	propagation constant, resonator
k_a	=	propagation constant, air
k_s	=	propagation constant, substrate
K_0, K_1	=	modified Bessel function of the second kind
L_m	=	mutual inductance
L_r	=	equivalent inductance of DR
M	=	magnetic dipole
NN	=	coupling plane [an integer]
Q	=	quality factor
Q_{ext}	=	external quality factor
Q_l	=	loaded quality factor
Q_u	=	unloaded quality factor
R_r	=	equivalent resistance of DR
S	=	scattering parameter
TEM	=	transverse electromagnetic
W	=	energy

W_m = magnetic energy
W_e = electric energy
Z_L = load impedance
Z_o = characteristic impedance

ε_0 = permittivity of free space $8.854*10^{-12}$ [F/m]
ε_s = dielectric constant of substrate
ε_r = dielectric constant of resonator
λ_0 = free space wavelength
μ_0 = permeability of free space $12.566*10^{-7}$ [H/m]
$\emptyset$ = angle

1

Communication systems requirements

In radio frequency systems, electromagnetic waves form the link between a transmitter and receiver. During their travel, these waves are subject to physical laws over which man has no control. Thus, in order to establish a high-quality radio link, it is important to have some understanding of the different factors which affect wave propagation. The physicist attempts to isolate the causes of the physical phenomena such as electromagnetic waves, whereas the engineer is more interested in their application and measurement.

For a given specification of a complete communication system, it would be ideal if one could predict the exact signal strength at any point along the path of propagation. Unforunately, there are a number of factors which interfere and make such a prediction difficult. To begin with, the medium through which the electromagnetic waves pass has properties which vary with temperature, water vapour and pressure, which in turn alter the direction, polarization and velocity of radio waves. Besides the transmission medium, it is also the characteristics of the station which have a bearing on the quality of the communication link.By characteristics one means the frequency used, the type of aerial system and the type of modulation employed. It is the task of the engineer to find and use the propagation conditions and electronic circuits which are most suitable to a high-quality link.

A table showing the position of microwaves in the electromagnetic spectrum is given in Fig. 1.1. From this it can be seen that communications, radar, navigation and measuring systems have allocations in the microwave band of frequencies between about 1 GHz and 100 GHz. In Fig. 1.2, a summary of the possible applications of microwaves is shown. In measuring systems, for example, microwaves are used for the measurement of mechanical quantities (distance, level, displacement, vibration, speed and acceleration), geometrical dimensions (width, thickness, length) and physical properties (degree of curing, moisture content, humidity). The advantages of microwaves in connection with the above-mentioned examples are that the system is free from radiation hazards,is non-contacting and possesses good accuracy.

In conclusion,it can be said that an understanding of the behaviour of electromagnetic waves helps to optimize electrical systems which process the information carried by these waves.

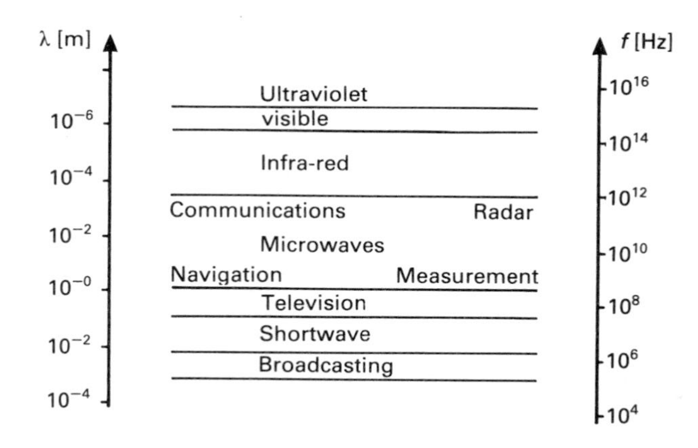

Fig. 1.1 — Position of microwaves in the electromagnetic spectrum.

Applications	Comments
Measuring systems	Distance, level, displacement, vibration, speed, acceleration, length
Communications	Wide bandwidth, directivity High channel capacity
Domestic	Burglar alarms, Cooking control

Fig. 1.2 — Some applications of microwaves.

BASIC REQUIREMENTS OF ELECTRICAL SYSTEMS

In the context of microwaves, electrical systems can be used either for communication or measuring purposes. Communication may be defined as a process whereby information is transferred from one point, called the transmitter, to another point, the receiver. The transmission medium for the information bearing signal may be a cable, a wave guide,an optical fibre or it may be free space. The fundamental requirement for a transmission medium is that a chosen frequency band is propagated with as little distortion and noise as possible. For free space transmission, efficient aerials are required since the signals received by these aerials are very small. For this reason, a communication receiver has to satisfy two criteria.In the first instance, a receiver must be able to process the weakest signal which may be comparable in magnitude to the noise of the receiving system. This requires low-noise, high gain amplifiers in order that the signals achieve levels which are usable for further processing. Secondly, owing to the type of signal received (e.g. pulses), wideband circuits are required with certain conditions regarding transient behaviour.

One of the two key stages in any communication receiver, especially at microwave frequencies, is the mixer which should have a low conversion power loss (CPL) and a low noise figure (F). The other stage is of course the radio frequency amplifier. Here again, it is mainly the low-noise performance which is of primary interest. A typical communication receiver block-diagram is shown in Fig. 1.3. Depending on

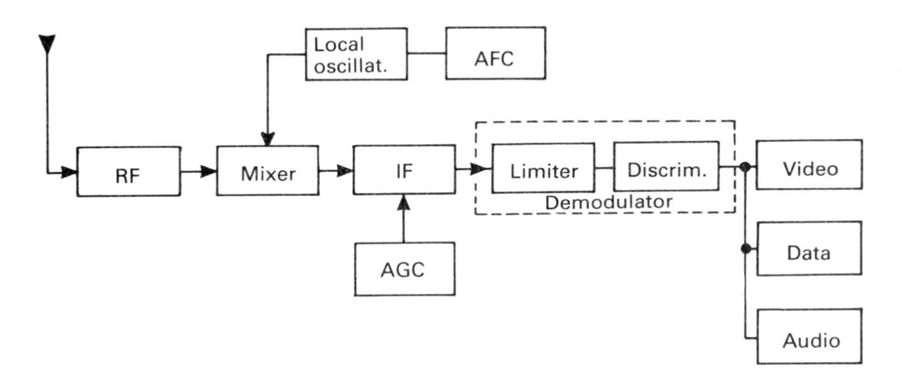

Fig. 1.3 — Block diagram of a communication receiver.

the frequency which is used, a radio frequency amplifier may not be employed because above a certain frequency the amplifier noise figure is no longer better than that of the mixer or the amplifier no longer exhibits gain. The number of amplifier stages required depends on the gain achievable per stage, the bandwidth and the noise figure required. The design of the front-end of the receiver can be divided into the design of amplifier, mixer, oscillator and coupling circuits. More complex systems include automatic frequency control (AFC) and/or automatic gain control (AGC) which result in a substantially constant intermediate frequency (IF) at the mixer output and a constant IF signal to the demodulator input.

Mixers in particular have been studied extensively. Mixing circuits can be constructed using both passive and active devices, the only requirement being that they possess a non-linear characteristic which is the key to the mixing process. For the most part, solid state devices have superseded thermionic devices, resulting in mixer stages of small dimensions, high reliability and reproducibility. Three terminal devices such as Ga As FET's are now more widely used for frequency conversion.

Diodes, however, owing to their low parasitics, are ideally suited to microwave circuits, e.g. mixers. A configuration of four diodes is known as a lattice, and mixers of this configuration have been extensively studied [1,2,3]. Most efforts have been directed towards a better understanding of the mechanisms for reducing the CPL. With the emergence of better three terminal active devices the diode has now received a rival. It seems that active mixers at microwave frequencies will give results similar to, or better than, those of diode mixers.

COMMUNICATION FREQUENCY BANDS

Scientists and engineers have divided the entire frequency spectrum into adjacent bands and have given to each an appropriate name. Whilst several methods of frequency sub-division have been proposed, the following method can be used as a guideline. The very high frequency (VHF) band occupies the region from 30–300 MHz, the ultra high frequency (UHF) band occupies the adjacent region from 300–3000 MHz and the super high frequency (SHF) band occupies the region from 3–30 GHz. This book concerns itself mainly with frequencies up to X-band, that is about 12 GHz.

The energy of propagating radio waves is confined principally to the shell between the earth and the ionosphere, and this space is frequently denoted as the terrestrial waveguide and the type of communication as terrestrial communication. Signal transmission which goes beyond the ionosphere is generally referred to as extra-terrestrial communication. Some of the advantages of high frequencies such as microwaves are to be found in wide bandwidth (high channel capacity) and relative secrecy of information transmission (directivity). A global satellite system is now in operation which uses microwaves for the transmission of meteorological data, telemetry and for international telephone and television channels. The useful operating range of communication systems is limited by the frequency in use and because of noise. It is prevalent in electronic processes but it is also due to extra-terrestrial radiation (sky noise) which is most intense between about 10 and 300 MHz and around the frequencies of 20–60 GHz. Figure 1.4 shows the sky noise as a function of frequency with a minimum near about 4 GHz. This is of course the reason

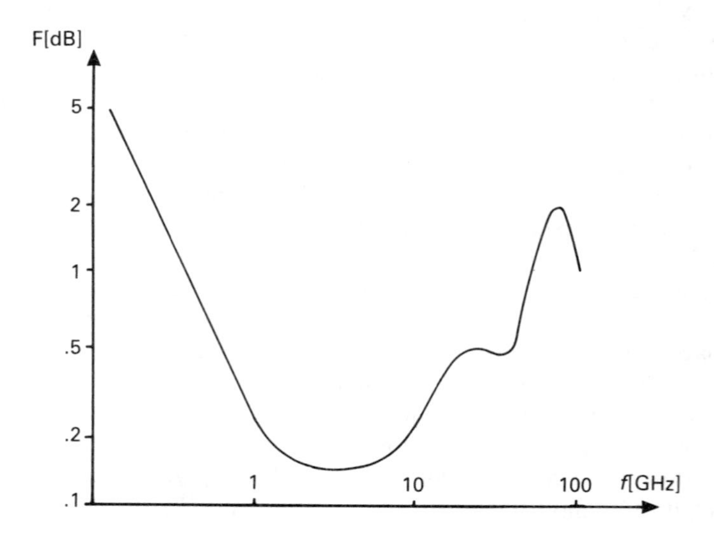

Fig. 1.4 — Skynoise characteristic.

why frequencies in the 1–10 GHz range are used for satellite communications. Additional frequency space requirements have necessitated the use of higher frequencies which experience higher attenuation and noise levels. One of the major benefits that can be derived from this is more secure communication,especially over short distances.

APPLICATION OF NON-LINEAR DEVICES

The high losses associated with the propagation of audio frequencies in free space has been one factor in the argument for the use of higher frequencies. These can be obtained either by harmonic generation or frequency multiplication. The process of shifting a lower frequency band to a higher frequency band is known as frequency translation.

In harmonic generators, a power at frequency f is applied to a non-linear device whose output produces frequencies which are harmonically related to f. This technique of frequency generation is employed whenever a direct method is not practicable. There are two classes of non-linear elements suitable for harmonic generation and frequency conversion:non-linear resistors which include diodes and non-linear reactances which include varactors. Manley and Rowe [4] have shown that for a lossless non-linear reactance the total harmonic power is equal to the power of the fundamental input frequency. In a practical situation, the desired harmonic is extracted from the non-linear device by means of a bandpass filter.

In frequency converters two signals of different frequency are applied to a non-linear element. One of the signal frequencies is much larger in power than the other and is usually referred to as the pump. The pump drives the non-linear element whose output contains a large number of frequencies which combine with the frequency of the small information-carrying signal. The resulting output contains the sum and difference of those frequencies. A network which shifts a higher frequency band to a lower frequency is referred to as a mixer.

NOISE IN SYSTEMS

In communication systems noise places a limit on the weakest detectable signal. If the noise is of the same order of magnitude as the signal then this can lead to an increase in error rate for a digital system or it can become irritating or incomprehensible in a voice or video communication system. For this reason an engineer must not only evaluate a system in terms of the signal, but also with respect to the noise.Other factors which have a bearing on circuits and systems are those of gain, bandwidth and sensitivity. Furthermore, attention must not only be focussed on the noise coming into a receiver, but also on the noise sources within, which contribute to the receiver output noise. A possible classification of noise is given in Fig. 1.5.

The gain in a system can be maximized by careful circuit design and component selection. For example, in some applications one will utilize discrete devices whilst in others integrated circuits are more beneficial. On the other hand, if noise is not the

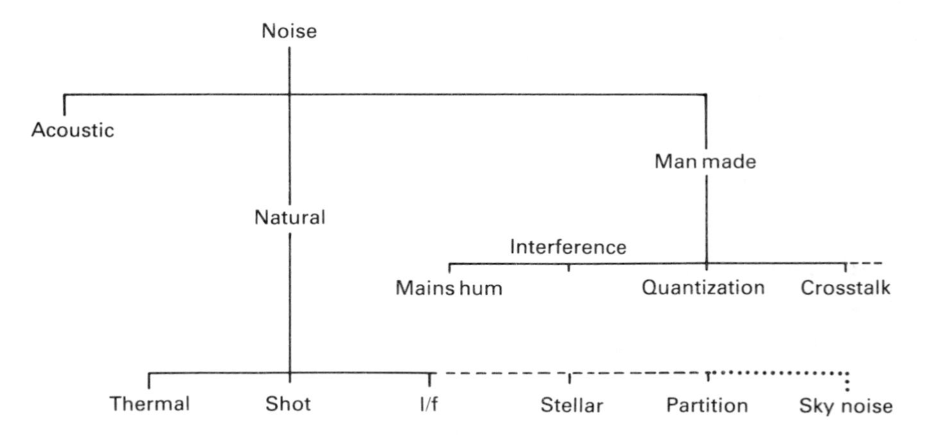

Fig. 1.5 — Possible classification of noise.

prime factor under consideration, one will use integrated circuits and lumped components.

With regard to bandwidth, the system performance can be optimized by making the bandwidth just large enough to accommodate the information. Any excess bandwidth only deteriorates the signal-to -noise ratio of the system. In certain circumstances, the system bandwidth may be made smaller if special signal processing techniques are used [5], such as frequency modulated feedback, the phase locked loop or a dynamic tracking filter.

REFERENCES

[1] Liechti, C. A., Down converters using Schottky barrier diodes, *IEEE Transact. on Electron Devices*, **ED-17**, No. 11, Nov., 1970, pp. 975–983.

[2] Howson, D. P., Minimum conversion loss in the broadband mixer, *Radio and Electronic Engineer*, **42**, May 1972, pp. 237–242.

[3] Kulesza, B. L. J., General theory of a lattice mixer, *Proc. IEEE*, **118**, No. 7, July 1971, pp. 864–870.

[4] Manley, J. M. and Rowe, H. E., Some general properties of non-linear elements: Part 1 — General energy relations, *Proc. IRE*, **44**, July 1956, pp. 904–913.

[5] Slater, J. and Trinogga, L. A., *Direct Broadcasting Satellites: Planning and Design*, Ellis Horwood, Chichester, 1985.

2

Planar transmission lines

In order to transmit energy or couple energy from one point to another a transmission medium is required. For RF/microwave communication links this may be air. In most cases, however, some kind of structure is required for the purpose of the transmission. For non-microwave frequencies, i.e. frequencies typically below 1 GHz, parallel wire lines and coaxial cables can be employed for signal/wave transmission. Owing to the overcrowding of the lower frequency spectrum higher and higher frequencies are being used. Obviously, it is not possible to use coaxial cable and lumped circuit techniques at those higher frequencies which are generally referred to as microwaves. Instead one uses distributed transmission line techniques.These can also be used to build complete microwave integrated circuits,also known as MICs. Planar transmission lines are now discussed in the following sections and in the context of their use in the following chapters.

THE COAXIAL LINE

In order to introduce some of the terminology used in planar line technology consider the coaxial cable in Fig. 2.1. Assume a current I to flow into the centre

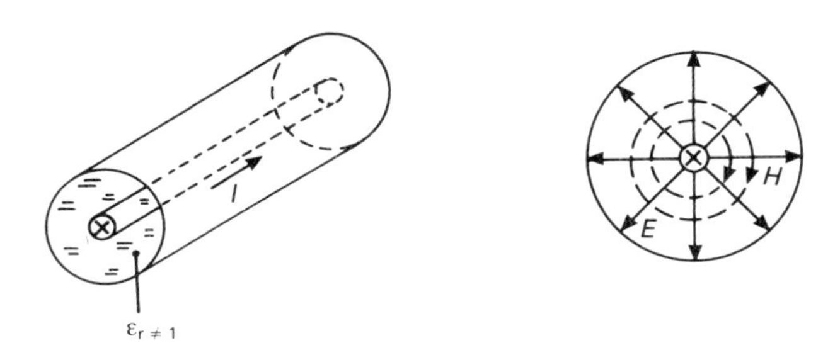

Fig. 2.1 — Coaxial cable filled with a dielectric and associated field lines.

conducter as indicated by the cross (X). The electric field lines are then emanating radially from the centre conducter and end on the outer conductor of the coaxial cable. The magnetic field lines H follow a concentric circle pattern around the centre conductor in clockwise direction. At all times the electric and magnetic field lines are perpendicular to each other, i.e. the field lines cross at right angles. This mode of the electromagnetic wave propagation is referred to as the transverse electromagnetic mode of propagation or simply TEM-mode.

Now consider the cable in Fig. 2.2(a) pressed into a rectangular shape, thus giving

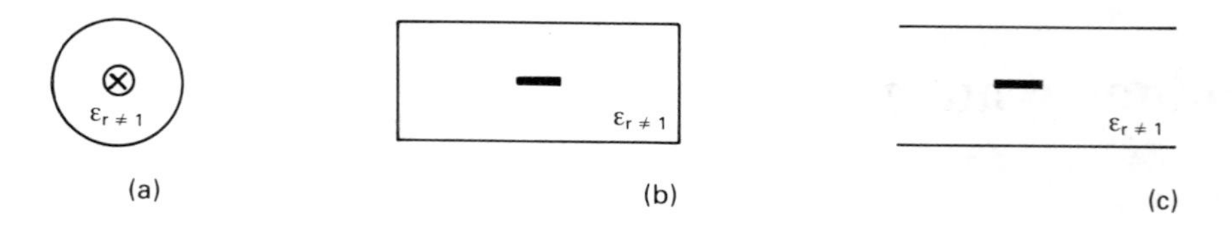

Fig. 2.2 — Evolution of a planar transmission line.

it a cross-section as shown in Fig. 2.2(b). If we now remove the sidewalls, which are not necessary for wave propagation, then we obtain the cross-section Fig. 2.2(c). A configuration of this nature is known as a planar structure since the elements which make up this kind of structure lie in a plane. Since in our case such a structure is used for the transmission of electromagnetic waves we call this transmission line a planar transmission line. In the following a selection of different planar transmission lines are discussed.

MICROSTRIP

Microstrip is one of the most frequently used planar transmission lines. The main reason is the ease with which components can be mounted on such a structure. The microstrip geometry and a cross-sectional view are shown in Figs 2.3 and 2.4(a). This

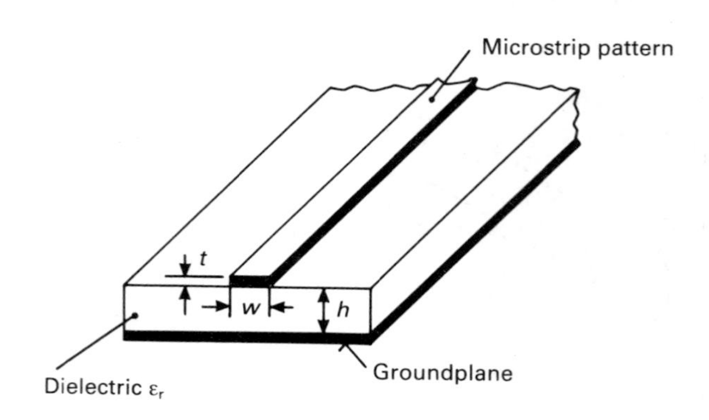

Fig. 2.3 — Microstrip geometry.

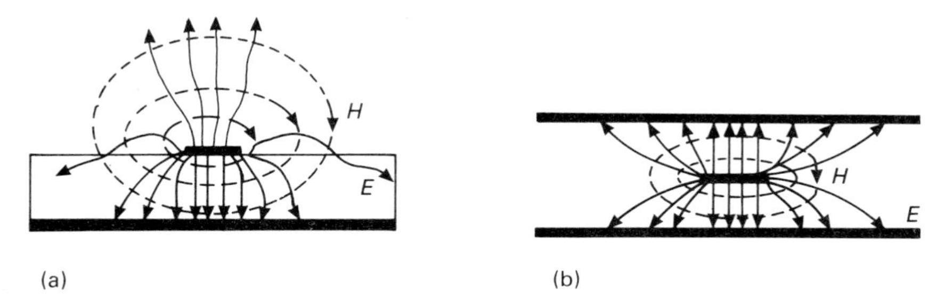

Fig. 2.4 — Field pattern in (a) a microstrip, (b) a stripline.

transmission line consists of a dielectric substrate which has a ground plane on one side and a strip pattern on the other, hence the name microstrip. As a result of this a microstrip has the following features:

(1) There is easy access to the surface of the microstrip. This makes it easy to mount both passive and active devices anywhere on the structure. Minor adjustments are possible after having fabricated the circuit.
(3) As is apparent from Fig. 2.4(a), the electromagnetic field distribution of a microstrip is asymmetrical, i.e. in the lower half the field is contained between the strip and groundplane, on the upper half it radiates into the air. Although this feature is exploited in microstrip aerials, radiation and its associated losses are considered a drawback of microstrip. Most of the electromagnetic energy may be confined to the vicinity of the microstrip if a substrate of high dielectric constant is selected and if the circuit is enclosed in a metal housing (box).
(3) Furthermore, because of the two dielectric constants associated with the strip (air and the dielectric) the mathematical analysis of microstrip is difficult, in other words, a microstrip represents a mixed dielectric structure.

STRIPLINE

The stripline is another form of a planar transmission line. A stripline is a three-conductor transmission line and looks like a sandwich. The central strip, as can be seen from Fig. 2.4(b),is sandwiched between two groundplanes. It is held in a central position by the dielectric material on either of its sides. A stripline is thus an inherently enclosed and symmetrical structure. There is literally no radiation loss. Thus it is the ideal structure for passive circuits such as filters, power dividers and couplers. A stripline is characterized by the following features:

(a) It has a minimum of radiation loss
(b) It is difficult if not impossible to mount surface components.Any drilling of holes or any other recesses would affect predicted stripline circuit performance.

SLOTLINE

The slotline is another form of planar transmission line. Its cross-sectional view and approximate field distribution are shown in Fig. 2.5(a)–(c). Please note that the

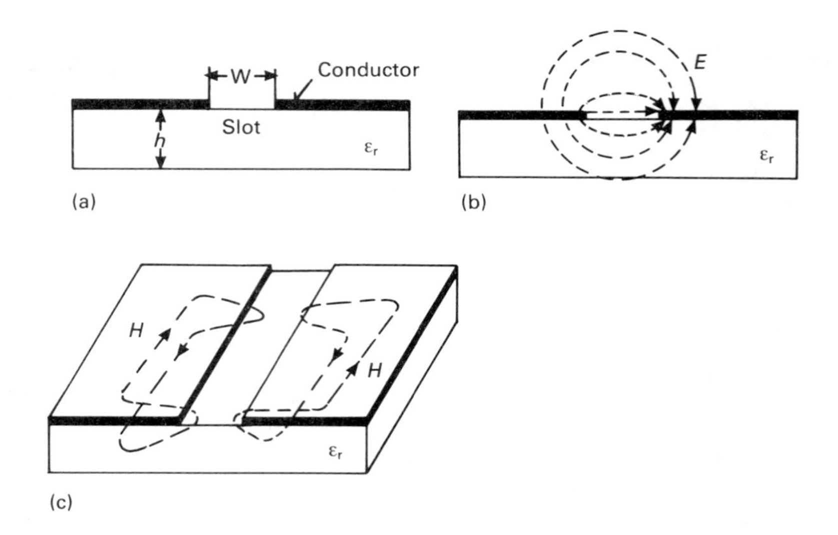

Fig. 2.5 — Cross-sectional view and field pattern of a slotline.

copper conductors lie on one side only. The impedance of a slotline increases as the slot width increases. Since it is difficult to make high impedances in microstrip a slotline may be used to achieve this. Since the two slotline conductors lie in one plane it offers itself for the shunt mounting of microwave components. The most suitable location for mounting any component is governed by the electromagnetic field configuration along the slot. Compared to microstrip, transitions to coaxial lines are more difficult to make with slotlines.

COPLANAR LINE

A coplanar line as shown in Fig. 2.6 consists of three copper conductors in one plane. The two groundplanes run parallel to the microstrip centre conductor. From the nature of the structure it appears obvious that a coplanar line combines some of the positive features of a microstrip and slotline structure.

The centre conductor lends itself for the series mounting of microwave components such as for example a transistor or chip capacitor. Shunt mounting of components can take place between the centre conductor and the groundplane. Coplanar lines can work in both, a balanced and unbalanced configuration. Thus, a coplanar line can be used to construct balanced or single ended (unbalanced) mixers or it can be used to change a balanced circuit into an unbalanced one or vice versa.

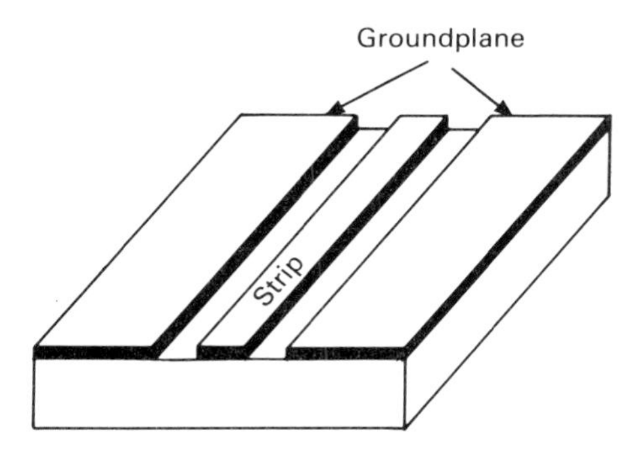

Fig. 2.6 — Geometry of a coplanar line.

There are other types of planar lines, some of which are being mounted into waveguides such as the fin line. There is almost always a transmission medium available to cater for specific requirements. It is not uncommon to use two or more types of planar transmission lines in a microwave subsystem for performance optimization. Nevertheless, microstrip is still the most commonly used transmission medium. All of the microwave circuit designs in the following chapters are based on microstrip only.

KEY ADVANTAGES OF PLANAR TECHNOLOGY

Compared with waveguides or coaxial techniques, planar technology or planar lines must have certain advantages, otherwise they would not be used. Some of these advantages may be summarized as follows:

— small size;
— light weight;
— good reliability;
— good reproducibility;
— performance improvement;
— low cost for large production runs.

Microwave circuits which are made in planar technology are often referred to as microwave integrated circuits or MICs. Owing to the permittivity of the substrate material, the physical wavelength within the circuit is reduced. This results in small compact circuits or subsystems. Also, the cross-section of a planar line is usually much smaller than a waveguide or coaxial cable. This results additionally in a light weight.On occasions circuit size can be even further reduced through the use of lumped elements rather than distributed elements.

A microwave subsystem employing planar technology is easier to construct than one with waveguides or coaxial cable. Considerable benefits are to be derived from the lack of necessary interconnections between circuits, in other words no connectors and cables are required to join circuits to make up a subsystem. Also, because of the subsystem approach, there is no relative motion between circuits, improving reliability even further. Furthermore, the circuit layout is very flexible. Corners and bends are easily made.

Reliability is good, especially for subsystems, owing to the lack of a large number of connectors, to give one example. The use of photolythographic techniques ensures good reproducability and hence a uniform product. The above features result in a performance improvement which need no further elaboration apart from one additional aspect. The microwave components to be used in circuits made in planar technology are of similar size to the lines to which they are soldered or bonded unlike in the case of a waveguide circuit. Better matches can thus be obtained. Further improvements in circuit performance can be achieved through the use of unencapsulated devices. These improvements do not only relate to size and weight, but also electrical performance owing to the absence of package and mounting parasitics.

A low cost, especially for large production runs, is the result of these advantages. In actual fact, cost is almost independent on the complexity of the circuit. Modern satellite TV converters bear evidence of this.

Naturally, there are also disadvantages and difficulties associated with planar transmission lines. These are in general the lack of any circuit adjustment possibilities once the circuit has been etched. In standard microwave circuits one can always use such techniques as variable shorts and tuning screws. It is thus clear that the design philosophy for planar transmission line circuits has to be different from conventional microwave circuit design.

PERMITTIVITY AND PERMEABILITY

The effectiveness with which an electric or magnetic field is produced within a medium depends on the latter's properties. In the context of microwaves the medium is also referred to as a guide or guiding structure.

For an electric field it is the permittivity of the dielectric material involved and for a magnetic field it is the permeability. Permittivity is quantitively expressed as ε and describes how easily a material can be polarized. The same applies to materials with magnetic properties where the symbol μ is used. The absolute constants for permittivity and permeability have the following well known values:

$$\varepsilon_0 = 8.855 \times 10^{-12} \text{ [F/m]}$$
$$\mu_0 = 4\pi \times 10^{-7} \text{ [H/m]}$$

Permeability and permittivity of different materials are usually expressed in relation to the above as

$$\varepsilon_r = \varepsilon/\varepsilon_0 \quad \text{or} \quad \varepsilon = \varepsilon_0\, \varepsilon_r \tag{2.1}$$
$$\mu_r = \mu/\mu_0 \quad \text{or} \quad \mu = \mu_0\, \mu_r \tag{2.2}$$

Table 2.1 gives useful typical relative permittivities, also known as dielectric constants, of materials used in microwave circuits.

PROPAGATION VELOCITY

The velocity with which an electromagnetic field propagates is finite. For light travelling through air this is $c = 3*10^{10}$ cm/s. For an electromagnetic wave travelling through a medium the velocity will be smaller than that above and will depend on the properties of the medium, namely ε and/or μ. Solving Maxwell's equations shows that the following relationships hold:

$$v = \frac{1}{\sqrt{\varepsilon\,\mu}} = \frac{1}{\sqrt{\varepsilon_0\,\varepsilon_r\,\mu_0\,\mu_r}} \tag{2.3}$$

For air $\mu_r = \varepsilon_r = 1$ and hence

$$c = v = \frac{1}{\sqrt{\varepsilon_0\,\mu_0}} \tag{2.4}$$

In microwave circuits the substrates frequently have no magnetic properties. Using eqns (2.3) and (2.4) and letting $\mu = 1$ one obtains

$$v = \frac{1}{\sqrt{\varepsilon_0\,\varepsilon_r\,\mu_0}} = \frac{c}{\sqrt{\varepsilon_r}} \tag{2.5}$$

where $c = \lambda \times f$. This shows that the velocity through the medium or guiding material in relation to light is reduced by the inverse of the square root of the dielectric constant. Thus a dielectric constant of nine will reduce the velocity by a factor of three. As a result we observe the important feature that the microwave circuit size will also reduce by the same factor. The wavelength or velocity through a medium is often accompanied by a suffix 'g' which refers to 'guide'. Hence eqn. (2.5) may be re-written as

$$v_g = \frac{c}{\sqrt{\varepsilon_r}} \tag{2.6}$$

Example 1

Calculate the velocity of propagation of an electromagnetic wave through Epsilam.

Table 2.1 — Dielectric constants of frequently used microwave substrate materials

Material	ε_r
air (vacuum)	1.0
alumina	9.6, 10.2, 10.5
quartz	6 to 8
beryllia	6.6
Duroid	2.2
Polyguide	2.32
Epsilam	10.0
3 M	10.3

Answer
From eqn. (2.6) and Table 2.1 we obtain

$$v_g = \frac{c}{\sqrt{\varepsilon_r}} = \frac{3 \times 10^{10}}{\sqrt{10}} = 0.948 \times 10^{10} \text{ cm/s}$$

Example 2
What is the physical wavelength of a 1 GHz signal travelling through (a) air; (b) Epsilam; (c) Polyguide?

Answer

From eqn. (2.6) and Table 2.1 we obtain

(a) $\lambda = c/f = 30$ cm

(b) $\lambda_g = v_g/f = \dfrac{c}{f\sqrt{\varepsilon_r}} = 0.948$ cm

(c) $\lambda_g = \dfrac{3 \times 10^{10}}{10 \times 10^9 \sqrt{2.32}} = 19.69$ cm

MICROSTRIP REALIZATION

As was indicated earlier on, the impedance of a microstrip depends on the properties of the substrate used, namely the dielectric constant ε_r, the substrate height h and the line width w. In a practical situation the impedance is given and one wants to know the width of the microstrip in order to realize that impedance. Many researchers [1,2,3] have looked into this problem and have derived closed loop equations of various degrees of accuracy. In order to realize microstrip circuits one is interested in a synthesis rather than an analysis.

When synthesizing microstrip, consideration must be given to the width to height ratio, *w/h*, of the strip or line. This is explained by means of Fig. 2.7(a) and (b). From

Fig. 2.7 — Definition of wide and narrow microstrip.

this it is evident that the dispersion of a narrow line is much greater than that on a wide line, hence affecting the accuracy of the synthesis. For this reason two sets of equations are frequently used in order to achieve a reasonable degree of accuracy. Thus one set is used for so called narrow lines with a small *w/h* ratio and the other set for a large *w/h* ratio. There is of course a grey area where it is difficult to decide whether one deals with a wide or narrow line. The ultimate decision lies then with the designer, or one has to make two circuits and choose that with the better performance.

From the above and Fig. 2.7 it is evident that microstrip calculations need to take into account its dispersion characteristics owing to the open structure on one side. This may be done by introducing the effective permittivity ε_{eff} as explained in the following.

The characteristic impedance of a transmission line is given by the well known equation

$$Z_o = \sqrt{\frac{L}{C}} = v_p \, L = \frac{1}{v_p \, C} \tag{2.7}$$

and hence

$$v_p = 1/\sqrt{L\,C} \tag{2.8}$$

where v_p is the phase velocity in the transmission line medium. Unlike the case of a stripline where analysis is undertaken with a single dielectric constant (field distribution is symmetrical with respect to the line), microstrip analysis involves two dielectric constants. Since a microstrip structure is open, the dielectric constant above the line is that of air and below the line it is that of the substrate. If we were now to remove the substrate material only as shown in Fig. 2.8, the microstrip would

Fig. 2.8 — Microstrip with and without substrate.

float in air and wave propagation between the microstrip and groundplane would be as that in free space, i.e. $3 * 10^{10}$ cm/s. As a result the microstrip impedance in air Z_{o1} can be found from eqn. (2.7) as

$$Z_{o1} = \sqrt{\frac{L}{c\, C_1}} = c\, L = \frac{1}{c\, C_1} \tag{2.9}$$

since $v_p = c$. The suffix 1 denotes air.

As the inductance L increases per microstrip unit length, the capacitance C increases proportionally thus keeping the ratio Z_{o1} constant. From eqns (2.7) and (2.9) we obtain

$$Z_o = \frac{1}{c\sqrt{C\, C_1}} \tag{2.10}$$

Thus knowing the unit capacitances with and without the substrate material present (Fig. 2.8) we can evaluate Z_o.

From eqn. (2.9) we have

$$\sqrt{\frac{L}{C_1}} = c\, L \text{ or } c = \frac{1}{\sqrt{L\, C_1}} \tag{2.11}$$

Using eqns (2.11) and (2.8) we can establish the relationship between the velocities in air and a medium, namely

$$\frac{C}{c_1} = \left(\frac{c}{v_p}\right)^2 = \varepsilon_{eff} \tag{2.12}$$

This ratio is defined as effective permittivity. Referring to Fig. 2.7, a wide microstrip behaves for all practical purposes like a parallel plate capacitor and hence the effective permittivity can be looked upon as that of the substrate, i.e. $\varepsilon_{eff} = \varepsilon_r$. For the narrow line the field distribution could be looked upon as being shared between the air and substrate dielectric, i.e. $\varepsilon_{eff} = (\varepsilon_1 + \varepsilon_r)/2$. Substituting $\varepsilon_1 = 1$ for the permittivity of air, we have $\varepsilon_{eff} = (1 + \varepsilon_r)/2$. In fact the permittivity depends on many more parameters which includes the operating frequency and substrate height.

The following states frequently used equations for the synthesis of a microstrip of width w to meet a required impedance Z_o, i.e. $w = f(Z_o, h, \varepsilon_r)$.

(a) Narrow line, when $Z_o > [44 - 2\,\varepsilon_r]$

$$\frac{w}{h} = \left[\frac{e^H}{8} - \frac{1}{4e^H}\right]^{-1} \tag{2.13}$$

where

$$H = \frac{Z_o\sqrt{2(\varepsilon_r + 1)}}{119.9} + \frac{(\varepsilon_r - 1)}{2(\varepsilon_r + 1)}\left[\ln\frac{\pi}{2} + \frac{1}{\varepsilon_r}\ln\frac{4}{\pi}\right] \tag{2.14}$$

If $Z_o > 63 - 2\,\varepsilon_r]$ it may be useful to replace ε_r by the effective ε_r, namely

$$\varepsilon_{eff} = \frac{\varepsilon_r + 1}{2}\left[\frac{\varepsilon_r - 1}{2H\,(\varepsilon_r + 1)}\left(\ln\frac{\pi}{2} + \frac{1}{\varepsilon_r}\ln\frac{4}{\pi}\right)\right]^{-2} \tag{2.15}$$

A microstrip may be considered a wide line if

(b) wide line, when $Z_o < 44 - 2\varepsilon_r]$

Hence

$$\frac{w}{h} = \frac{2}{\pi} - (d - 1) - \ln\,(2d - 1)] + \frac{\varepsilon_r - 1}{\pi\,\varepsilon_r}$$

$$\times\left[\ln\,(d - 1) + 0.293 - \frac{0.517}{\varepsilon_r}\right] \tag{2.16}$$

where

$$d = \frac{59.95\,\pi^2}{Z_{op}\sqrt{\varepsilon_r}} \tag{2.17}$$

As for the narrow line, we can use for the case $Z_o > [63–2\varepsilon_r]$

$$\varepsilon_{eff} = \frac{\varepsilon_r}{0.96 + \varepsilon_r\,(0.109 - 0.004\,\varepsilon_r)\,[\log(10 + Z_o) - 1]} \tag{2.18}$$

It is easy to write a computer program using the above or similar equations for the quick evaluation of the microstrip width. The following is an example of how computer program POLSTRI can be used to do this.

As can be seen, a 100 ohm line results in a narrower line than the 50 Ohm line. Apart from the full wavelength, fractions of a wavelength such as a half and quarter wavelength can easily be incorporated in the program. This proves convenient when designing circuits which employ $\lambda/4$ and $\lambda/2$ line lengths. It should also be noted that the dielectric constant ε_{eff} is used when calculating the physical wavelength. It is a useful exercise for the reader to modify a single parameter of the inputted data in order to see the effect it has on the other parameters and in particular a change in substrate height h or dielectric constant ε_r. A listing of program POLSTRI is given in the following.

```
RUN


            Microstrip width calculation

 ENTER Zo [Ohm]?50

 ENTER Er?2.3

 ENTER height h [mm]?1.27

 ENTER frequency f [GHz]?1



 ratio of w/h =3.02623521

 given h =1.27,thus w =3.84331871 [mm]

 free space wavelength =300 [mm]

 Eeff =2.01999472

 effective wavelength =211.079539 [mm]

 effective half wavelength =105.53977 [mm]

 effective quarter wavelength =52.7698848 [mm]



     Do you wish to run the

     programme again  (Y/N)?

  THE END
```

```
RUN



                    Microstrip width calculation

  ENTER Zo [Ohm]?100

  ENTER Er?2.3

  ENTER height h [mm]?1.27

  ENTER frequency f [GHz]?1



 ratio of w/h =0.860288924

 given h =1.27,thus w =1.09256693 [mm]

 free space wavelength =300 [mm]

 Eeff =1.82317428

 effective wavelength =222.181125 [mm]

 effective half wavelength =111.090563 [mm]

 effective quarter wavelength =55.5452814 [mm]
```

```
L.
   10 REM Name of this programme is POLSTRI
   20 REM POLSTRI calculates microstrip width from Zo , Er , h and f
   30 CLS
   40 VDU 19,1,0,0,0,0
   50 VDU 19,2,2,0,0,0
   60 PRINT""''''''""
   70 INPUT "                    Microstrip width calculation

          ENTER Zo [Ohm]",ZO
   80 PRINT
   90 INPUT "   ENTER Er",ER
  100 PRINT
  110 INPUT "   ENTER height h [mm]",HI
  120 PRINT
  130 INPUT "   ENTER frequency f [GHz]",F
  140 H=(ZO*SQR(2*(ER+1))/119.9)+((ER-1)/(2*(ER+1)))*(LN(PI/2)+(1/ER)*LN(4/PI))
  150 IF ZO<=44-2*ER GOTO 180
  160 RATIO=1/(EXP(H)/8-1/(4*EXP(H)))
  170 GOTO 200
  180 D=59.95*(PI)^2/ZO/SQR(ER)
  190 RATIO=2/PI*(D-1-LN(2*D-1))+(ER-1)/PI/ER*(LN(D-1)+0.293-0.517/ER)
  200 IF ZO<=63-2*ER GOTO 240
  210 RD=1/(1-((1/2/H)*(ER-1)/(ER+1)*(LN(PI/2)+(1/ER)*(LN(4/PI)))))^2
  220 EF=(ER+1)/2*RD
  230 GOTO 250
  240 EF=ER/(0.96+ER*(.109-.004*ER)*(LOG(10+ZO)-1))
  250 W=RATIO*HI
  260 LO=300/F
  270 LEFF=LO/SQR(EF)
```

```
280 LQ=LEFF/4
290 LH=LEFF/2
300 CLS
310 PRINT""''''""
320 PRINT"  ratio of w/h =";RATIO
330 PRINT
340 PRINT"  given h =";HI;",thus w =";W;" [mm]"
350 PRINT
360 PRINT"  free space wavelength =";LO;" [mm]"
370 PRINT
380 PRINT"  Eeff ="EF
390 PRINT
400 PRINT"  effective wavelength =";LEFF;" [mm]"
410 PRINT
420 PRINT"  effective half wavelength =";LH;" [mm]"
430 PRINT
440 PRINT"  effective quarter wavelength =";LQ;" [mm]"
450 PRINT""''''""
460 PRINT"      Do you wish to run the "
470 PRINT
480 INPUT"      programme again  (Y/N)",A$
490 IF A$="Y" GOTO 30 ELSE PRINT "
                                        THE END"
```

PARALLEL TRANSMISSION LINES

During the design of a microwave circuit or subsystem it is inevitable that on occasions two or more copper tracks lie parallel to each other. These parallel tracks, lines or strips, especially if they are close to each other, will couple energy to each other. This undesired coupling will deteriorate circuit performance. The coupling between the tracks can be considered as the parallel coupling between transmission lines. Apart from the above one can make a problem into a virtue and exploit the coupling between adjacent lines for other purposes.

Parallel-coupled transmission lines are used or encountered, amongst others, in microwave couplers and filters and some examples are shown in Fig. 2.9. The gap or

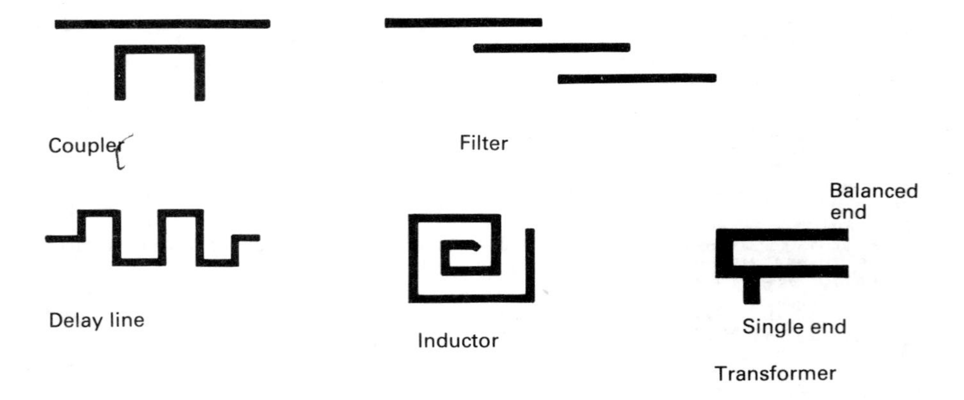

Fig. 2.9 — Examples of coupled transmission lines.

spacing between adjacent lines is usually denoted by the symbol s. Two different field distributions can exist between two microstrips depending on the relative current flow in the strips. In Fig. 2.10(a) the current flow in each conductor is in the same

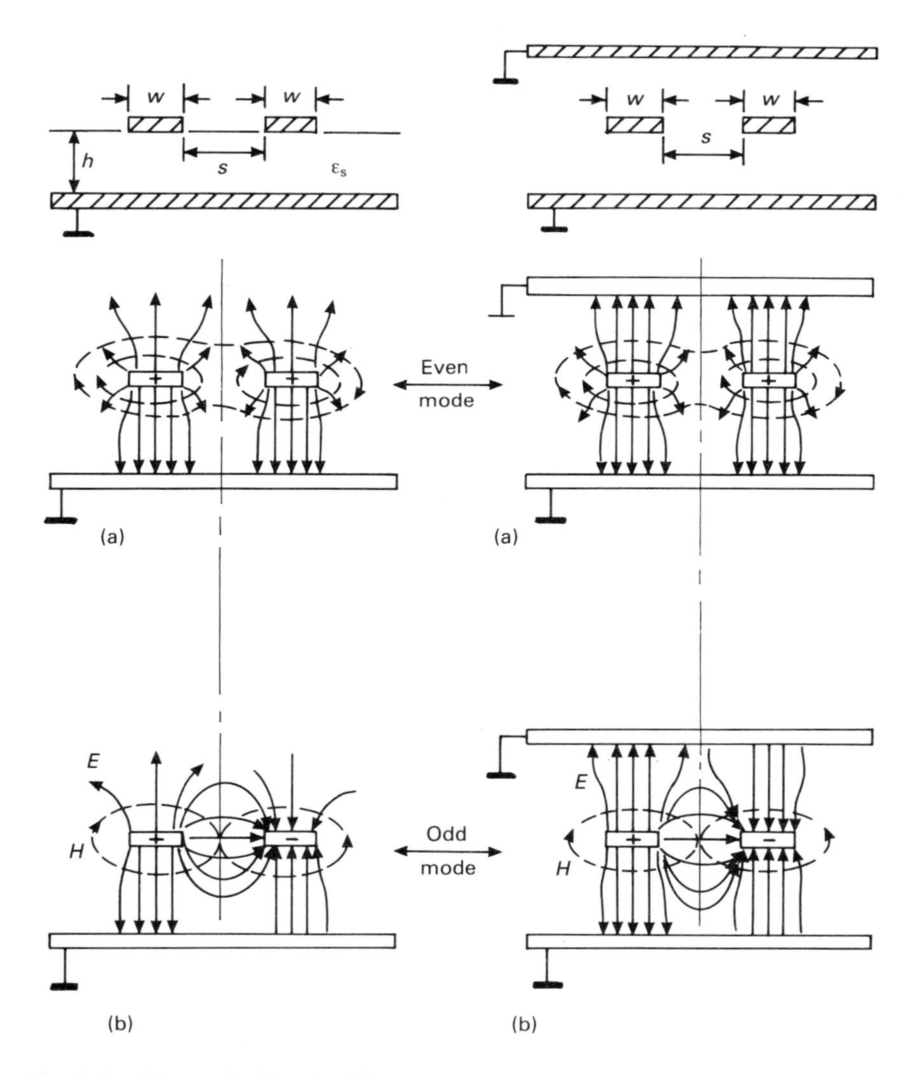

Fig. 2.10 — Even and odd mode field distribution in a microstrip.

Fig. 2.11 — Even and odd mode field distribution in a stripline.

direction and the field distribution is known as even mode. Owing to the dielectric constant of the microwave substrate ($\varepsilon_r > 1$) the electric field is more concentrated in the substrate than in the air. The same applies to the magnetic field lines shown in

Fig. 2.10. Both modes have an axis of symmetry. The even mode axis of symmetry is known as magnetic wall and the odd mode as electric wall. The names stem from the fact that in the case of even mode operation no tangential magnetic field can exist between the two microstrips. The same applies to the electric field in odd mode symmetry, i.e.there is no electric tangential component.The field distribution of a stripline is shown in Fig. 2.11.

Clearly,the characteristic impedances for either mode will be different. The even and odd mode impedances are denoted by Z_{oe} and Z_{oo} respectively, where the last suffix identifies the mode. It is these two impedances which can be used to specify the intentional or unintentional coupling between two adjacent microstrips as will be seen later. The associated phase velocities are also different, namely v_{pe} and v_{po}.

COUPLED MICROSTRIP

Diverse microwave circuits exist which do not only employ single lines, but also parallel or coupled lines. Coupled lines are used to achieve particular aims in terms of electric performance of the circuit in question. Basic examples of coupled microstrip were given in Fig. 2.9.

In order to predict circuit performance or to assess any circuit performance deterioration resulting from unwanted coupling, one needs to derive closed form mathematical expressions for the fundamental TEM modes which exist on a pair of parallel conducting lines against a ground plane. The appropriate physical structure is shown in Fig.2.12. If the parallel lines were sandwiched between two ground

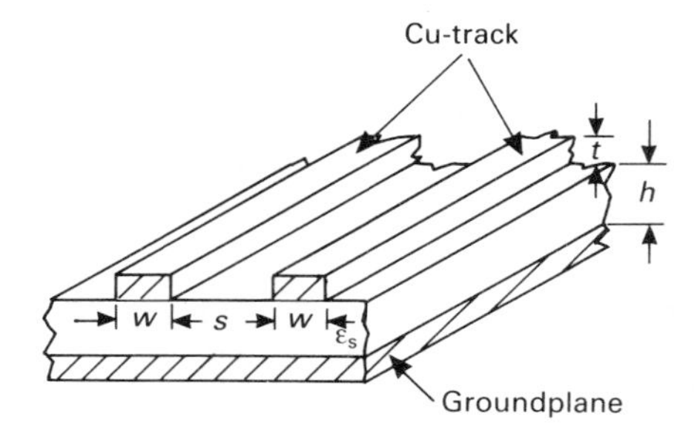

Fig. 2.12 — Geometry of coupled microstrip.

planes (stripline structure) then a true TEM situation would exist. This is easier to solve than the case of a microstrip structure which is only quasi-TEM. Edwards [4] presents an analysis for coupled microstrip. He assumed the lines to be sandwiched between two layers of a uniform dielectric, resulting in even and odd mode impedances, but only one phase velocity value v_p.

The parallel coupled line analysis is based on Fig. 2.13 where all ports have been terminated in the characteristic impedance of the line. This includes the source at port 1 which has a driving point impedance equal to the characteristic impedance, usually 50 Ω. Under those special conditions we can re-draw Fig. 2.13 as shown in

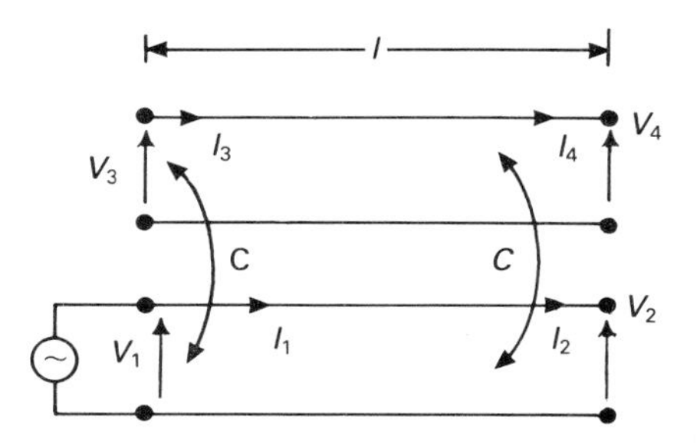

Fig. 2.13 — Parallel coupled lines.

Fig. 2.14, where l is the length over which two adjacent lines couple. For the purpose of analysis Fig. 2.14 can be separated into an even mode and odd mode case as given in Fig. 2.15. Since the four circuits are all symmetrical only one set of figures need be analysed, either the primary arm or the secondary arm. Here we use the primary arm for ease of comparison with results in other literature.

The voltage and current relationships existing on the primary arms for the even and odd mode can thus be expressed as follows:

$$\begin{bmatrix} V_{1e} \\ I_{1e} \end{bmatrix} = \begin{bmatrix} \cos\theta & jZ_{oe}\sin\theta \\ jY_{oe}\sin\theta & \cos\theta \end{bmatrix} \begin{bmatrix} V_{2e} \\ I_{2e} \end{bmatrix} \tag{2.19}$$

and

$$\begin{bmatrix} V_{1o} \\ I_{1o} \end{bmatrix} = \begin{bmatrix} \cos\theta & jZ_{oo}\sin\theta \\ jY_{oe}\sin\theta & \cos\theta \end{bmatrix} \begin{bmatrix} V_{2o} \\ I_{2o} \end{bmatrix} \tag{2.20}$$

From this, eight expressions can be obtained for all currents and voltages for both even and odd modes. A significant mathematical simplification can be obtained by assuming all ports as matched. As a result the following important equality is obtained

$$Z_o^2 = Z_{oe}\, Z_{oo} \tag{2.21}$$

which says that the characteristic impedance Z_o in a matched parallel line system is the geometric mean of Z_{oe} and Z_{oo}. Furthermore, by defining the coupling parameter C such that

$$C = \frac{Z_{oe} - Z_{oo}}{Z_{oe} + Z_{oo}} \tag{2.22}$$

or as a voltage ratio in dB as

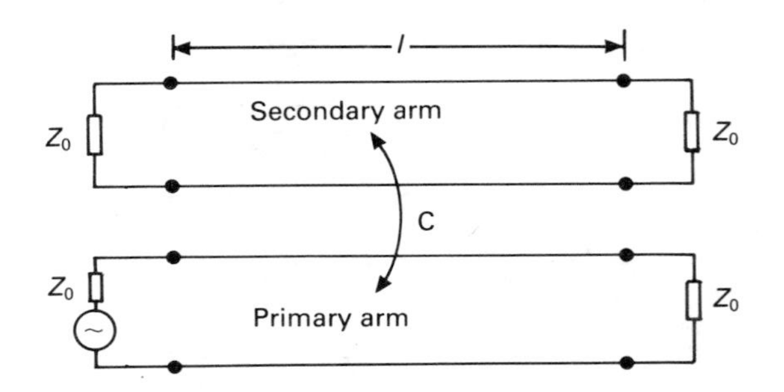

Fig. 2.14 — Terminated parallel coupled lines.

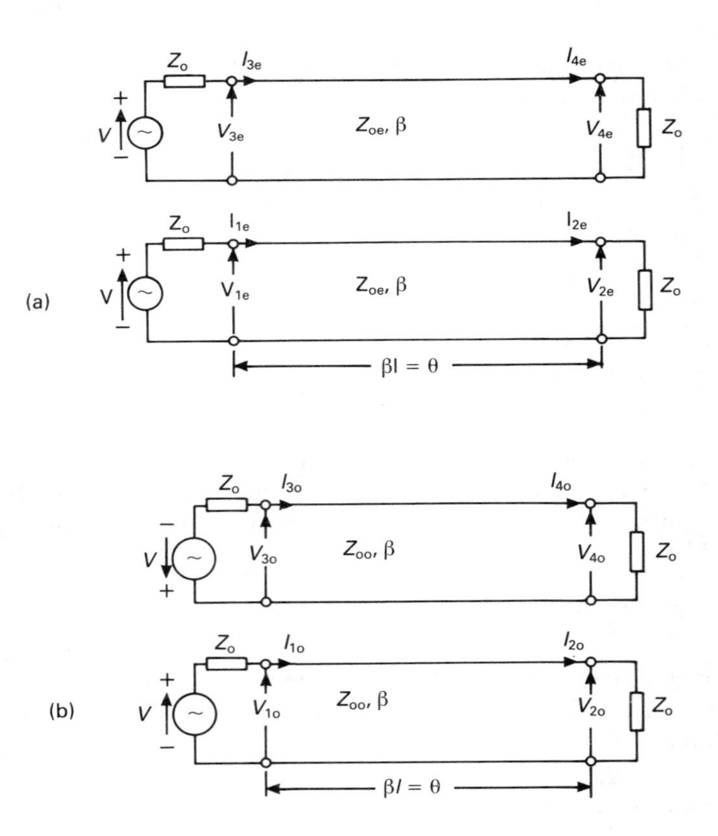

Fig. 2.15 — Separated even- and odd-mode equivalent circuits for parallel-coupled microstrip. (a) even mode; (b) odd mode.

$$C\,[\text{dB}] = 20 \log \left| \frac{Z_{oe} - Z_{oo}}{Z_{oe} + Z_{oo}} \right| \tag{2.23}$$

results in the following simple expressions for the terminal voltages of a pair of coupled microstrip:

$$V_1 = 1 \tag{2.24}$$

$$V_2 = \frac{\sqrt{1 - C^2}}{\sqrt{1 - C^2}\cos\theta + j\sin\theta} \tag{2.25}$$

$$V_3 = \frac{j\,C\sin\theta}{\sqrt{1 - C^2}\cos\theta + j\sin\theta} \tag{2.26}$$

$$V_4 = 0 \tag{2.27}$$

From this we see that the coupled output voltages (V_2, V_3) or powers depend on the coupling C and the electrical coupling length θ of the two lines. Maximum coupling takes place when V_2 and V_3 are a maximum or when the denominator of the respective equation is a minimum. This minimum exists for a coupling length of $\lambda/4$ or when $\theta = 90^{\circ} = \pi/2$.

Thus eqns (2.24) to (2.27) become:

$$V_1 = 1 \tag{2.28}$$

$$V_2 = -\mathrm{j}\sqrt{1 - C^2} \tag{2.29}$$

$$V_3 = C \tag{2.30}$$

$$V_4 = 0 \tag{2.31}$$

A comparison of eqns (2.29) and (2.30) shows the quadrature phase relationship as evidenced by $-j$. Also, for $C = 0.7$, i.e. 3 dB coupling, we have an equal powersplit between ports 2 and 3, namely

$$V_3 = -j\,V_2 \tag{2.32}$$

Finally we observe $V_4 = 0$ since this is the isolated port. Remembering that

$$\theta = \frac{2\,\pi\,l}{\lambda_g} = \frac{2\,\pi\,f\,l}{v} \tag{2.33}$$

we can plot the frequency response $|V_2| = f(\theta)$ and $|V_3| = f(\theta)$. Rearranging eqn. (2.23) and substituting eqn. (2.21) give the following useful initial design equations:

$$Z_{oe} = Z_o\sqrt{\frac{1 + 10^{C/20}}{1 - 10^{C/20}}} \tag{2.34}$$

$$Z_{oo} = Z_o \sqrt{\frac{1 - 10^{C/20}}{1 + 10^{C/20}}} \tag{2.35}$$

where C must be entered in dB as shown in eqn. (2.23). Results from this section will be used later in the design of some couplers and filters.

DESIGNING COUPLED MICROSTRIP

The following concerns itself with the design aspects of parallel coupled microstrip. Most of the aspects discussed here are relevant where coupled structures are employed in microwave circuits such as for example the obvious parallel line coupler (Ch. 9) and filters (Ch. 6). Because of the presence of two or more microstrips, the even and odd-mode impedances need to be taken into account when synthesizing. These impedances are usually expressed as the mean of Z_{oe} and Z_{oo},namely:

$$Z_{ose} = \frac{Z_{oe}}{2} \tag{2.36}$$

$$Z_{oso} = \frac{Z_{oo}}{2} \tag{2.37}$$

where the suffix s refers to the single microstrip equivalent. Z_{ose} is thus the characteristic impedance of a single microstrip in the even mode.

Amongst others [5,6], Akhtarzad *et al.* [7] have addressed the problem of coupled microstrip and derived formuli which offer themselves readily for computer-aided design (CAD). The basic problem in the design of coupled microstrip is the establishment of a relationship between the even and odd mode impedances (Z_{oe}, Z_{oo}) and the physical dimensions of the coupling structure in terms of the shape ratios w/h and s/h. These relationships can be established or, in other words, the problem can be solved in two ways; graphically or purely numerically (CAD).

A graphical approach requires two sets of curves with Z_{oe} and Z_{oo} plotted against the shape ratios w/h and s/h. The problem is then solved by looking for those w/h and s/h values on the graph which simultaneously give Z_{oe} and Z_{oo}. The numerical design solution used here starts by finding the even and odd mode shape ratios of a single line (eqns 2.36 and 2.37) or more specifically $(w/h)_{se}$ and $(w/h)_{so}$. The shape ratio $(w/\mathrm{h})_{se}$ is thus the w/h ratio of a single microstrip in the even mode. Wheeler [8] has produced curves for these single lines based on the following equations:

$$(w/h)_s = \frac{2(D-1)}{\pi} - \frac{2}{\pi} \ln (2D-1) + \frac{\varepsilon_r - 1}{\pi\, \varepsilon_r}\left[\ln (D-1) + 0.293 - \frac{0.517}{\varepsilon_r}\right] \tag{2.38}$$

where

$$D = \frac{60\ \pi^2}{Z_o\ (\varepsilon_r)^{0.5}} \tag{2.39}$$

In short, we observe that for a single microstrip

$(w/h)_s = f\ (\varepsilon_r,\ Z_o)$

The coupled line shape ratios may now be found through the simultaneous solution of eqns (2.40) and (2.41). Hence

$$(w/h)_{se} = \frac{2}{\pi} \cosh^{-1}\left(\frac{2H - G + 1}{G + 1}\right) \tag{2.40}$$

and

$$(w/h)_{so} = \frac{2}{\pi} \cosh^{-1}\left[\frac{2H - G - 1}{G - 1}\right] + \left[\frac{4}{\pi\ (1 + \varepsilon_r/2)}\right] \cosh^{-1}\left[1 + \frac{2w/h}{s/h}\right] \tag{2.41}$$

It should be noted that the upper case letter H denotes a constant as defined in eqn. (2.44) whereas the lower case letter h is the substrate height.

Equation (2.41) is for $\varepsilon_r <= 6$ where it gives best results. For $\varepsilon_r >= 6$ we use for the single microstrip odd-mode shape ratio

$$(w/h)_{so} = \frac{2}{\pi} \cosh^{-1}\left[\frac{2H - G - 1}{G - 1}\right] + \frac{1}{\pi} \cosh^{-1}\left[1 + \frac{2w/h}{s/h}\right] \tag{2.42}$$

where

$$G = \cosh\left(\frac{\pi\ s}{2\ \mathrm{h}}\right) \tag{2.43}$$

and

$$H = \cosh\left(\frac{\pi\ w}{h} + \frac{\pi\ s}{2\ h}\right) \tag{2.44}$$

For the purpose of computation cosh$x$ may be represented in its exponential form:

$$\cosh x = \frac{e^x + \quad e^{-x}}{2} \tag{2.45}$$

If the term $\cosh^{-1}[1+\{(2w/h)/(s/h)\}]$ in eqn. (2.41) is neglected, then the solution of eqns (2.40) and (2.41) becomes simpler, resulting in a s/h shape ratio of

$$\frac{s}{h}=\frac{2}{\pi}\cosh^{-1}\left\{\frac{\cosh\,[\pi/2)(w/h)_{se}]\;+\cosh\,[(\pi/2)(w/h)_{so}]-2}{\cosh\,[(\pi/2)(w/h)_{so}]\;-\cosh\,[(\pi/2)(w/h)_{so}]}\right. \tag{2.46}$$

Substitution of the above s/h ratio into eqns (2.40) and (2.41) and the use of an iterative process will result in an optimum w/h shape ratio. This techniques is used when for example, designing a parallel line coupler in Chapter 9.

The program ZOE calculates the space and width of two microstrips which couple over a certain length. An example is given in the following as well as the program listing.

```
 RUN
Enter the even mode impedance Zoe [Ohm]?69.5
Enter the odd mode impedance Zoo [Ohm] ?36
Enter the substrate permittivity Er     ?9
Enter the substrate thickness h [mm]    ?1

This programme calculates the s/h and w/h ratio
for a given Zoe and Zoo

R E S U L T S


space to height ratio (s/h)                                        = 0.204633069

width to height ratio (w/h)                                        = 0.913083435

Hence for the given substrate thickness h   :

the WIDTH of the microstrip in the coupled region [mm]            = 0.913083435

the SPACE between the microstrips in the coupled region [mm] = 0.204633069
```

```
L.
  10 REM This programme is called ZOE
  20 REM This programme calculates the s/h and w/h ratio for
       a given Zoe and Zoo
  30 MODE 3
  40 INPUT"Enter the even mode impedance Zoe [Ohm]";ZOE
  50 INPUT"Enter the odd mode impedance Zoo [Ohm] ";ZOO
  60 INPUT"Enter the substrate permittivity Er     ";ER
  70 INPUT"Enter the substrate thickness h [mm]    ";H
  80 ZO=ZOE/2
  90 PROCcalwh
 100 WHSE=WHS
 110 ZO=ZOO/2
 120 PROCcalwh
 130 WHSO=WHS
 140 PROCcalsh
 150 TEMP1=SH*H
 160 PROCleng
 170 ZO=50
 180 PROCcalwh
 190 WX=WHS*H
 200 PRINT""
 210 PRINT"This programme calculates the s/h and w/h ratio
          for a given Zoe and Zoo"
 220 PRINT""
 230 PRINT"R E S U L T S "
 240 PRINT""
 250 PRINT""
 260 PRINT"space to height ratio (s/h)                                     = ";SH
```

```
  270 PRINT""
  280 PRINT"width to height ratio (w/h)                                    = ";WH
  290 PRINT""
  300 PRINT"Hence for the given substrate thickness h  :"
  310 PRINT""
  320 PRINT"the WIDTH of the microstrip in the coupled region [mm]        = ";WH*
H
  330 PRINT""
  340 PRINT"the SPACE between the microstrips in the coupled region [mm] = ";SH*
H
  350 PRINT""
  360 END
  370 MODE 128
  380 DEF PROCimp
  390 ENDPROC
  400 DEF PROCcalwh
  410 H1=ZO*SQR(2*(ER+1))/119.9+.5*((ER-1)/(ER+1))*(LN(PI/2)+(1/ER)*LN(4/PI))
  420 WHS=((EXP(H1))/8-1/(4*EXP(H1)))^-1
  430 GOTO 460
  440 D1=59.95*PI*PI/(ZO*SQR(ER))
  450 WHS=(2/PI)*((D1-1)-LN(2*D1-1))+((ER-1)/(PI*ER))*(LN(D1-1)+.293-.517/ER)
  460 ENDPROC
  470 DEF PROCcalsh
  480 MN1=(PI*WHSE/2):NM1=(PI*WHSO/2)
  490 MN=(EXP(MN1)+EXP(-MN1))/2:NM=(EXP(NM1)+EXP(-NM1))/2
  500 TEMP=(MN+NM-2)/(NM-MN):SH=(2/PI)*LN(TEMP+SQR(TEMP^2-1)):IF SH<.005 THEN 51
0 ELSE 520
  510 ENDPROC
  520 g=(EXP(PI*SH/2)+EXP(-PI*SH/2))/2
  530 d=(g-1+((g+1)*MN))/2
  540 WH=(LN(d+SQR(d^2-1))-(PI*SH/2))/PI
  550 TEMP=(2*d-g-1)/(g-1):z=(2/PI)*LN(TEMP+SQR(TEMP^2-1)):TEMP=1+2*(WH/SH)
  560 TEMP2=LN(TEMP+SQR(TEMP^2-1)):IF (ER<6) THEN 570 ELSE 590
  570 WHSO2=z+(4*TEMP2/(PI*(1+(ER/2))))
  580 GOTO 600
  590 WHSO2=z+(TEMP2/PI)
  600 IF (WHSO2-WHSO)>.1 OR (WHSO-WHSO2)>.1  THEN 610 ELSE 650
  610 IF (WHSO2<WHSO) THEN 620 ELSE 640
  620 SH=SH-.01
  630 GOTO 650
  640 SH=SH+.01
  650 IF (WHSO2-WHSO)>.1 OR (WHSO-WHSO2)>.1 THEN 520 ELSE 660
  660 IF (WHSO2-WHSO)>.01 OR (WHSO-WHSO2)>.01 THEN 670 ELSE 710
  670 IF (WHSO2<WHSO) THEN 680 ELSE 700
  680 SH=SH-.001
  690 GOTO 710
  700 SH=SH+.001
  710 IF (WHSO2-WHSO)>.01 OR (WHSO-WHSO2)>.01 THEN 520 ELSE 720
  720 ENDPROC
  730 DEF PROCleng:LOCAL ER,C
  740 EEFF=1:ER=1:EO=8.85E-12:C=3E8
  750 g=(EXP(PI*SH/2)+EXP(-PI*SH/2))/2
  760 d=(EXP(PI*WH+PI*SH/2)+EXP(-(PI*WH+PI*SH/2)))/2
  770 x=(2*d-g+1)/(g+1)
  780 PROCikosh:WHSE=2*TEMP1/PI:x=(2*d-g-1)/(g-1):PROCikosh
  790 x=(1+WH/SH)
  800 PROCikosh2
  810 WHSO=2*TEMP1/PI+(4/(PI*(1+ER/2)))*TEMP2
  820 IF (WH<3.3) GOTO 920
  830 TEMP3=LN(EXP(1)*PI*PI/16)/(2*PI)*(ER-1)/(ER^2)
  840 TEMP1=WHSO/2+LN(4)/PI+TEMP3
  850 TEMP4=(ER+1)/(2*PI*ER)
  860 TEMP2=TEMP4*(LN(PI*EXP(1)/2)+LN(WHSO/2+.94))
  870 ZOSO=(119.9*PI/(2*SQR(ER)))*(1/(TEMP1+TEMP2))
  880 TEMP1=WHSE/2+LN(.4)/PI+TEMP3
  890 TEMP2=TEMP4*(LN(PI*EXP(1)/2)+LN(WHSE/2+.94))
  900 ZOSE=(119.9*PI/(2*SQR(ER)))*(1/(TEMP1+TEMP2))
  910 GOTO 970
  920 TEMP1=LN(4*(1/WHSO)+SQR(16*(1/WHSO)*(1/WHSO)+2))
  930 TEMP2=(.5*(ER-1)/(ER+1))*(LN(PI/2)+LN(4/PI)/ER)
  940 ZOSO=(119.9/SQR(2*(ER+1)))*(TEMP1-TEMP2)
  950 TEMP1=LN(4*(1/WHSE)+SQR(16*(1/WHSE)*(1/WHSE)+2))
  960  ZOSE=(119.9/SQR(2*(ER+1)))*(TEMP1-TEMP2)
  970 ZO1=SQR(ZOSE*ZOSO)*2
  980 CP=EO*ER*WH
  990 CF=SQR(EEFF)/(C*ZO1*2)-CP/2
 1000 TEMP1=EXP(-.1*EXP(2.33-2.53*WH))
```

```
1010 TAH=(EXP(8*SH)-EXP(-8*SH))/(EXP(8*SH)+EXP(-8*SH))
1020 CF1=CF*SQR(ER/EEFF)/(1+(TEMP1*(1/SH)*TAH))
1030 K=SH/(SH+2*WH)
1040 K1=SQR(1-K^2)
1050 IF (K^2<.5) GOTO 1080 ELSE 1060
1060 TEMP1=PI/(LN(2*(1+SQR(K))/(1-SQR(K))))
1070 GOTO 1090
1080 TEMP1=LN(2*(1+SQR(K1))/(1-SQR(K1)))/PI
1090 CGA=EO*TEMP1
1100 TAH1=(EXP(PI*SH/4)-EXP(-PI*SH/4))/(EXP(PI*SH/4)+EXP(-PI*SH/4))
1110 TEMP1=LN(1/TAH1)
1120 TEMP2=.02*SQR(ER)/SH+1-1/(ER^2)
1130 CGD=EO*ER*TEMP1/PI+.65*CF*TEMP2
1140 CE=CP+CF+CF1
1150 CO=CP+CF+CGA+CGD
1160 Z01E=1/(C*CE)
1170 Z010=1/(C*CO)
1180 ENDPROC
1190 DEF PROCikosh
1200 TEMP1=LN(x+SQR(x*x-1))
1210 ENDPROC
1220 DEF PROCikosh2:TEMP2=LN(x+SQR(x*x-1)):ENDPROC
```

REFERENCES

[1] Bahl, I. J. and Garg, R., Simple and accurate formulas for microstrip with finite strip thickness, *Proc. IEEE*, No. 65, 1977, pp. 1611–1612.

[2] Owens, R. P., Accurate analytical determination of quasi-static microstrip line parameters, *Radio and Electronic Engineer*, **46**, No.7, July 1976, pp. 360–364.

[3] Yamashita, E. and Atsuki, K., Design of transmission line dimensions for a given characteristic impedance, *IEEE Trans.* **MTT 17**, Aug. 1969, pp. 638–639.

[4] Edwards, T. C., *Foundations for microstrip circuit design*, John Wiley, 1981.

[5] Bryant T. G. and Weiss, J. A., Parameters of microstrip transmission lines and of coupled pairs of microstrip lines, *IEEE Trans.* **MTT 16**, No. 12, Dec. 1968, pp. 1021–1027.

[6] Cohn, S. B., Shielded coupled-strip transmission line, *IRE Trans.* **MTT,** No.3, Oct. 1955, pp. 29–38.

[7] Akhtarzad, S., Rowbotham, R. and Johns P. B. , The design of coupled microstrip lines, *IEEE Trans.*, **MTT 23**, No. 6, June 1975.

[8] Wheeler, H. A., Transmission-line properties of parallel strips separated by a dielectric sheet, *IEEE Trans.* **MTT 13**, No.3, March 1965, pp. 172–185.

3

Practical tools for microstrip design

This chapter concerns itself with the practicalities of making a microwave substrate and mounting of appropriate components. To start with it is described how to prepare the artwork for a microwave circuit and what equipment is available to execute the artwork. This is then followed by a discussion of substrates and their etching, since its choice will have an effect on the final circuit performance as does the housing into which the circuit or subsystem is to be incorporated. Finally, a selection of components peculiar to microwave circuits and their assembly are treated.

ARTWORK

As has been pointed out on many occasions, the lack of frequency space led to the use of higher frequencies, technology permitting. Thus, microwave circuits up to at least X-band are widely used and technology and production techniques may be considered well understood. The more extensive demand for microwave circuits lead not only to the advancement of CAD for those circuits, but also the artwork which will now be considered.

Originally the drawings required for making microwave circuits in planar transmission line technology were done by hand. Usually drafting tape was used or a masking film was cut by hand to arrive at a layout. It is evident that the above two techniques have their limitations when it comes to fine lines, narrow gaps, repeatability, and curves of diverse radii or very dense or complex circuits. Also, it is not easy to make changes. Moreover, the benefit derived from CAD is negated by manual layout. What is the use of the speed of CAD to design the microwave circuit if it takes days if not weeks to produce the artwork? Clearly, CAD of microwave circuits and their artwork generation need to be coordinated in a commercial environment. This then brings us to the techniques available for modern artwork generation.

Modern artwork generation is usually automated and uses a fraction of the time required for manual layout. The resulting drawing or mask is of high quality and accuracy. There are three methods which may be used in the automatic generation of artwork: coordinatographs, photoplotters and pattern generators. These are now discussed.

COORDINATOGRAPH

The automated coordinatograph represents a more sophisticated version of its well-known manually operated counterpart. It can be looked upon as an *x–y* plotter where the drawing pen is replaced by a cutting tool such as a knife. In essence, the coordinatograph consists of a base plate above which a cutting tool is mounted, and which can be moved in both the *x* and *y* coordinates. A film consisting of two layers, i.e. a transparent layer with an opaque upper layer is usually placed on the base plate. A special knife is then lowered onto the film so that it just penetrates the opaque film layer. In an automated version the coordinatograph is usually interfaced with a computer which then controls the knife position. Clearly, as the knife is moved along it cuts a pattern into the film. At the end of the computer program the knife is lifted up again. The cut out opaque film layer is then peeled off, leaving a mask with a transparent pattern. In summary, the advantages of an automatic coordinatograph in relation to its manual layout lie in the considerable speed of mask production (mask throughput), resolution, precision and repeatability.

PHOTOPLOTTER

Photoplotters, like coordinatographs, are precision machines. In a photoplotter a light source directly exposes a film in order to produce a pattern as input into the *x–y* control system via a computer. After exposure the film is developed to give a negative which can then be used to produce a microwave circuit pattern. Unlike a coordinatograph, a photoplotter needs to be accommodated in a darkroom. Ferranti, for example, manufacture a photoplotter which carries out the above operation.

PATTERN GENERATOR

Pattern generators are similar to photoplotters and use a microprocessor as a data base. Rectangular shapes of varying dimensions are directly flashed onto a photosensitive glassplate to create a one-to-one size pattern. The main use of pattern generators is, however, reserved for high density artwork such as required for integrated circuits. Here, rectangular elements are most common. Smooth contours cannot be obtained. The tolerances in a pattern generated by this machine are extremely low.

Naturally the question arises as to which technique or equipment is best suited to microwave circuits. Pattern generators produce square patterns (pixels) and are thus not ideal for producing smooth contours, bends or radii. Owing to their precision they are also very expensive. Thus they are infrequently used for microwave circuit artwork. This leaves the photoplotter and coordinatograph.

The choice between the two is a difficult one since several factors need to be taken into account, such as price, skill of operating personnel, quality of artwork and operational environment. A photoplotter is usually much more expensive, requires a controlled environment such as a darkroom and requires skilled personnel, i.e. those who can cope with the associated photographic processing. In contrast, a coordinatograph can be placed in a drawing office. Also, a coordinatograph can quickly produce large square areas whilst a photoplotter needs to define the boundaries and then fill in the area which requires time. On the other hand, a photoplotter can produce higher circuit density. Microwave circuit can be affected significantly by the sharpness of line edges. A knife cut is sharper than the grey line edge produced by a photoplotter.

Either system has thus about the same number of advantages and drawbacks. It must thus be left to the microwave engineers concerned to make a judgement on the basis of a sound understanding of their main product line and their financial budget.

SCHEMATIC CAPTURE AND LAYOUT

Barnard Microsystems Ltd of St Albans has created software under the trade name of 'WaveMaker' which provides the microwave engineer with a most appropriate graphics system for editing microwave circuit layouts and circuit schematics. In addition to providing schematic capture and layout autoprocessing of the net lists of the well known Touchstone and Super compact microwave programs, this software will also assist in the creation of the layout of electronic devices, printed circuit boards, MMICs and hybrid circuits. There are two factors which point at the importance of layout and schematic capture software:

(1) The layout of microwave circuits is intensive in terms of human interaction. The better the software, the less time it will take to create the optimum microwave circuit layout.
(2) The design 'first time right' objective is becoming more essential as the cost of failure increases. A good software will thus support the accurate and comprehensive net list to layout transformation.

WaveMaker can transform a hirarchical Super Compact or Touchstone net list to an exact graphical equivalent layout. Cells containing the user-defined layout or schematic of active and passive elements are incorporated into the autoprocessed circuit. The coordinate values are held exactly in integer units of nanometers up to a maximum of one meter. All calculations are done using double precision floating point numbers. The layout can be done in metric or imperial units of length.

The software has an open architecture to allow the user to enhance it for his own applications. Once the final circuit layout has been obtained, a hard copy can be made for circuit realization. WaveMaker can also be accessed by many other graphics systems such as Aristomat, Wild Aviotab, HP-GL in outline and filled polygon mode, IEDS line and polygon mode but to name a few.

PHOTOGRAPHY

If the artwork does not directly result in a film negative then photographic techniques need be used to arrive at this. The Agfa-Gevaert REPROMASTER MK3 camera system is most suitable for undertaking this task. The Repromaster handles copies up to A3 size with a standard lens at 1:1 exposure. The copyboard measures 48 × 66 cm and the maximum copying area is 38.5 × 47.5 cm when using back lighting. The two lenses, a 21 cm G-Claron and a 15 cm Comparon, are of the highest precision and enable the production of enlargements or reductions of up to four times linear. Any intermediate size up to the above can of course be obtained. The sharpness of the resulting negative is extremely high. The sturdy all-steel frame ensures the necessary stability. Two solidly constructed guide shafts accurately control the vertically travelling copyboard and lens holder. Total mechanical precision is guaranteed.

The cranks, percentage scales and matt registration sheet permit correct size adjustment and focusing. The vacuum back of the camera ensures perfect contact for the film which is necessary for razor-sharp reproductions. Control is by means of a vacuum gauge. Four 500 W halogen lamps ensure uniform illumination of the original artwork which is deposited on the copyboard. These lamps are integral with the copyboard so that the lamp position needs never be adjusted.

In addition to diverse camera systems Agfa-Gevaert offers a wide range of photographic materials for use with the repromaster camera. For microwave artwork COPYPROOF CPN, where N stands for negative, has proved most useful since its resolving power guarantees faithful detail rendering. The paper is sensitive to blue and blue-green lights only and can thus be processed in the darkroom in bright yellow light. Additionally, CPN film material is compatible with regular Copyproof receiving positives. This allows a positive to be made for recording and report writing purposes.

Having taken a photograph of the microwave artwork the film needs to be developed in order to obtain the necessary negative. This may be easily done using a COPYPROOF processor such as for example the CP380. In a darkroom the exposed negative film CPN is fed with an acetate sheet of CPF through the processor. Both sheets then emerge from the unit in contact. After about 60 seconds the positive is peeled from the negative which is then dried and ready for use in conjunction with the microwave substrate to be prepared. The etching procedure is outlined in a later section of this chapter.

SUBSTRATES

The potential of planar transmission lines, i.e. flat transmission line structures, became apparent to the microwave industry in early 1950. Particularly attractive was the ease of mass production. At the beginning problems were encountered with the substrate materials in that they lacked the mechanical and electrical characteristics required for reliable microwave operation and microwave circuit fabrication. This then brings us to the point that microwave circuits require laminates or substrates with special properties, which are not normally found in or required from the more familiar printed circuit boards.

Microwave substrates have now been used for designs up to mm-wave frequencies, for aerial systems such as phased arrays and spiral aerials, high precision microwave components, cooled receiver front ends, high speed digital systems, data link applications. They are also being used for space applications. There are a number of manufacturers of microwave substrates known under names such as

Duroid, 3M, Polyguide, Rexolite, OAK. Over the years materials research into microwave substrates has resulted in their having most of the following desirable properties:

(1) low loss tangent;
(2) uniform dielectric constant;
(3) high sheet flatness;
(4) no post etch warpage;
(5) good peel strength;
(6) high batch uniformity;
(7) inert to chemicals;
(8) high mechanical strength;
(9) easily machined, cut and drilled;
(10) low thickness variation;
(11) long term stability;
(12) high heat resistance.

The above points are now briefly enlarged upon in the following sections.

From a manufacturing point of view the machinability of substrates is very important. Most of the laminates may be drilled using well known printed circuit techniques and tools, since the sheets are relatively thin. Sharp new drills, preferably made from carbide, should be used. The latter is imperative if the laminate contains ceramics to increase the dielectric constant. This reduces the risk of localized heat development and guarantees hole size and shape. In order to prevent any burrs suitable backup boards should be used. Most microwave laminates can be cut or sheared, provided the cutting tool is sharp.

Microwave substrates should have long term stability. Naturally, there will be physical variations, temperature being one cause of it. When copper is removed, the changes to which the substrate may be subjected in the directions of width and length increase. The degree of change depends on the amount of copper removed for a given temperature change. Thermal expansion of microwave substrates does not take place in a linear fashion. Indeed, expansion in the z-direction can be significant within certain temperature limits. It is well known that the thermal expansion coefficient of certain dielectrics is high. When constraint is in the direction of length and width, as in the case of woven substrate, expansion in the z-direction becomes more pronounced. The above factors also influence any post etch warpage.

It is important that there is as little thickness variation in the substrate as possible, in other words the two copper planes on either side of the dielectric should be parallel. This will minimize impedance variations and changes in phase velocity. Additionally, sheet flatness should be high to prevent any curvature and hence physical copper track deformation.

The peel strength of a substrate refers to the forces required to remove a copper track from the substrate. Clearly, it is more difficult to peel off a wide than a narrow track. The peel strength also influences the heat that may be applied when soldering components to the copper track. Indeed, solder pads may need to be provided in order to realize certain microwave features such as high impedance bias lines.

A low loss tangent is essential to keep the losses down, that is it keeps the attenuation along the line to a minimum and reduces power dissipation in the dielectric of the substrate. This is most important where planar transmission lines are used for signal power transmission.

Naturally, high batch uniformity is desirable in order to allow for consistent performance of the final product. Finally, it is almost superfluous to say, the laminates should be inert to chemicals and any other fluids in order to preserve laminate characteristics.

Microwave substrates may be manufactured by casting the substrate dielectric material or by including a weaving process. Rexolite for example uses a cast dielectric material which is then cured. A sheet is thus formed which is ground to the desired thickness and coated, usually copper. Copper deposition can be either by rolling a copper foil onto the substrate or by elecro-deposition. The copper foil is obtained by roller milling an ingot to the desired copper weight which may be 0.5, 1, or 2 ounces per square foot. At present copper weight is still used to denote thickness. One ounce of copper spread over an area of 1 ft^2 results in a copper sheet thickness of about 35 μm (1.4 mil). A two ounce copper weight thus means a microstrip or stripline thickness of about 70 μm which is the maximum copper thickness used. If copper is rolled onto a substrate, adhesive is used to provide adhesion.

An alternative to employing copper foil is to deposit the copper by electrolysis. This is done by using a rotating drum as the cathode in the electrolysis process. The rotational speed of the drum controls the copper foil thickness on the substrate. As a result of the depositing techniques used, the surface of rolled copper laminates is smoother than that of electro-deposited copper. This can have a bearing on microwave circuit performance and depends on substrate thickness.

Weaving offers an alternative to cast dielectrics. The weave can either be fine, medium or of composite structure. A finely woven glassfibre/PTFE laminate is useful for applications where a high hole density is required or where good results are expected when punching. Medium woven glassfibre/PTFE laminates offer excellent repeatable electrical parameters. Woven substrates offer good dimensional stability as requried for larger boards. They lack, however, the desirable property of being more isotropic with respect to their dielectric characteristics. As in textiles, there is a warp and woof (filling) thread in a woven substrate. The pattern can be observed on the substrate especially if a magnifying glass is being used. This is because high pressure is applied when making the laminate and bonding the copper to it. The warp direction is the direction in which the foil was deposited and is of greater uniformity than the woof direction. This is shown in Fig. 3.1. Optimum microwave circuit performance is obtained by fabricating the bulk of the circuit along the warp axis.

Although we are mainly concerned with microstrip, striplines, as will be discussed later, form an important part in filter design. It might thus be opportune to contrast in this section the conventional clamped stripline assembly with the bonded stripline assembly.

The cross-sectional area of a conventional clamped stripline assembly is shown in Fig. 3.2(a). The thickness of the stripline or centre conductor is negligible resulting in an almost non-existent airgap. A clamping structure typically consisting of two

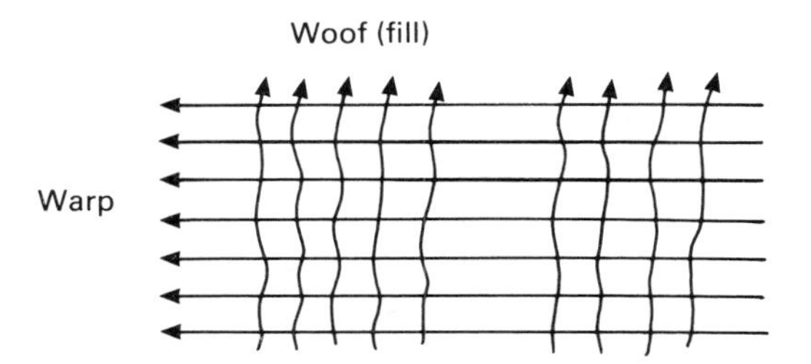

Fig. 3.1 — Warp/woof direction of a woven laminate.

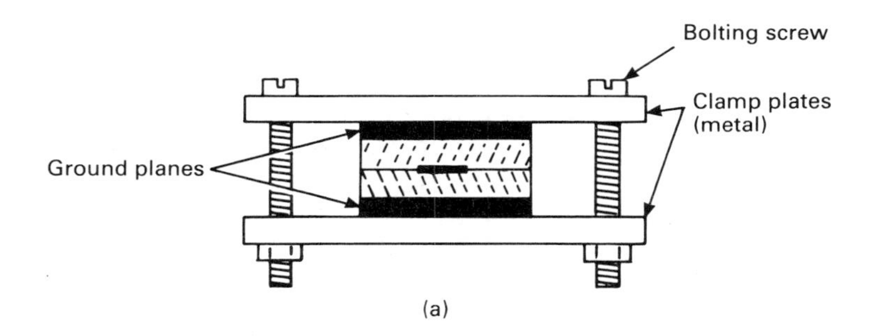

Fig. 3.2 — (a) Clamped stripline assembly.

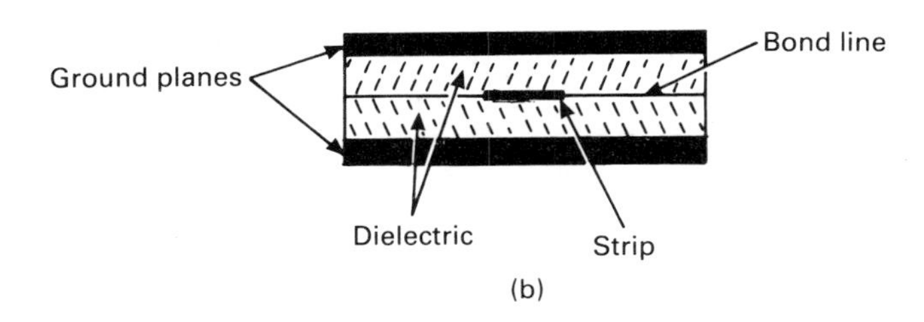

Fig. 3.2 — (b) bonded stripline assembly.

clamping plates and four or more clamping screws hold the stripline structure together. Although flexible when designing circuits, this method of assembly has inherent drawbacks. The performance of the circuit concerned will depend on the torque with which the screws have been tightened. This will affect line couplings and line impedances. In time and owing to temperature changes, stress relaxation is likely to take place with an accompanying change in circuit performance. Also, any clamping assembly adds weight and volume. An alternative to the above is a bonded assembly as shown in Fig. 3.2(b). This automatically results in a space and weight saving. There are basically three techniques available for bonding two substrates together and certain procedures need to be followed for optimum results. Once bonded, the layer between the two substrates, the so called 'bondline', is impervious to liquids and improves thermal conduction which is beneficial for higher power applications.

The above is only an outline of the properties which are desirable for a good microwave substrate and the factors which may affect the final circuit for which it is made. It is always best to obtain details from the individual substrate manufacturer with regard to technical specification, machinability, methods of bonding and handling.

CERAMIC SUBSTRATES

For very high quality microwave circuits or microwave integrated circuits (MICs) ceramic materials are used for the substrate. Some of the facts, properties and characteristics quoted here for substrates are also applicable to dielectric resonators (DRs) as discussed in Chapter 7. An example of such a substrate is known by its brand name 'Superstrate' and is manufactured by the Material Research Corporation (MRC) in America. The substrates consist of alumina of 99.6% purity and are produced by a taping process. The most important factors which determine substrate quality are thickness, size, surface quality, camber (flatness), grain size and density. A good substrate possesses these features. As a result fine line widths and high line densities can be obtained with ceramic substrates. They may also be coated with conductive metal layers such as chromium–copper or chromium–gold. This assists the bonding of components to the circuit pattern and reduces line attenuation. Another feature of ceramic substrates is its well controlled dielectric property owing to substrate flatness, and hence a uniform line impedance.

An example of a substrate specification is given in Table 3.1. The finished substrate has its two surfaces marked A-face and B-face. The A-face has the superior surface smoothness and it is on this side where the final circuit is made.

The ultimate microwave circuit or microwave integrated circuit (MIC) performance depends strongly on the perfection of the substrate. The latter is assessed with respect to surface defects. These are now briefly characterized, adopting the terminology used by MRC for its superstrates. The same approach could be used for dielectric resonators (DRs).

Green scratch. This is a fine scratch incurred by the handling of an unfired (also known as 'green') substrate. Scratch length is up to 7 mm and scratch depth to about 0.008 mm.

Blister. This is caused by small air bubbles in the green tape which expand during firing.

Colour spot. This is caused by an impurity particle in the green tape which fires with the substrate and thereby contaminates and colours the alumina. Such defects normally blister the substrate surface, and are frequently associated with a small fissure.

Table 3.1 — Substrate specification

Composition	99.6% aluminium oxide
ε_r	9.8 at 10 GHz, $t_{amb} = 25°C$
Surface finish	A-face 70–100 μm
	B-face 100–130 μm
Standard sizes	25.4 × 25.4 mm
	25.4 × 50.8 mm
	50.8 × 50.8 mm
Coating	
Chromium thickness	100–200 A sputtered
Gold thickness	Initial sputtered layer, balance electroplated
Gold purity	99.99%

Firing scratch. A scratch generated by a ceramic which lodges between the substrate and the surface on which it rests during firing. The shrinkage which occurs during firing causes this defect to generate a relatively deep (0.02 mm) scratch. These scratches are easily felt with the fingernail;green scratches are not.

Pit. A small depression in the surface.

Burr. A piece of excess substrate material which has become fired onto the substrate surface. These defects are flake-like in character, and are typically 0.013 mm in height and 0.13–0.25 mm in width or length. A scanning electronmicroscope shows burrs to be fine grained, high alumina particles.

Abrasion. This is a network of very fine scratches typically 0.005 mm in depth. It gives the appearance of a dull spot on the surface. An abrasion covers typically an area of 3 by 3 mm.

Table 3.2 lists some of the microwave substrate manufacturers and the most frequently required information. The thickness of the dielectric of the substrate is given in inches and in the brackets in millimetres.

COATING AND ETCHING

Having produced a suitable negative and selected the relevant substrate material the microwave circuit or subsystem must now be coated and etched. Different processes are required for etching standard copper laminates and ceramic substrates. For copper laminates the etching process follows well-known printed circuit techniques. The following is a typical sequence of the process involved.

(1) Choose laminate and cut to size.
(2) Clean laminate mechanically if necessary using very fine abrasive powder or metal polish.
(3) Clean laminate chemically using acetone or similar substance.
(4) Coat substrate with photoresist, either by spinning or dipcoating. This depends on the substrate size. Maximum size for spinning is typically 50.8 × 50.8 mm.

Table 3.2 — Names and properties of some commonly used microwave substrates

Brand name	ε_r	Thickness available	Manufacturer
RT/Duroid			Rogers Corporation, Microwave Materials Division, Box 3000 Chandler, AZ85224, USA
5870	2.33	0.025 (0.6350) 0.05 (1.2700) 0.1 (2.5400)	
5880	2.2	as 5870	
5500	2.5	0.015 (0.3810) 0.031 (0.7874) 0.062 (1.5748) 0.093 (2.3622) 0.125 (3.1750)	
6006	6.0	see 5870	
OAK		0.004 (0.1016) to 0.5 (12.7)	OAK Materials Group Inc., Laminates Division, Hoosick Falls, New York 12090, USA
OAK 601	2.54		
OAK 602	2.50		
OAK 605	2.20 2.33	0.01 (0.254) & 0.06 (1.524)	
Rexolite			
1422	2.54	0.031 (0.7874) to 0.25 (6.35)	
2200	2.64	0.031 (0.7874) to 0.25 (6.35)	
Polyguide			Electronized Chemicals Corp. Burlington, Mass., USA
	2.32	0.031 (0.7874) 0.062 (1.5748) 0.125 (3.1750)	
Epsilam			3M Company, Electronic Prod. Division, 3M Center, Saint Paul, Minnesota 55101, USA
	10.3	0.025 (0.635) 0.05 (1.270)	
Superstrate			Materials Research Corp., Orangeburg, New York 10962, USA
960	9.2	0.01 (0.254)	
996	9.8	to 0.05 (1.270)	
CuFlon	2.1	0.005 (0.127)	Polyflon Company, 35 River Street, New Rochelle, NY 10801, USA

(5) Dry photoresist.
(6) Place negative onto resist and expose to UV.
(7) Place substrate in photoresist developer. This will remove resist which has been exposed to UV and leaves the desired resist coated microwave pattern.
(8) Etch substrate in ferric chloride or similar copper etchant. This will remove exposed copper.
(9) Rinse in water.
(10) Place substrate in photoresist stripper. This will remove remaining photoresist on copper pattern.
(11) Rinse in running water for about 30 minutes to remove any etchant, otherwise after-etching can occur.
(12) Trim substrate if necessary and fit in suitable or purpose milled box, fit connectors and/or supply leads, test.

When the circuit has been etched there is a tendency for the copper to oxydize and it is advisable to solder connectors and components to the copper pattern as soon as possible. In case oxydation has taken place, those copper tracks which need soldering must be again mechanically cleaned. The substrate may of course need drilling in order to fit components such as transistors which need a connection to the ground plane.

It is possible to purchase ready-coated substrates. This eliminates the often messy process of resist coating. If a ready coated substrate is available then first remove the black plastic sheet which protects the laminate against daylight UV and mechanical damage. Then progress with step (6) above. More details on etching and coating may be obtained from the many books on printed circuit techniques.

Unless one has specialized facilities, alumina substrates are best purchased in coated form, i.e. with a layer of copper, chromium and gold. This combination of metals guarantees good adhesion to the substrate and to each other. Gold is not always necessary, but it helps component solderability and good electrical conduction. The etching process thus involves several stages owing to the different metals involved. Also, alumina substrates are usually not larger than 50.8×50.8 mm and hence require a spinner for photoresist deposition. A thin photoresist film gives better resolution than a thicker film. Film thickness depends mainly on spinner acceleration, spinning speed and lengths of the spinning process. Details are provided by the manufacturer of these spinners. A typical alumina preparation process is illustrated in Fig. 3.3.

The following is an example of etchants which may be made up in any chemistry department. Alternatively gold and chromium etches may be purchased.

Gold etch:
5 g of potassium iodine KI
0.5 g of iodine I_2
100 ml of deionized or distilled water
The solution may be heated to 70°C for faster etching.

Chromium etch:

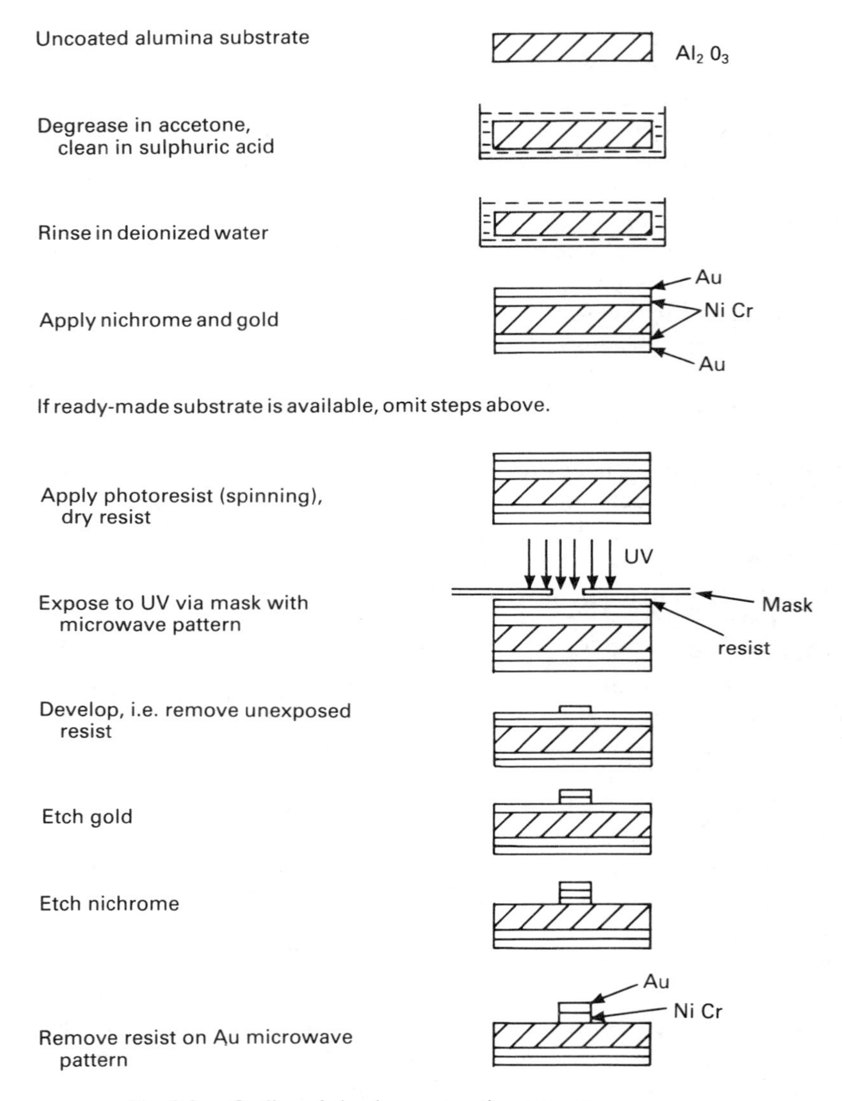

Fig. 3.3 — Outline of alumina preparation process.

Chromium etch consists of two solutions, part A and B, which are mixed just prior to use.
Solution A: 25 g of sodium hydroxide NaOH; 50 ml of deionized or distilled water.
Solution B: 10 g of potassium ferricyanide K_3 Fe (CN_6); 30 ml of deionized or distilled water.
This solution may also be heated to 70°C for faster action.

Once the substrate has been etched and trimmed it is ready to accept components. Some typical microwave components and means of attachment are discussed in the next sections.

MICROSTRIP COMPONENTS

Capacitors

In order to optimize RF and microwave circuits it is essential not only to choose the appropriate active device, but also the most appropriate passive component such as capacitors and inductors. This is frequently overlooked, or little relevance is attached to it. The choice of capacitor, for example, is very important.

Capacitors are used for coupling, tuning or matching purposes. The choice becomes more difficult as frequency and power levels increase. Capacitors for high frequency applications may be best characterized in terms of their dielectric loss or tan δ. The losses are due to the dielectric material of the capacitor. The higher the loss the larger the loss tangent, and the larger the equivalent resistance represented to the applied high frequency or microwave frequency. Larger losses result in larger heat generation within the capacitor which is clearly undesirable. Usually small chip capacitors are used in microwave circuits. Poor capacitor quality, especially if used in medium to high power applications, will result in overheating and circuit failure. If the same capacitor were, for example, used in a small signal amplifier then this would most likely result in poor matching or interstage coupling. Signal loss of several dB can result from bad capacitor selection. A capacitor with a loss tangent of 0.004 will generate four times as much heat as one with a loss tangent of 0.001. Ideally tan δ would be zero. In order to illustrate this point further, consider Table 3.3 which compares a glass epoxy substrate as used below GHz with a good microwave substrate such as alumina. Note that this is an illustrative example only. Hence, the higher the current the more dramatic the power dissipation.

Apart from heat dissipation, capacitor design and hence properties which depend

Table 3.3 — Capacitor power loss/dissipation at 500 MHz

Parameter	Glass epoxy	Alumina
$\tan\delta$	0.05	0.001
$Q = \dfrac{1}{\tan\delta}$	20	1000
C	10 pF	10 pF
$X_{500} = \dfrac{1}{2\pi f C}$	32 Ω	32 Ω
$R_{diss} = \dfrac{X_{500}}{Q}$	1.6 Ω	0.032 Ω
I	10 mA (1A)	10 mA (1A)
$P = I^2 R_{diss}$	0.16 mW (1.6W)	3.2 μW (32 mW)

on that design affect the useful upper frequency at which the capacitor may be used without having a detrimental effect on the circuit in which it is incorporated. Although chip capacitors are most frequently used (capacitors without leads) for microwave circuits, these capacitors have some inductance by virtue of their physical length. Thus, at a particular frequency, the capacitor will display self resonance. Furthermore, the larger the capacitor the lower the self resonance. Best performance is obtained at an operating frequency below self-resonance.

If the equivalent circuit of a capacitor is looked upon as a series L/C circuit, then it becomes predominantly inductive as the frequency increases, especially if it has a high quality factor Q, in other words, if the resistive losses are negligible. In certain circumstances the capacitor may thus be used as an inductive element. Use has been made of this in impedance matching networks. In actual fact a capacitor operated in this mode represents a low-cost high Q inductor. An additional advantage is the built-in dc block by virtue of its capacitance.

If we relate the previous discussion, for example, to microwave amplifier designs, then this may give a clue why sometimes the expected gain does not come up to expectation, perhaps up to several dB. Chip capacitors at low-cost are used in large quantities for thick film circuits. The properties of these capacitors are measured at low frequencies, typically 1 kHz. The information is of little use for microwave designs and only results in inferior circuit performance. The only way to circumvent this is to use capacitors specifically made and characterized for microwave frequencies. American Technical Ceramics (ATC) multilayer UHF/microwave porcelain and ceramic capacitors are most suitable for many applications.

A typical structure of an ATC self-encapsulating capacitor construction is shown

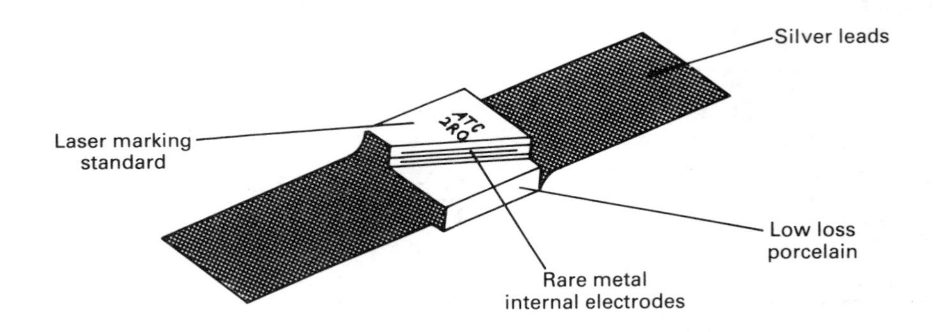

Fig. 3.4 — Low-loss interdigitated capacitor construction.

in Fig. 3.4 and an example of outline dimensions and termination styles is given in Table 3.4.

On occasions it may be desirable to have a limited adjustment or tuning facility on a microwave circuit. For this diverse surface mount precision trimmers of various styles are on the market, for example from Voltronics. They meet the demanding performance requirements of modern communication and microwave subsystems up to about 10 GHz. The choices of the dielectric are usually glass, air or sapphire. The quality factor at 100 MHz lies typically between 3000 and 6500.

Table 3.4 — Examples of ATC capacitor outline dimensions and termination styles

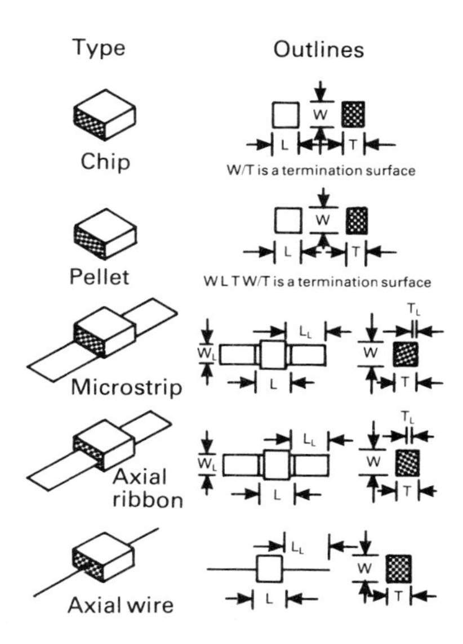

Another interesting range of capacitors for microwave and mm-wave applications is available from Dielectric Laboratories Inc. and some outline drawings are shown in Fig. 3.5.

Lumped components

Owing to the high frequencies involved in microwave circuits, inductors and capacitors can be made to form an integral part of the microwave circuit or system. There are of course limitations as to the physical values which can be usefully realized. For large values of capacitance and inductance discrete devices need to be used.

A straight section of wire or transmission line, especially if it is very narrow in width, will exhibit inductance. In the same way, a wide microstrip will exhibit capacitance. For a straight copper track the inductance is given approximately by

$$L\ [\mathrm{H}] = 5.08 \times 10^{-2}\, l \left\{ \ln \frac{l}{w+t} + 0.224\, \frac{w+t}{l} + 1.19 \right\}$$

$$L\ [\mathrm{mH}] = 50.8\, l \left\{ \ln \frac{l}{w+t} + 0.224\, \frac{w+t}{l} + 1.19 \right\} \tag{3.1}$$

Example

Let us assume we have a 100 Ohm transmission line on a substrate of thickness $h = 1.27$ mm, $\varepsilon_r = 2.4$ and $f = 10$ GHz. The thickness of the track is 0.035 mm. What is the inductance of this line?

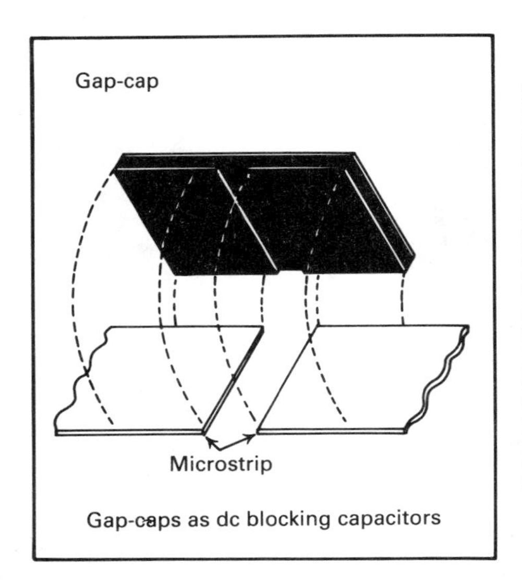

(a)

Parts	Designator	Description
	P	Gold over nickel barrier, Gold, 100μ in. standard thickness
	L	Single beam lead
	A	Axial beam lead
	R	Radial beam lead
	S	Standing axial beam lead (maintains tight capacitance tolerance)

(b)

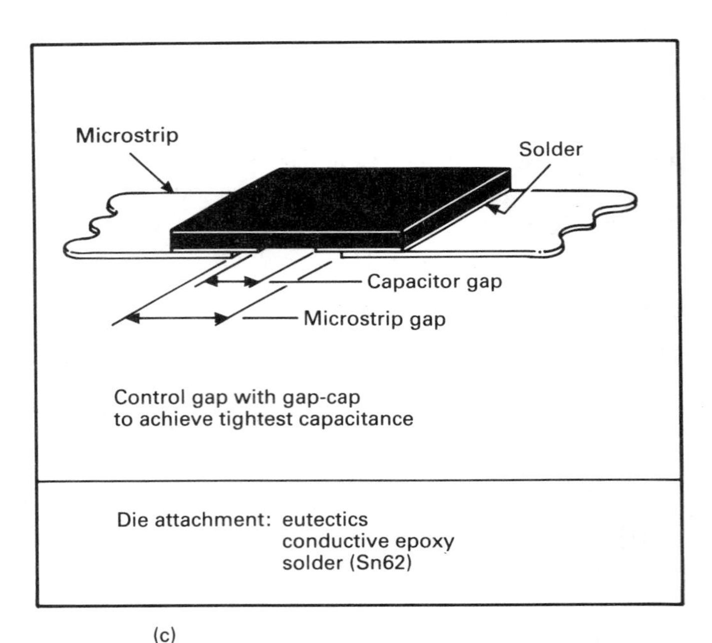

(c)

Fig. 3.5 — (a) Special version of a chip capacitor, from Dielectric Laboratories Inc. (b) Termination styles. (c) Mounting method.

Using the program POLSTRI in Chapter 2 we obtain for the microstrip width $w = 1.05$ mm. Substituting the above values into eqn. 3.1 we obtain an inductance value of 68 mH.

Inductors can be made more compact as shown in the examples of Fig. 3.6.

Referring to the specific case of microwave amplifier biasing circuits, it is shown

Fig. 3.6 — Examples of lumped inductors.

in Chapter 5 how an amplifier may be biased by using a $\lambda/4$ transmission line terminated at one end with a capacitor. The arrangement is shown here again in Fig. 3.7(a). The capacitor in Fig. 3.7(a) can be replaced by a wide microstrip as shown in Fig. 3.7(b). The capacitance is given by

$$C = \frac{\varepsilon_0 \varepsilon_r A}{h} \tag{3.2}$$

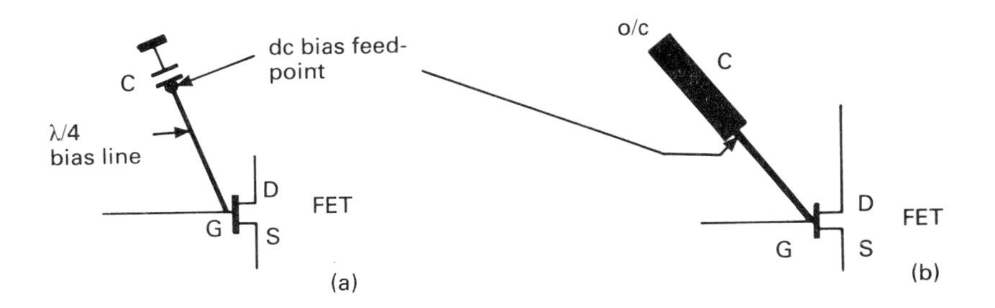

Fig. 3.7 — Biasing circuit using (a) discrete, (b) lumped capacitor for decoupling.

where h is the thickness of the substrate and A the area of the wide microstrip. Thus, any microwave frequency still reaching the d.c. feedpoint will be shorted to ground. This process can be further assisted by making the wide microstrip length $l = \lambda/4$. Seen from the feedpoint the wide microstrip is a transmission line terminated in an open circuit, i.e. $Z_r = \infty$. Hence, from the special case for transmission lines in Chapter 4, eqn. 4.71, we obtain

$$Z_s = \frac{-j\,Z_o}{\tan \beta l} = \frac{-j\,Z_o}{\tan\,(2\,\pi\,l/\lambda)} = -j\,Z_o/\tan 90° = j0\,\Omega$$

This acts as an additional microwave signal short circuit.

For in-line capacitors, as for example for signal coupling purposes, the capacitance can be considerably increased through the use of digitated structures, an example of which is given in Fig. 3.8. The capacitance of an interdigital capacitor depends on the finger length and spacing. Values of a few pF can be obtained.

It is obvious that it is difficult to manufacture pure reactances and any capacitor

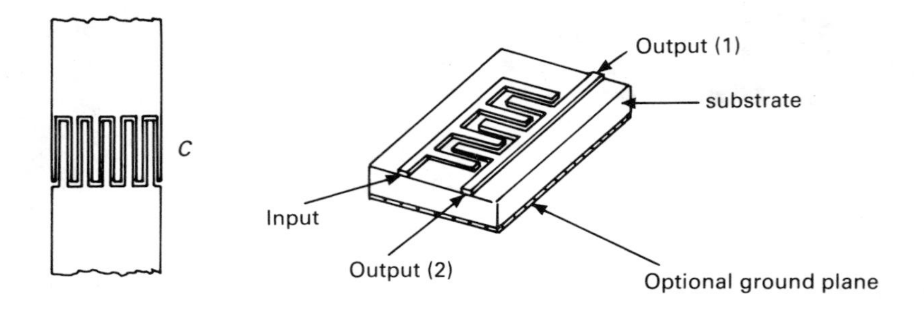

Fig. 3.8 — Lumped capacitors; either output can be used.

has an inductance associated with it and vice versa. Lumped circuits may thus be designed or capacitors and inductors may be combined to form lumped resonance circuits as shown in Fig. 3.9. Simple filter sections using this technique are shown in Fig. 3.10.

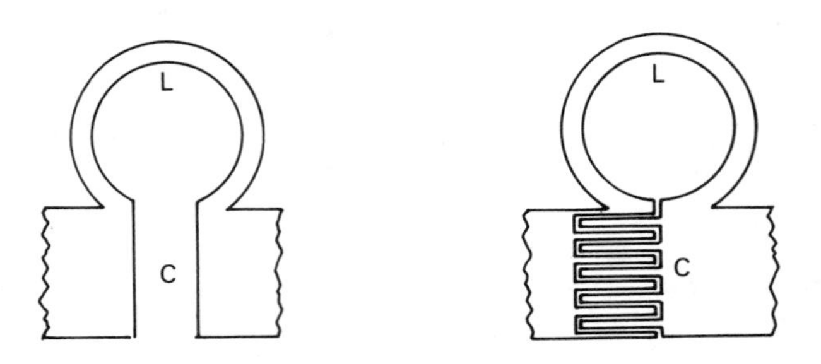

Fig. 3.9 — Lumped resonance circuits.

Transistors and diodes

Transistors and diodes are available in styles specially suited for microwave applications. The outline drawing of a typical packaged and unpackaged microwave transistor is given in Fig. 3.11. Many of these transistors are also available in unpackaged form in so called flip chip version. Sometimes they are referred to as LID or lead inverted devices.

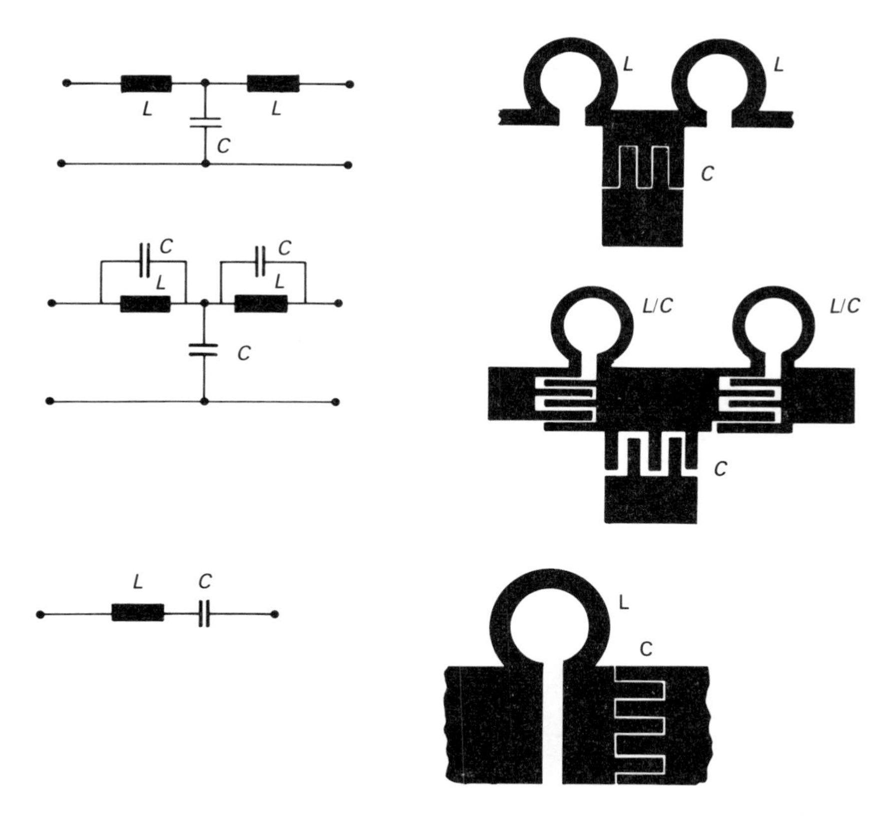

Fig. 3.10 — Application examples of L/C lumped elements.

Diodes also come in packaged and unpackaged form as pill, beam lead or LID structures as shown in Fig. 3.12. The most suitable style depends on the application and it is best to consult the manufacturer's data sheets.

ATTACHMENT TECHNIQUES

Owing to their small size, many microwave components require special attachment techniques. What is described here for capacitors applies in a similar way to other microwave components, be they active or passive. As a general rule every attachment problem should be solved on its own merit.

Soldering

A soldering operation requires that the parts to be soldered together must be free from any contamination. These contaminations stem most frequently from metal oxides and finger oils. Handling of the components should be minimized and gloves, finger cots or plastic tweezers should be used. Any metal oxides may be removed by

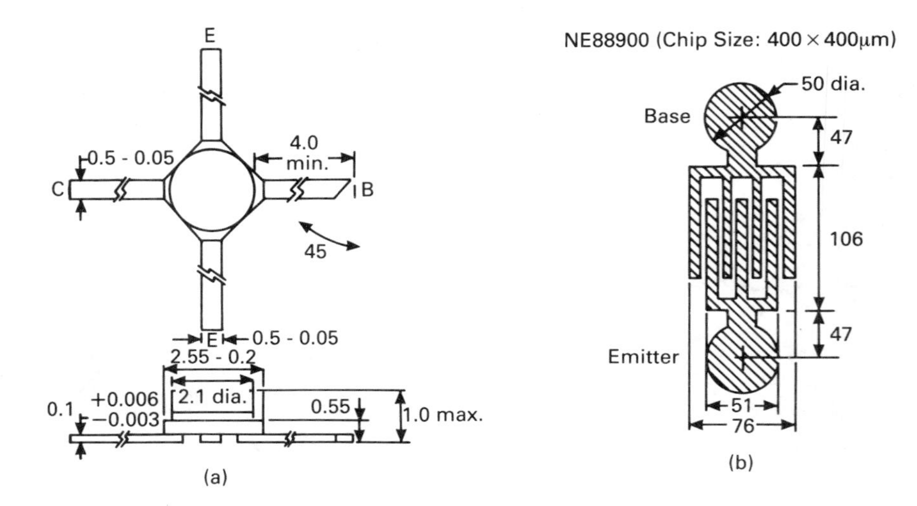

Fig. 3.11 — Microwave transistor (a) with leads and (b) in chip form.

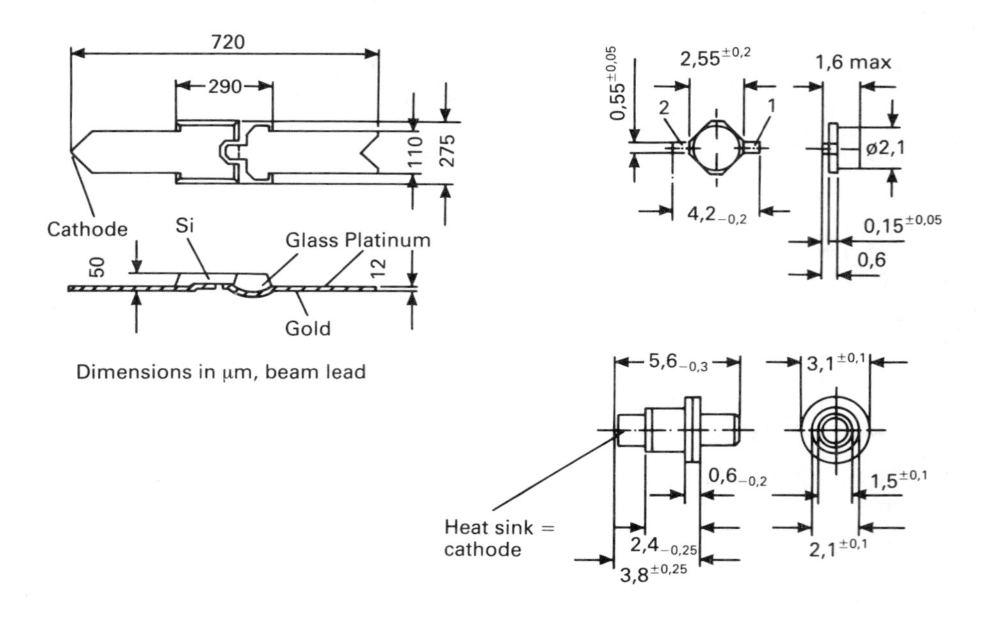

Fig. 3.12 — Microwave diode styles.

burnishing and finger oils by a suitable solvent such as a chlorinated hydrocarbon. Contamination of the parts to be joined together results in poor solder wetting and reflow operations.

For one-off laboratory assemblies a 10-W solder iron with a fine tip is suitable. It helps considerably if the components and microwave tracks are pre-tinned. The choice of solder composition and solder wire diameter is also important. The solder must contain a suitable flux. It is best to select the solder to the component manufacturers specification. Solders with a silver constituent, for example, should not be used if silver is contained in the component attachment pads since this will result in silver 'scavenging' and hence poor bonds.

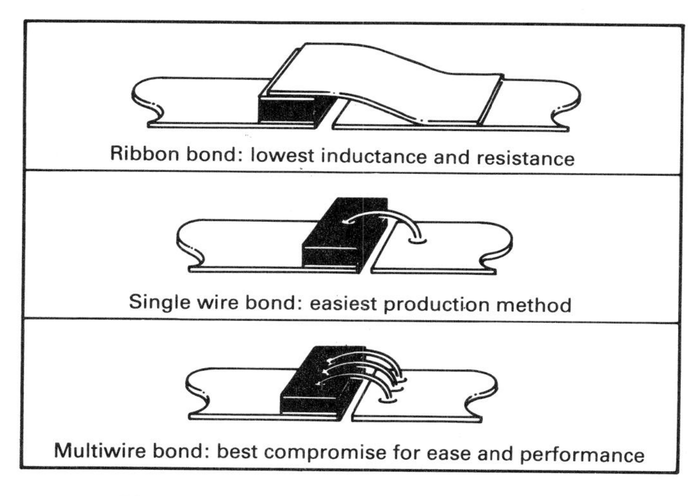

Fig. 3.13 — Die attachment and bonding methods.

Epoxy bonding

An alternative to soldering is epoxy bonding. Usually silver- or gold-based epoxies are used. Epoxy bonding requires more space between the component and microwave tracks or attachment pads since it will spread more easily. Gold terminations should be bonded with a gold-based epoxy and silver-based terminations with a silver-loaded epoxy. For best results the epoxy should be applied with a syringe although a fine needle can be used to transfer the epoxy from its small glass container to the bonding area. In the latter case care must be taken that not too much epoxy collects at the tip of the needle and hence creating a joint with a blob.

Wire bonding

The two most often used techniques are thermo compression and ultrasonic bonding. The bond relies entirely on the close atomic interface at the joint interface. This technique requires, of course, an appropriate bonding machine which is not cheap. To obtain a good bond the surfaces need to be clean both mechanically and chemically. Some bonding joints are illustrated in Fig. 3.13.

4

Transmission lines

INTRODUCTION

Transmission lines are used extensively at microwave frequencies both as low loss circuit interconnections and as circuit elements in microwave components. The most common transmission line used for signal transmission is the coaxial cable which is screened and free from interference and radiation effects (Fig. 4.1). Transmission

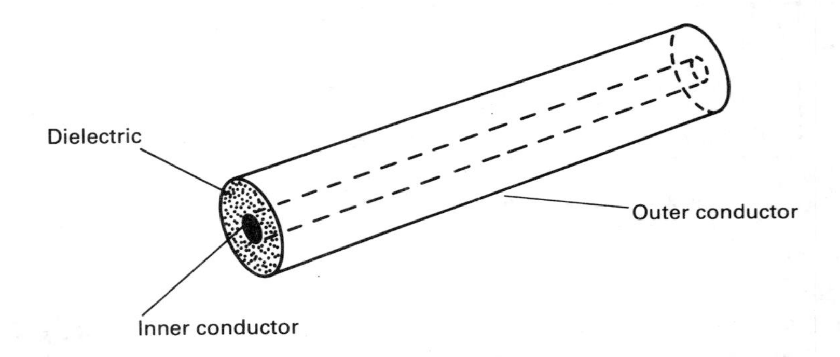

Fig. 4.1 — Coaxial cable.

lines are used as circuit elements in microwave integrated circuit (MIC) realization of amplifiers, filters, oscillators and couplers (to name but a few components). In this application transmission lines replace lumped inductors, capacitors and transformers with simple printed patterns on MIC substrates. These transmission line elements perform better than their lumped equivalent (Fig. 4.2). In order to use transmission lines correctly it is necessary to have a quantitative understanding of their electrical behaviour; this is the basis of transmission line theory and also of this chapter.

It is assumed in the following analysis that we restrict ourselves to the class of uniform transmission lines. That is, lines whose dimensions and hence electrical behaviour are identical at all planes normal to the direction of signal propagation

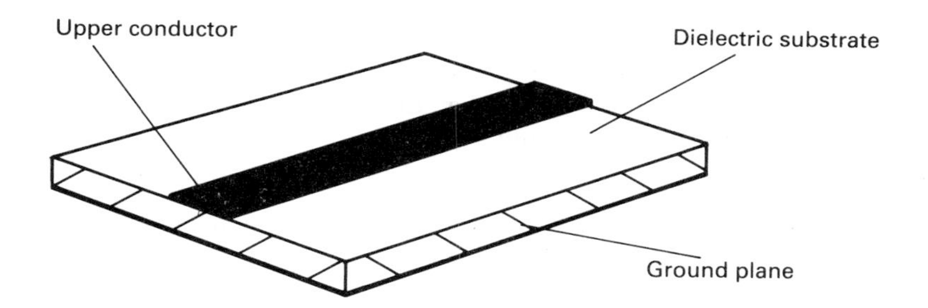

Fig. 4.2 — M.I.C. transmission line.

along the line. Although non-uniform lines are used, their analysis is more complex and beyond the scope of this book.

Transmission lines are distributed circuits, that is they possess series inductance and resistance, and shunt capacitance and conductance which is distributed uniformly along the length of the line. This is different to lumped circuits where the inductance or capacitance is located at a single point. Distributed circuits are more complicated than lumped circuits because the voltage and current measured at any point along the line is a function of the distance along the line [1,2].

DERIVATION OF TRANSMISSION LINE EQUATIONS

The transmission line can be analysed by considering it as a cascade of elemental lumped sections of length δx (see Fig. 4.3).

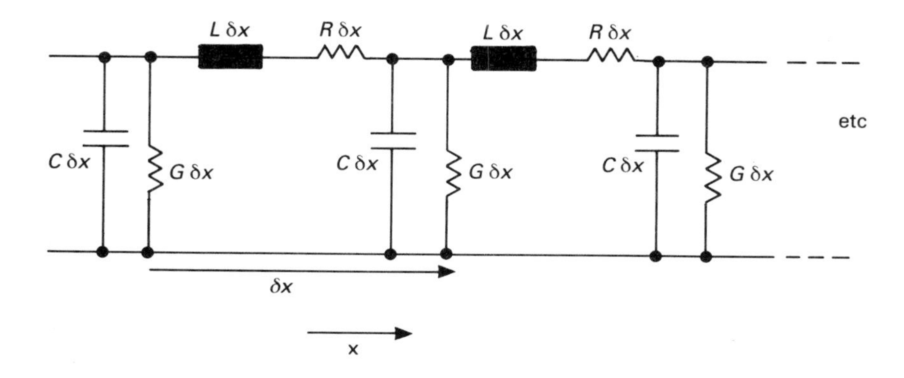

Fig. 4.3 — Transmission line equivalent circuit based upon a cascade of elemental sections.

R = series resistance per unit length (conductor loss)
L = series inductance per unit length
G = shunt conductance per unit length (dielectric loss)
C = shunt capacitance per unit length

We then solve Kirchhoff's laws along a basic section of line of length δx (see Fig. 4.4).

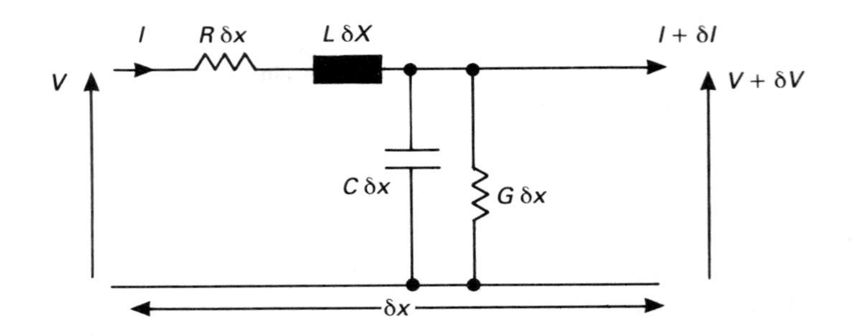

Fig. 4.4 — A single elemental section of line.

Provided δx is short with respect to the wavelength of the applied signal then the analysis will be reasonably accurate. An exact analysis is then obtained by letting δx tend to zero and obtaining differential equations for the line.

Applying a voltage V at the input of the section produces a current I. The voltage at δx along the line will then change by δV, and the current will change by δI. Kirchhoff's laws yield the following equations:

$$V = R \times \delta x \times I + L.\delta x.\frac{\mathrm{d}I}{\mathrm{d}t} + V + \delta V \tag{4.1}$$

$$I = G \times \delta x \times (V + \delta V) + C\delta x.\frac{\mathrm{d}}{\mathrm{d}t}(V + \delta V) + I + \delta I \tag{4.2}$$

Rearranging, we obtain

$$-\frac{\delta V}{\delta x} = RI + \frac{L\mathrm{d}I}{\mathrm{d}t} \tag{4.3}$$

$$-\frac{\delta I}{\delta x} = GV + \frac{C\mathrm{d}V}{\mathrm{d}t} \tag{4.4}$$

Letting δx tend to zero, we obtain the following differential equations:

$$-\frac{dV}{dx} = RI + \frac{LdI}{dt} \tag{4.5}$$

$$-\frac{dI}{dx} = GV + \frac{CdV}{dt} \tag{4.6}$$

Now differentiating eqn. (4.5) with respect to x and eqn. (4.6) with respect to t, we obtain:

$$-\frac{d^2V}{dx^2} = \frac{RdI}{dx} + \frac{Ld^2I}{dxdt} \tag{4.7}$$

$$-\frac{d^2I}{dxdt} = \frac{GdV}{dt} + \frac{Cd^2V}{dt^2} \tag{4.8}$$

Then substituting eqns (4.5) and (4.8) into eqn. (4.7) to eliminate derivatives of I, we obtain:

$$-\frac{d^2V}{dx^2} = -RGV - \frac{RCdV}{dt} - \frac{GLdV}{dt} - \frac{LCdV}{dt^2} \tag{4.9}$$

or

$$\frac{d^2V}{dx^2} = LC\frac{d^2V}{dt^2} + (RC + GL)\frac{dv}{dt} + RGV \tag{4.10}$$

Similarly, we obtain

$$\frac{d^2I}{dx^2} = LC\frac{d^2I}{dt^2} + (RC + GL)\frac{dI}{dt} + RGI \tag{4.11}$$

These are second order coupled partial differential equations and are recognizable as wave equations. The solution of these equations should enable us to find the voltage and current at any point x along the line at any time t.

These equations are now simplified by letting R and G equal zero, corresponding to a lossless line. This is a valid technique as in practice we tend to use lines which are nearly lossless.

Thus we obtain:

$$\frac{d^2V}{dx^2} = LC\frac{d^2V}{dt^2} \tag{4.12}$$

$$\frac{d^2I}{dx^2} = LC\frac{d^2I}{dt^2} \tag{4.13}$$

A general solution of these equations for V or I is:

$$V(x,t) = f(t \pm \sqrt{LC}x) \tag{4.14}$$

This solution implies a wave travelling along the line in the plus or minus x directions with a velocity v where

$$V = \frac{1}{\sqrt{LC}} \tag{4.15}$$

f is a function which depends upon the applied signal, for example if we apply a sinusoidal voltage to the line then f will be sinusoidal. That eqn. (4.13) is indeed a solution of eqn. (4.12) can be shown by substituting the solution back into the equations to see if they equate. This is left to the reader.

Next we must find the relationship between the voltage and current at any point along the transmission line. As we shall see, the relationship is a simple one.

We apply a voltage V at the input of the line to generate a current I where:

$$V = f_1(u) \tag{4.16}$$

$$I = f_2(u) \tag{4.17}$$

and

$$u = t - \sqrt{LC}x \tag{4.18}$$

Now

$$\frac{dV}{dx} = \frac{df_1}{du} \times \frac{du}{dx} = -\sqrt{LC}\frac{df_1}{dU} \tag{4.19}$$

and

$$\frac{dI}{dt} = \frac{df_2}{du} \times \frac{du}{dt} = \frac{df_2}{du} \tag{4.20}$$

Substituting eqn. (4.5) (with $R = 0$) into eqn. (4.20)

$$-\frac{dV}{dx} = L\frac{dI}{dt} = L\frac{df_2}{du} \tag{4.21}$$

and from eqns (4.19) and (4.21)

$$\sqrt{LC}\frac{\mathrm{d}f_1}{\mathrm{d}u} = L\frac{\mathrm{d}f_2}{\mathrm{d}u} \tag{4.22}$$

Rearranging and integrating

$$\frac{f_1}{f_2} = \sqrt{\frac{L}{C}} = \frac{V}{I} = Z_0 \tag{4.23}$$

Z_0 is the ratio of voltage to current along the line and is a constant, simply determined by the inductance and capacitance and hence the dimensions of the line.

SINUSOIDAL EXCITATION

Most microwave signals are sinusoids so it is instructive to analyse the transmission line with sinusoidal excitation. Remembering the use of complex notation for sinusoidal signals we can write:

$$-\frac{\mathrm{d}V}{\mathrm{d}x} = (R + j\omega L)I = ZI \tag{4.24}$$

$$-\frac{\mathrm{d}L}{\mathrm{d}x} = (G + j\omega C)V = YV \tag{4.25}$$

Repeating the earlier procedure, we obtain the partial differential equation

$$\frac{\mathrm{d}^2V}{\mathrm{d}x^2} = ZYV \tag{4.26}$$

with a solution

$$V = V_o^+ e^{-\gamma x} + V_o^- e^{-\gamma x} \tag{4.27}$$

where

$$\gamma = \sqrt{ZY} = \sqrt{(R + j\omega L)(G + j\omega l)} \tag{4.28}$$

γ is the propagation constant of the transmission line. If the line is lossy either R or G are finite and γ will be complex, thus the sinusoidal signal will be attenuated as it travels along the line.

In general

$$\gamma = \sqrt{(R + j\omega l)(G + j\omega l)} = \alpha + j\beta \tag{4.29}$$

α is the attenuation constant of the line measured in Nepers/m and β is the phase constant.

For a lossless line, $R = G = 0$ and

$$\beta = \pm j\omega\sqrt{LC} \tag{4.30}$$

or

$$\beta = \pm \frac{j\omega}{\upsilon} \tag{4.31}$$

and

$$\upsilon = \frac{1}{\sqrt{LC}} \tag{4.32}$$

where υ is the velocity of signal propagation along the line.

Now, since $\upsilon = f\lambda$ and $\omega = 2\pi f$ then

$$\beta = \frac{2\pi}{\lambda} \tag{4.33}$$

And for sinusoidal excitation with angular frequency ω, we can write

$$V(x,t) = Ve^{-\alpha x} \times e^{j(\omega t - \beta x)} \tag{4.34}$$

and

$$I(x,t) = \frac{V(x,t)}{Z_o} \tag{4.35}$$

TWO PORT PARAMETER REPRESENTATION OF A LINE

It is useful to be able to represent a transmission line in terms of its two port parameters. In Fig. 4.5 we represent a line as a two port network where V_1, I_1 are the voltage and current at the input of the network and V_2, I_2 are those at its output.

The voltage at point x along the line is then given by

$$V(x) = Ae^{-\gamma x} + Be^{\gamma x} \tag{4.36}$$

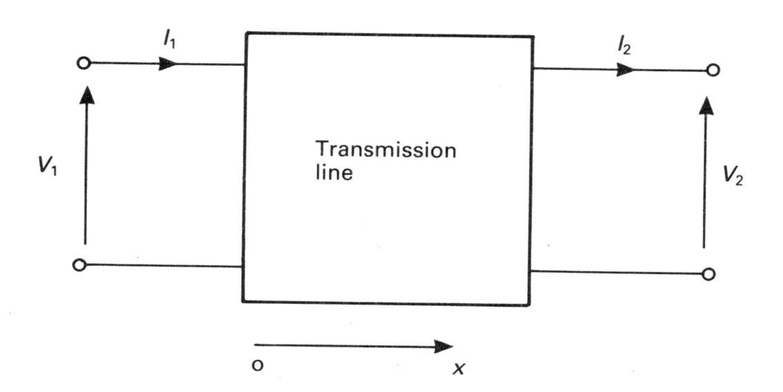

Fig. 4.5 — Two port representation of a line.

and the current at point x is

$$I(x) = \frac{Ae^{-\gamma x}}{Z_o} - \frac{Be^{\gamma x}}{Z_o} \tag{4.37}$$

Now let

$$V_1 = A + B \tag{4.38}$$

Then

$$I_1 = \frac{A}{Z_o} - \frac{B}{Z_o} \tag{4.39}$$

or

$$I_1 Z_o = A - B$$

So

$$A = \frac{V_1 + I_1 Z_o}{2} \tag{4.40}$$

and

$$B = \frac{V_1 - I_1 Z_o}{2} \tag{4.41}$$

Now substituting eqns (4.40) and (4.41) into eqns (4.36) and (4.37) we obtain

$$V(x) = e^{-\gamma x}\frac{[V_1 + I_1 Z_o]}{2} + e^{\gamma x}\frac{[V_1 - I_1 Z_o]}{2} \tag{4.42}$$

and

$$I(x) = e^{-\gamma x}\frac{[V_1 + I_1 Z_o]}{2Z_o} - e^{\gamma x}\frac{[V_1 + I_1 Z_o]}{2Z_o} \tag{4.43}$$

or

$$V(x) = V_1 \cosh(\gamma x) - I_1 Z_o \sinh(\gamma x) \tag{4.44}$$

$$I(x) = I_1 \cosh(\gamma x) - \frac{V_1}{Z_o}\sinh(\gamma x) \tag{4.45}$$

Thus in matrix form the relationship of V_2 and I_2 at the output of the line to V_1 and I_1 is given by:

$$\begin{bmatrix} V_2 \\ I_2 \end{bmatrix} = \begin{bmatrix} \cosh(\gamma x) & -Z_o \sinh(\gamma x) \\ \dfrac{-\sinh(\gamma x)}{Z_o} & \cosh(\gamma x) \end{bmatrix} \begin{bmatrix} V_1 \\ I_2 \end{bmatrix} \tag{4.46}$$

Inverting (4.46), we obtain the familiar transfer [ABCD] matrix equation for the line

$$\begin{bmatrix} V_1 \\ I_2 \end{bmatrix} = \begin{bmatrix} \cosh(\gamma x) & Z_o \sinh(\gamma x) \\ \dfrac{\sinh(\gamma x)}{Z_o} & \cosh(\gamma x) \end{bmatrix} \begin{bmatrix} V_2 \\ I_2 \end{bmatrix} \tag{4.47}$$

For a lossless line $\gamma = j\beta$ and we obtain

$$\begin{bmatrix} V_2 \\ I_2 \end{bmatrix} = \begin{bmatrix} \cos(\beta x) & jZ_o \sin(\beta x) \\ \dfrac{j\sin(\beta x)}{Z_o} & \cos(\beta x) \end{bmatrix} \begin{bmatrix} V_1 \\ I_2 \end{bmatrix} \tag{4.48}$$

where

$$\beta x = \frac{\omega x}{v} \tag{4.49}$$

let

$$a = \frac{\omega}{v} \tag{4.50}$$

then

$$\beta x = a\omega \tag{4.51}$$

and

$$\begin{bmatrix} V_1 \\ I_1 \end{bmatrix} = \begin{bmatrix} \cos(a\omega) & jZ_o \sin(a\omega) \\ \dfrac{j\sin(a\omega)}{Z_o} & \cos(a\omega) \end{bmatrix} \begin{bmatrix} V_2 \\ I_2 \end{bmatrix} \tag{4.52}$$

Eqn. (4.52) is a useful result as it expresses the voltages and currents as a function of frequences.

TERMINATED TRANSMISSION LINES

Transmission lines used as circuit elements normally have some form of termination on the end of the line. First we shall consider a termination in an arbitrary load Z_L (also often referred to as terminating impedance Z_t) as shown in Fig. 4.6(a).

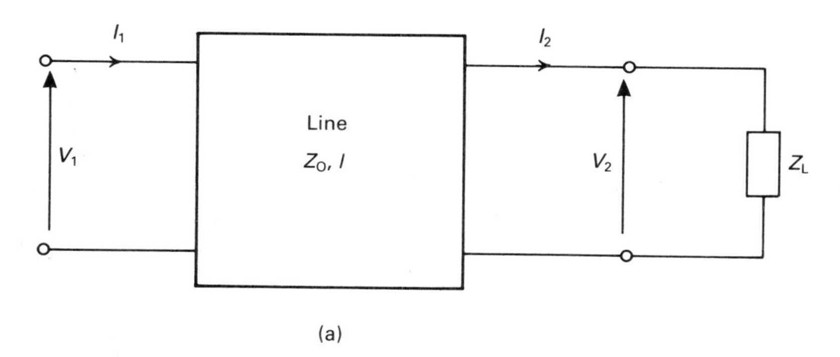

Fig. 4.6(a) — A line with arbitrary termination.

From eqn. (4.52) the input impedance (also often referred to as sending end impedance Z_s) of a line is given by

$$Z_{in} = \frac{V_1}{I_1} = \frac{\cos(a\omega)V_2 + jZ_o \sin(a\omega)I_2}{j\dfrac{\sin(a\omega)}{Z_o} + \cos(a\omega)I_2} \tag{4.53}$$

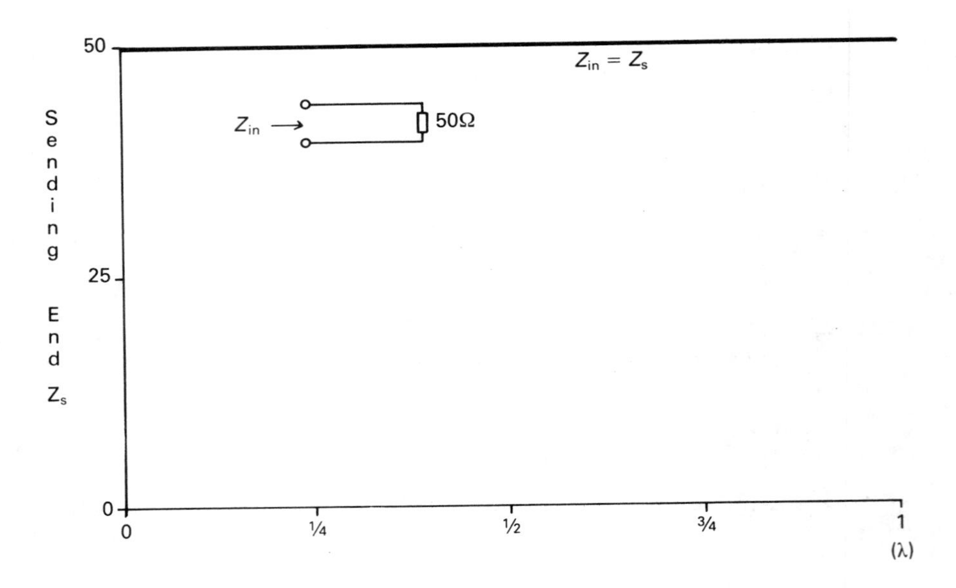

Fig. 4.6(b) — Input impedance of a 50 Ω matched transmission line. Input: $Z_o = 50\,\Omega$, $f = 1$ Hz, $\alpha = 0$ dB/cm, $\varepsilon_r = 1$, $l = 7.5$ cm, $R_L = 50\,\Omega$, $X_L = 0\,\Omega$.

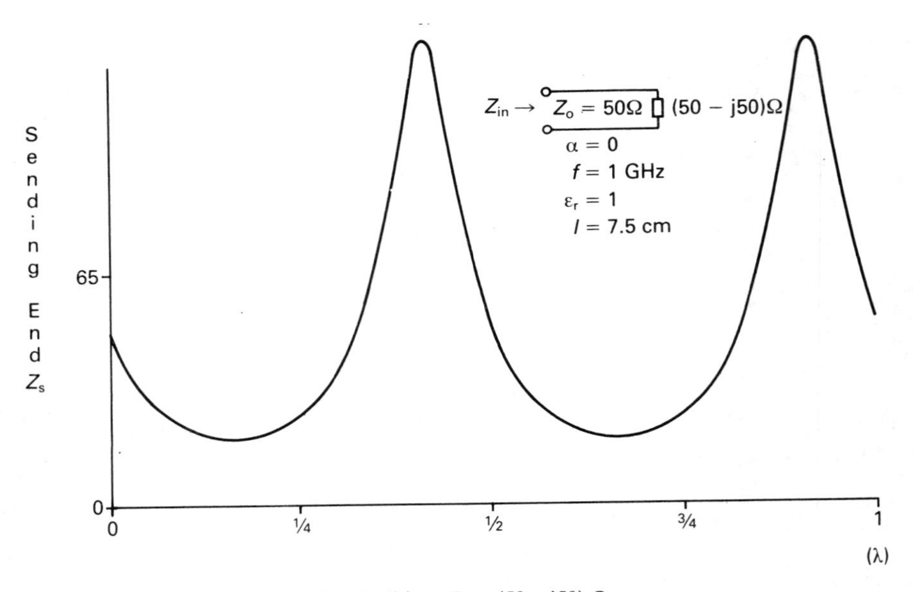

Fig. 4.6(c) — $Z_L = (50 - j50)\,\Omega$.

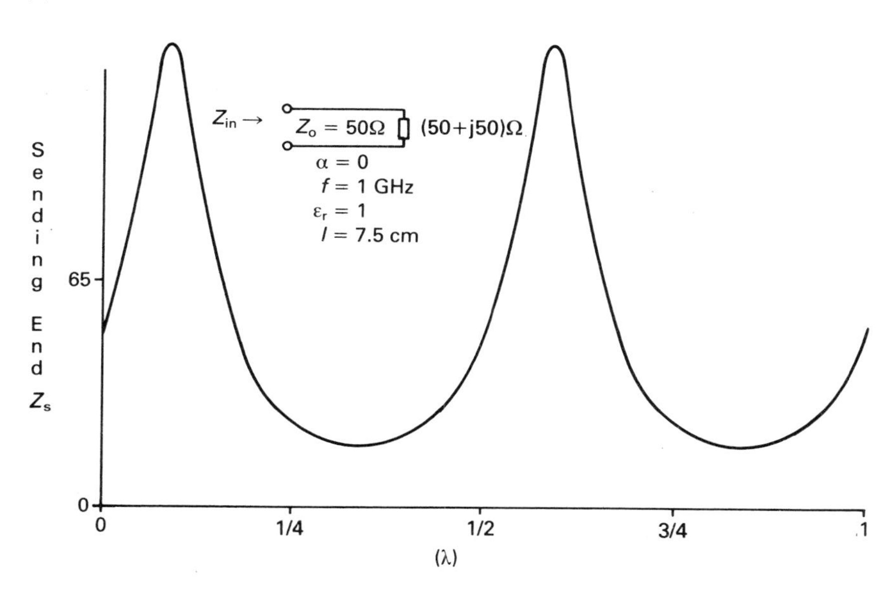

Fig. 4.6(d) — $Z_L = (50 + j50)\Omega$.

Dividing top and bottom of the right-hand side of eqn. (4.53) by I_2 and recognizing that

$$\frac{V_2}{I_2} = Z_L \tag{4.54}$$

We obtain

$$Z_{in} = \frac{Z_L \cos(a\omega) + jZ_o \sin(a\omega)}{\cos(a\omega) + \dfrac{jZ_L}{Z_o} \sin(a\omega)} \tag{4.55}$$

Eqn. (4.55) is a very useful general result.

An important class of termination is when the termination impedance is equal to the characteristic impedance of the line. In this case when $Z_L = Z_o$ from eqn. (4.55) we obtain

$$Z_{in} = Z_o = Z_L \tag{4.56}$$

This is known as the matched condition, since it states that the input impedance of lossless line terminated in a resistive load of value Z_o is purely resistive and of value

Z_o. Under these conditions there will be no reflected signal from the line and maximum energy will be transferred to the load; Fig. 4.6.(b).

If a line is terminated in a short circuit then obviously $Z_L = 0$ and from eqn. (4.49)

$$Z_{in} = jZ_o \tan(a\omega) \tag{4.57}$$

If a line is termianted in an open circuit then obviously $Z_L = \infty$ and from eqn. (4.55)

$$Z_{in} = -jZ_o \tan(a\omega)\ . \tag{4.58}$$

THE SMITH CHART

The termination line equations can be used to solve a variety of problems associated with RF and microwave circuits. A quick way to arrive at the solution of a particular problem is to use a graphical method, namely the Smith chart, which consists of two sets of circles, a concentric set and an orthogonal set as shown in Fig. 4.7.

The reflection coefficient Γ as shown in Fig. 4.7(a) is represented by the radius of the concentric circles and varies from zero to unity. Its associated phase angle φ varies from 0 to + 180° and 0 to − 180°. The concentric circles can also be used to

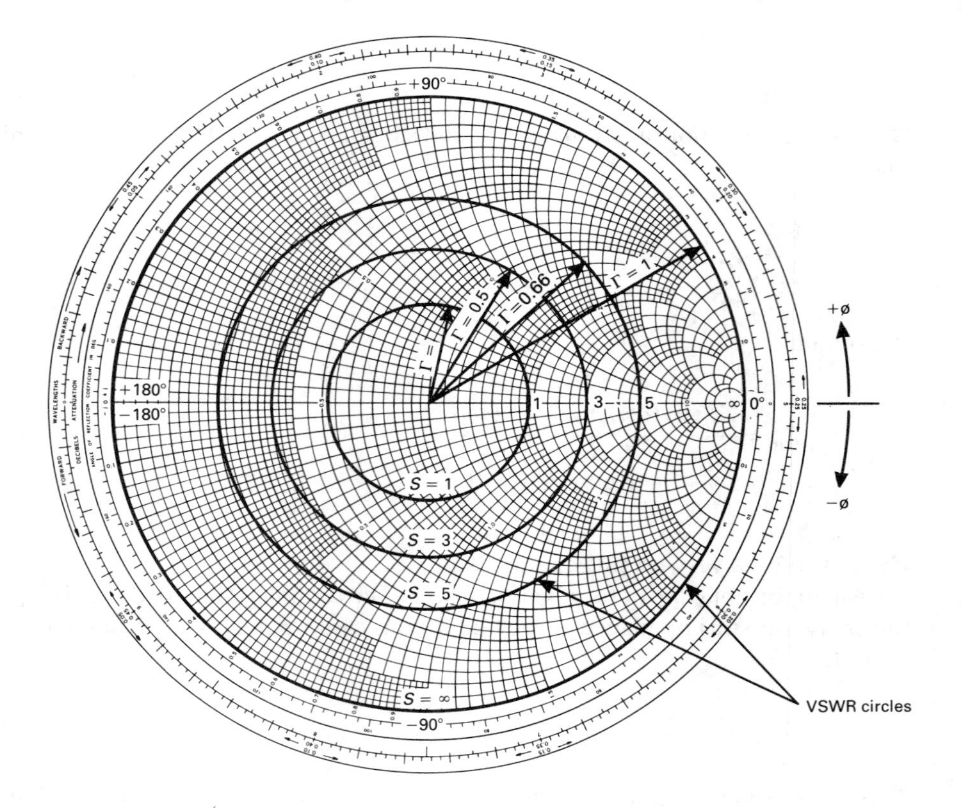

Fig. 4.7(a) — VSWR circles and Γ on a Smith chart.

read off the voltage standing wave ratio and therefore they are also referred to as the VSWR circles. Referring to Fig. 4.7(b), we have a set of circles which are orthogonal to each other and which represent the resistive and reactive components of an impedance. The inductive reactance lies in the top half and the capacitance reactance

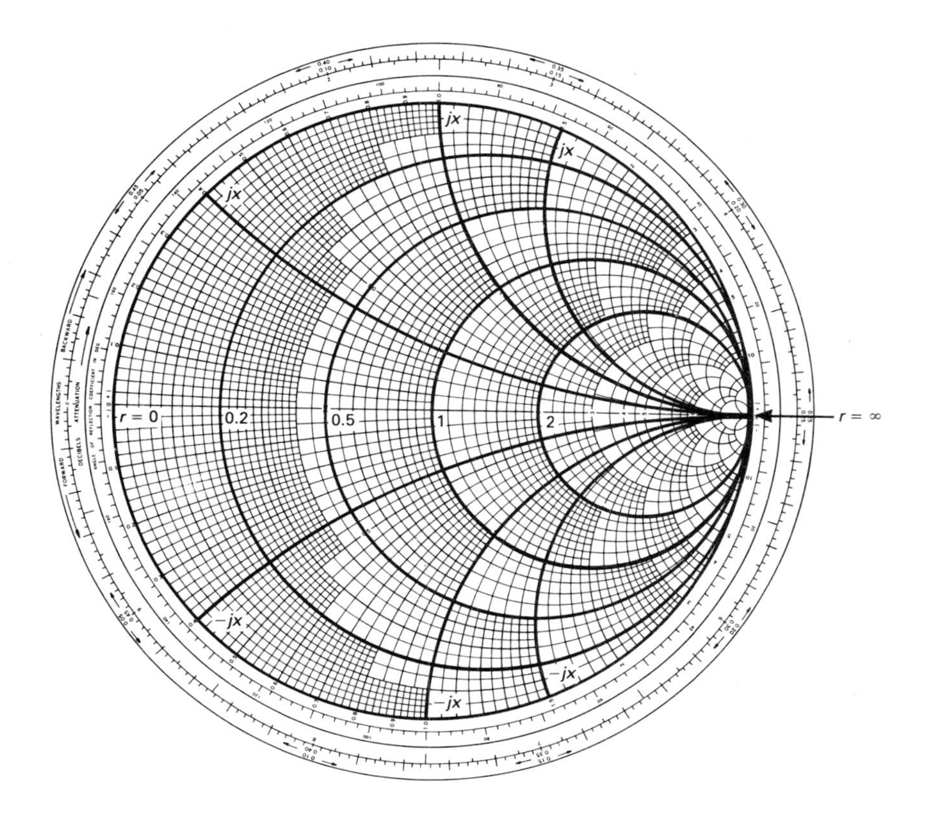

Fig. 4.7(b) — Resistance and reactance circles of a Smith chart.

in the lower half of Fig 4.7(b). The circles are normalized with respect to the characteristic impedance of the transmission line system, which is normally 50 Ω. The direction of signal flow is shown on the circle perimeter as BACKWARD (or towards the GENERATOR) and FORWARD (or towards the LOAD). Different chart producers use slightly different notations. In the following the equations for the construction of a Smith chart are derived. These equations can also be used for programming a computer in order to draw a Smith chart.

The normalized resistance and reactance are given by

$$r = \frac{R}{Z_o} \quad \text{and} \quad x = \frac{X}{Z_o} \tag{4.59}$$

where R and X represent the physical values of the impedance. Since R and X can lie within the range of zero to infinity, r and x lie within the same range. The intersection of a resistance circle with a reactance circle represents a normalized impedance point. Impedance points in the upper half of Fig. 4.7 represent an inductive impedance and in the lower half a capacitance impedance.

Thus

$$z = \frac{Z}{Z_o} = r \pm jx \qquad \text{and} \qquad y = \frac{Y}{Y_o} = g \pm jb \tag{4.60}$$

For an arbitrary load impedance Z_L and VSWR S the normalized impedance z may be expressed as

$$z = \frac{Z_L}{Z_o} = S = \frac{1+\Gamma}{1-\Gamma} = r + jx \tag{4.61}$$

For a complex reflection coefficient $\Gamma = a + jb$ we obtain

$$z = r + jx = \frac{1+a+jb}{1-a-jb} \tag{4.62}$$

Multiplication by the conjugate complex gives

$$r = \frac{1-a^2-b^2}{(1-a)^2+b^2} \tag{4.63}$$

and

$$x = \frac{2b}{(1-a)^2+b^2} \tag{4.64}$$

These equations result in circles if one can complete the squares. Hence

$$\left[a - \frac{r}{r+1}\right]^2 + b^2 = \frac{1}{(r+1)^2} \tag{4.65}$$

and

$$(a-1)^2 + \left[b - \frac{1}{x}\right]^2 = \frac{1}{x^2} \tag{4.66}$$

The above two equations result in two sets of circles as shown in Fig. 4.8 and if, merged along the appropriate axis as shown, will result in the Smith chart. Only a quarter of the reactance circles is made use of.

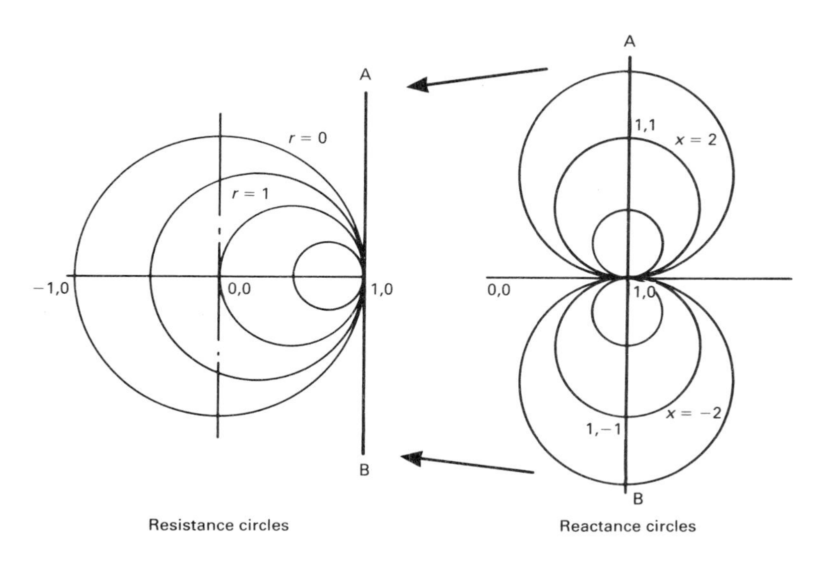

Fig. 4.8 — Evolution of Smith chart by merging lines marked AB. The circle coordinates are obtained from eqns (4.65) and (4.66).

The transformation equations for the Smith chart as given by eqns (4.65) and (4.66) can also be used to generate a Smith chart on the computer. At the same time the computer can be used to display and calculate any transmission line behaviour on that chart.

Using the computer program S-CHART, the Smith chart as shown in Fig. 4.9 can be generated. The operator has complete control over the number of normalized

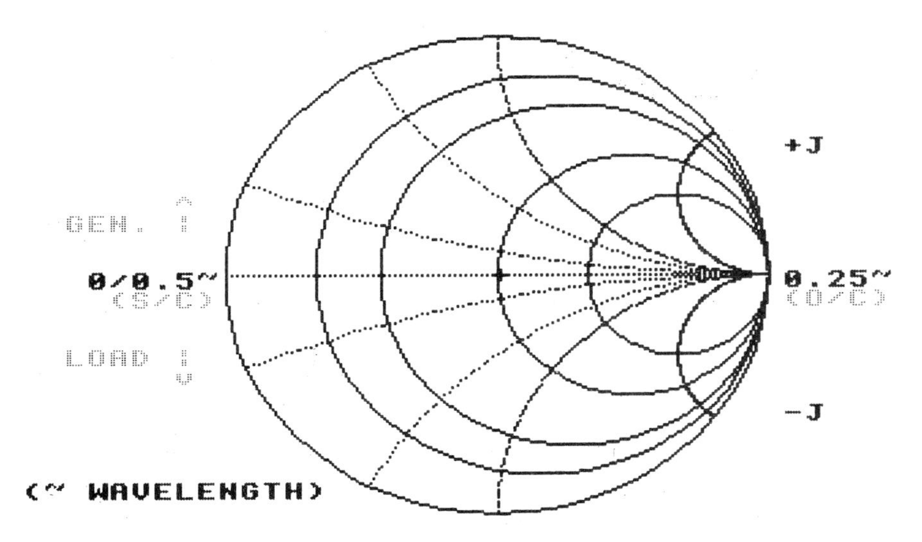

Fig. 4.9 — Generation of a Smith chart with program SMITH.

resistance and reactance circles to be drawn. The following parameters may be displayed with the S-CHART program:

(a) resistance points;
(b) reactance points;
(c) impedance points;
(d) VSWR circles.

The Smith chart in Fig. 4.10 shows an example of a resistance point $r = 2 + j0 = 2$, a reactance point $x = 0 + j1 = j1$, an impedance point $z = 2 - j1$ and two VSWR circles for 2 and 5 entered on the Smith chart. The program S-CHART is easy to use, especially if one follows the prompts carefully.

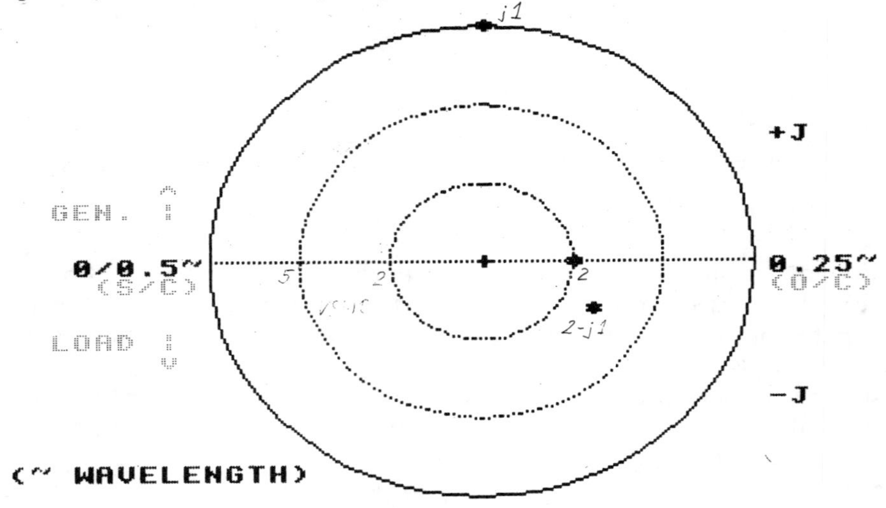

Fig. 4.10 — Smith chart with values entered.

On the next few pages a number of examples are given for familiarization with the Smith chart. It is left to the reader to implement them with the computer.

Example 1
Using the Smith chart, plot the following assuming a 50 Ω transmission line system:

(a) VSWR circle of 5;
(b) resistance of 100 Ω;
(c) reactance of 100 Ω;
(d) impedance of $(200 + j250)\,\Omega$;
(e) the admittance of (d);
(f) the reflection coefficient of (d);
(g) verify the admittance by calculation;
(h) a normalized impedance of $(3 - j2)\,\Omega$.

Solution

Referring to Fig. 4.11, the VSWR of any point on the Smith chart is found by drawing a circle, through that point using the normalized resistance point $r = 1.0$ as the centre. The value of the VSWR is marked as 'a' along the normalized resistance line, i.e. it starts as 1.0 in the centre of the chart and it stops on the right as infinity.

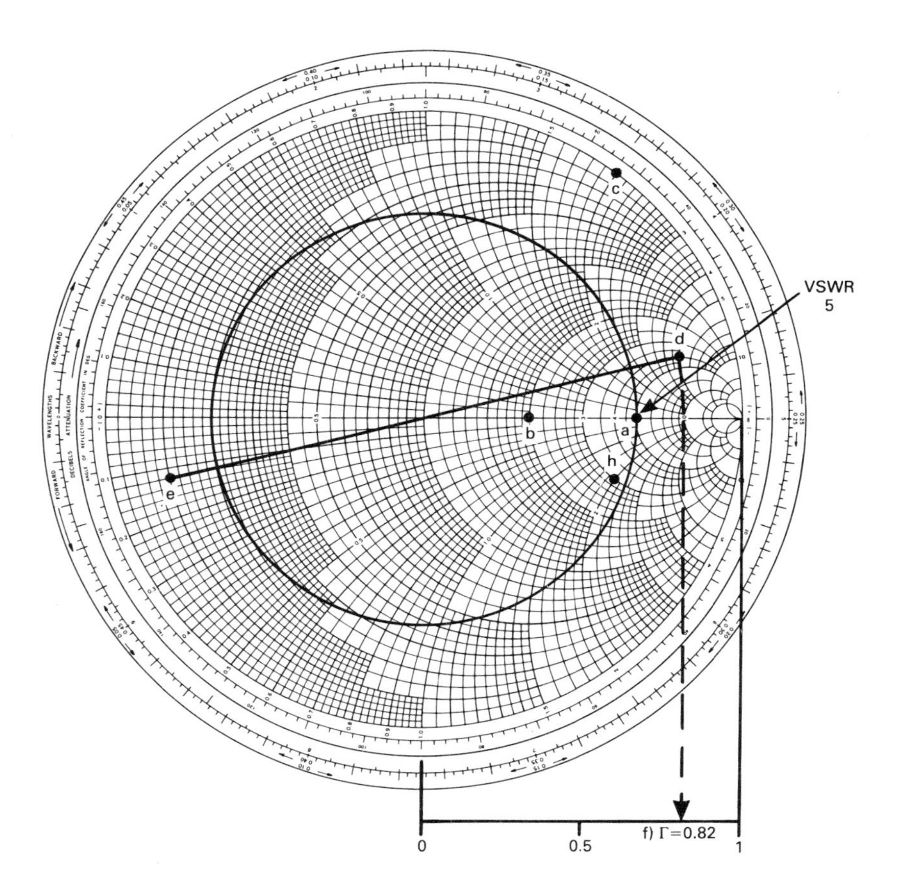

Fig. 4.11 — Solution for example 1.

If we want to plot the resistance of 100 Ω, then we must first normalize this by dividing it by 50 Ω, since we deal with a 50 Ω system. This gives a normalized resistance of 2 Ω indicated by 'b'.

Reactance implies no resistance. Hence any pure reactance must lie on the zero resistance circle. Normalization of the 100 Ω reactance results in its location at point 'c' on the Smith chart.

Upon normalization one obtains $z = Z/50 = (4 + j5)\,\Omega$. The impedance is thus located at point 'd' at the intersection of the normalized resistance circle $r = 4\,\Omega$ and normalized reactance circle $jx = 5\,\Omega$.

Admittance is the reciprocal of impedance and is obtained by drawing a straight line from point 'd' through the centre of the chart to point 'e'. Points 'd' and 'e' are equidistant from the Smith chart centre. The admittance at 'e' can be read off as $y = (0.1 - j0.12)S$, where S stands for Siemens.

The magnitude of the reflection coefficient is given by the distance between the Smith chart centre and point 'd'. It may be directly read off as 0.82 from the line of radially scaled parameters.

The normalized admittance may be calculated as $(0.097–j0.122)S$ using the relationhsip $y = 1/z$.

Since the impedance is already normalized it may simply be plotted as point 'h'. Note that the negative sign in front of j denotes a capacitive reactive component.

Example 2

(a) Using a Smith chart, draw the reflection coefficient circles for $\Gamma = 0.3, 0.5$ and 1.
(b) Draw the VSWR circles for 2.2, 4 and 6 on the same chart.
(c) Show mathematically that a VSWR circle of two corresponds to a reflection coefficient circle of 0.333.

Solution

(a) and (b): see Fig. 4.12
(c):

$$\Gamma = \frac{\text{VSWR} - 1}{\text{VSWR} + 1} = \frac{2 - 1}{2 + 1} = 0.333$$

Example 3

Determine

(a) the input impedance of the lossless transmission line circuit shown in Fig. 4.13 using a Smith chart;
(b) its reflection coefficient;
(c) its VSWR;
(d) its input impedance for 0.018λ.

Solution

(a):

1. Normalize load impedance $z_l = Z_l/50 = (3 + j1)$.
2. Plot as 'A' on chart.
3. Travel $\lambda/3$ wavelength backwards, i.e. towards generator.

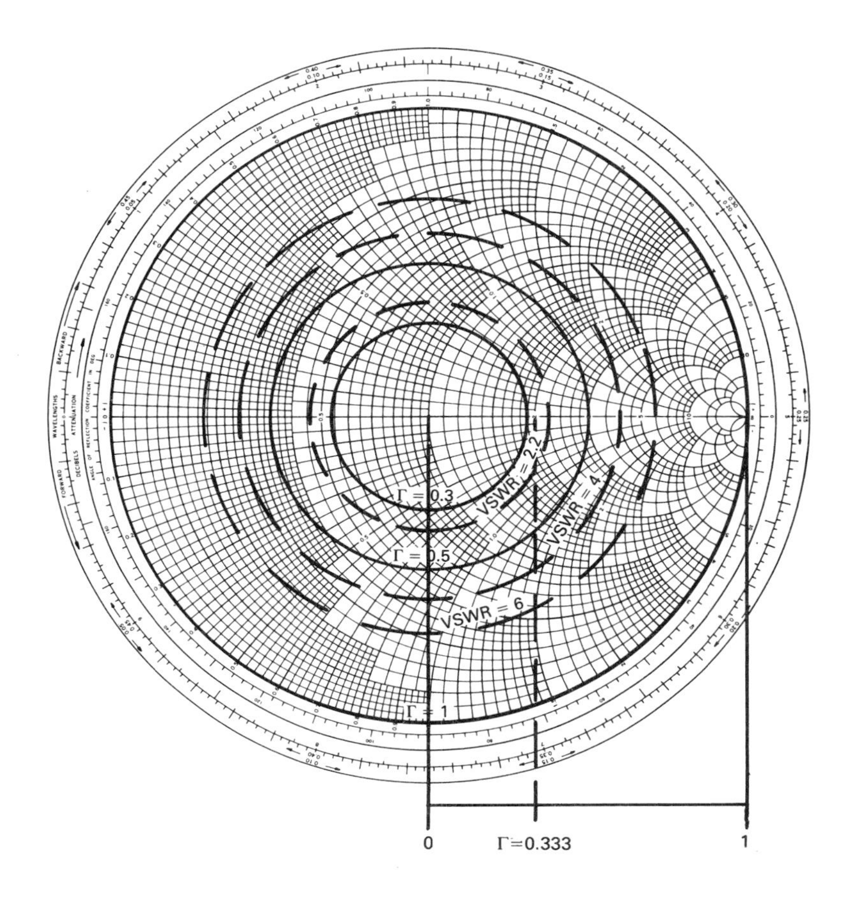

Fig. 4.12 — Solution for example 2.

4. Draw circle around 'O' with radius OA. Intersection is 'B'.
5. Read off $z_{in} = (0.33 + j0.33)$.
6. Denormalize $Z_{in} = z_{in} \times 50 = (16.5 + j16.5)\,\Omega$.

(b): $\Gamma = 0.55$.
(c): VSWR = 3.4 as read off at 'C' on the Smith chart.
(d): Travel 0.018λ backwards from 'A' which gives 'C'. From this we obtain $Z_{in} = z_{in} \times 50 = 3.4 \times 50 = 170\,\Omega$ which is a pure resistance.

Example 4

Match a $(100 + j100)\,\Omega$ load to a $50\,\Omega$ transmission line using a single short circuit stub as shown in Fig. 4.14.

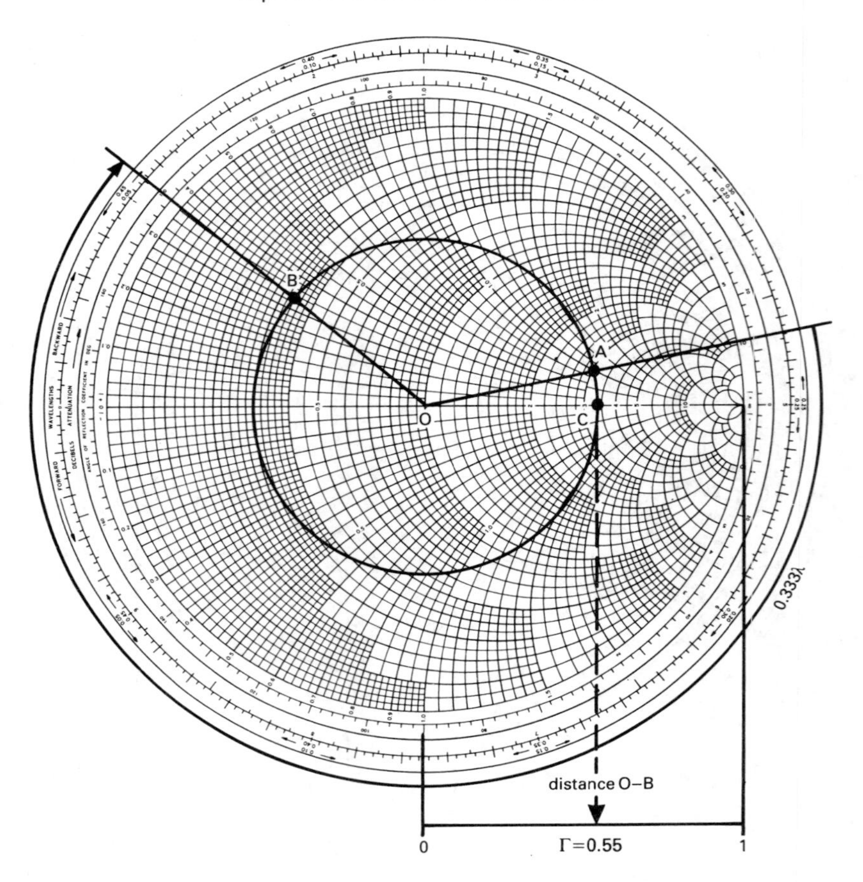

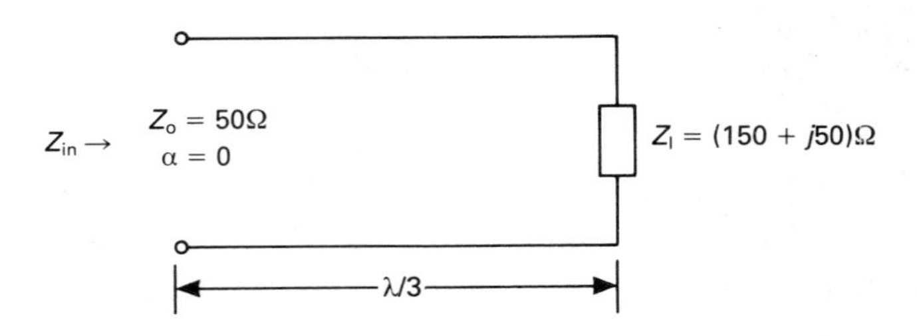

Fig. 4.13 — Solution for example 3.

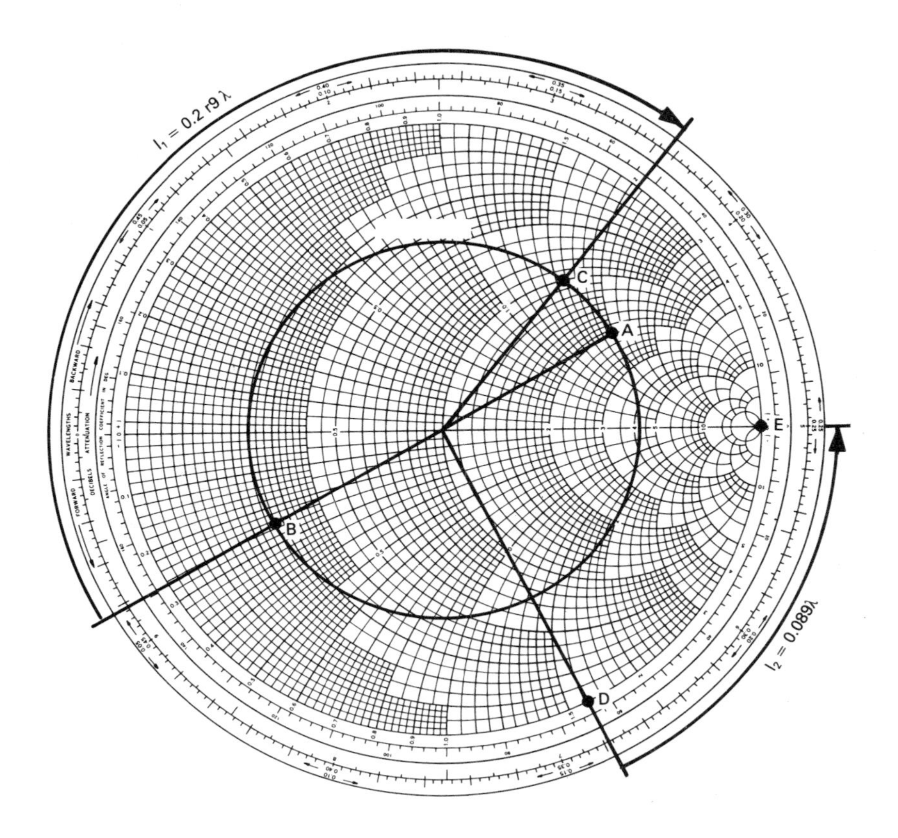

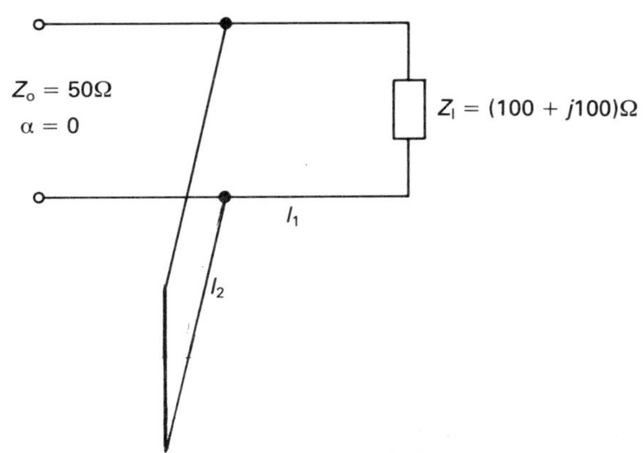

Fig. 4.14 Solution for example 4.

Solution

(1) Use Smith chart in admittance mode.
(2) Normalize to $z_l = 2 + j2$ and plot as 'A'.
(3) Plot admittance y_l and mark as 'B'.
(4) Draw circle towards generator from 'B' to intersect unity resistance circle at 'C'. Distance BC $= l_1 = 0.219\lambda$.
(5) Normalized admittance at point 'C' is $1 + j1.6$. Stub length of l_2 is now required to produce susceptance of $-j1.6$.
(6) Plot $-j1.6$ at 'D' and travel towards the load up to 'E' since the load is a short circuited stub and $Y = \infty$ at 'E'. Remember, in this case we use the Smith chart in its admittance mode.
(7) Read off distance DE $= l_2 = 0.089\lambda$.

Example 5

A load has a normalized impedance of 0.4–j0.4 at 3 GHz as shown in Fig. 4.15.

(a) Calculate line length l_1 and l_2 if a short circuit stub is used for matching.
(b) Calculate the physical line lengths l_1 and l_2.

Solution

(a)

(1) Use the Smith chart in admittance mode.
(2) Plot z_l as 'A'.
(3) Plot y_l as 'B'.
(4) Draw circle from 'B' towards the generator until it intersects with the unit resistance circle at 'C'. Angle BC $= \mathrm{l}_1 = 0.153\lambda$.
(5) At 'C' we have $y_l = 1 - j1.15$. A susceptance of $+j1.15$ is required to make $y_l = 1$.
(6) Thus plot $+j1.15$ at 'D'.
(7) Owing to point 1 above (admittance mode), point 'E' is $Y = \infty$ which represents a short circuit (a short circuited stub was specified). Move from 'D' to 'E' which is 0.385λ. This is the length l_2 required to obtain the conjugate susceptance of $+j1.15$.

(b) $\lambda = c/f = (3 \times 10^{10})/(3 \times 10^9) = 10\,\text{cm} = 100\,\text{mm}$ $l_1 = 15.3\,\text{mm}$ and $l_2 = 38.5\,\text{mm}$

Example 6

Referring to Fig. 4.16, determine

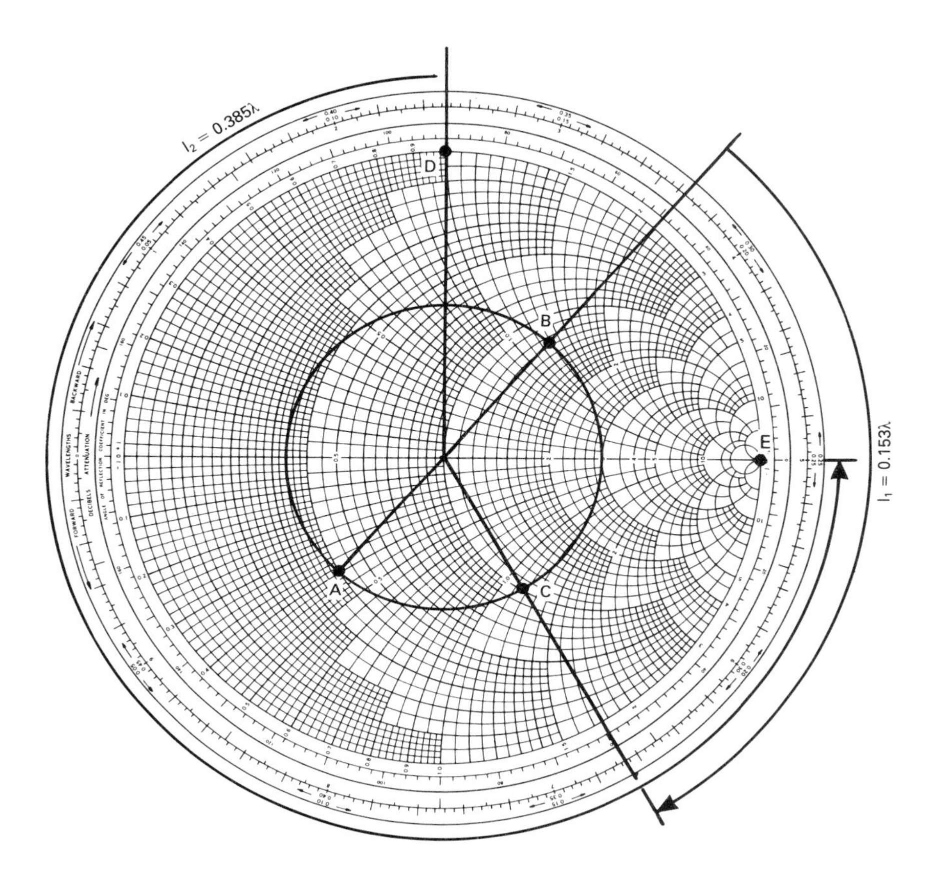

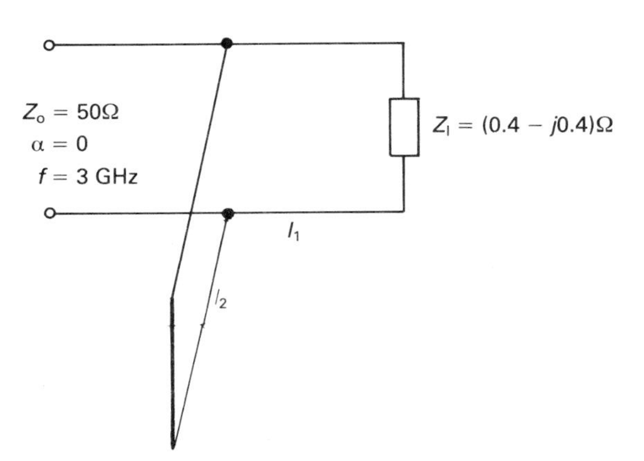

Fig. 4.15 — Solution for example 5.

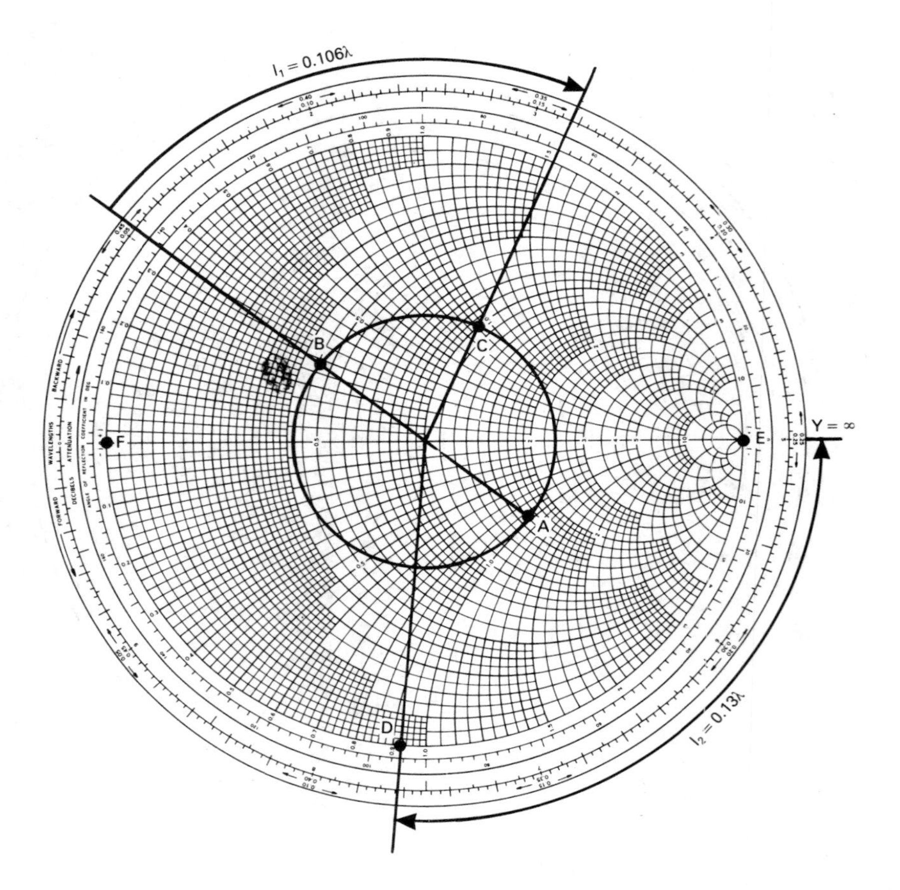

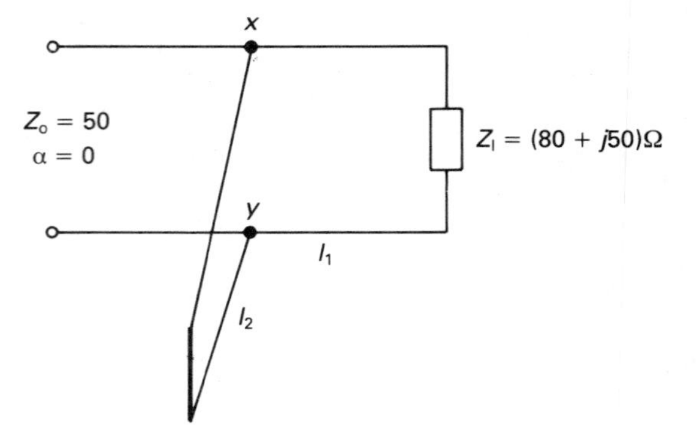

Fig. 4.16 — Solution for example 6.

(a) The location l_1 and length l_2 of a single short circuit stub to match a 50 Ω line to a load of $(80 - j50)$ Ω.
(b) What is the length of l_2 if the short circuit stub is replaced by an open circuit stub? What conclusions can be drawn from this?

Solution
(a)

(1) Normalize to $z_l = 1.6 - j1$ and plot as 'A'.
(2) Plot y_l as 'B'. This gives $0.44 + j0.28$.
(3) Move from 'B' toward generator to intersect unit resistance circle at 'C'.
(4) Angle BC $= l_1 = 0.106\lambda$.
(5) 'C' has an impedance coordinate of $1 + j0.94$. A negative susceptance of magnitude 0.94 is required to make the admittance real.
(6) Thus plot $-j0.94$ as 'D'. From the terminals x and y the short circuit is seen as the load. Hence move from 'D' in direction towards the load up to 'E' which represents a short circuit. Remember, we use the chart in its admittance mode.
(7) $l_2 = 0.13\lambda$.

(b) To obtain l_2 for an open circuit stub continue from 'D' via 'E' to 'F' where $y = 0$. $l_2 = 0.13\lambda + 0.25\lambda = 0.38\lambda$.

The conclusion which can be drawn is that the difference between an open circuit stub and a short circuit stub is always $\lambda/4 = 0.25\ \lambda$.

Example 7
With the aid of a Smith chart find the load impedance at the end of a $\lambda/8$ and $\lambda/4$ line as shown in Fig. 4.17. Assume an input impedance of $(50 + j10)$ Ω.

Solution:
1. Normalize to $z_{in} = 1 + j0.2$ and plot as 'A'.
2. Travel from 'A' $\lambda/8$ and $\lambda/4$ towards the load.
3. Obtain 'B' and 'C'.
4. Read off normalized load impedances as $(0.81 - j0.01)$ and $(0.95 - j0.2)$.
5. After normalization $Z_l = (40.5 - j0.5)$ and $(47.5 - j10)$ Ω respectively.

Example 8
The Smith chart may also be used for characteristic line impedances which are not 50 Ω. Thus, with reference to Fig. 4.18 calculate the length l_1 and l_2.

Solution
1. Normalize the load impedance by dividing it by 200. Thus we have $z_l = 0.25 - j0.75$. Plot as 'A'.
2. Plot admittance as 'B'.

Impedance or admittance coordinates

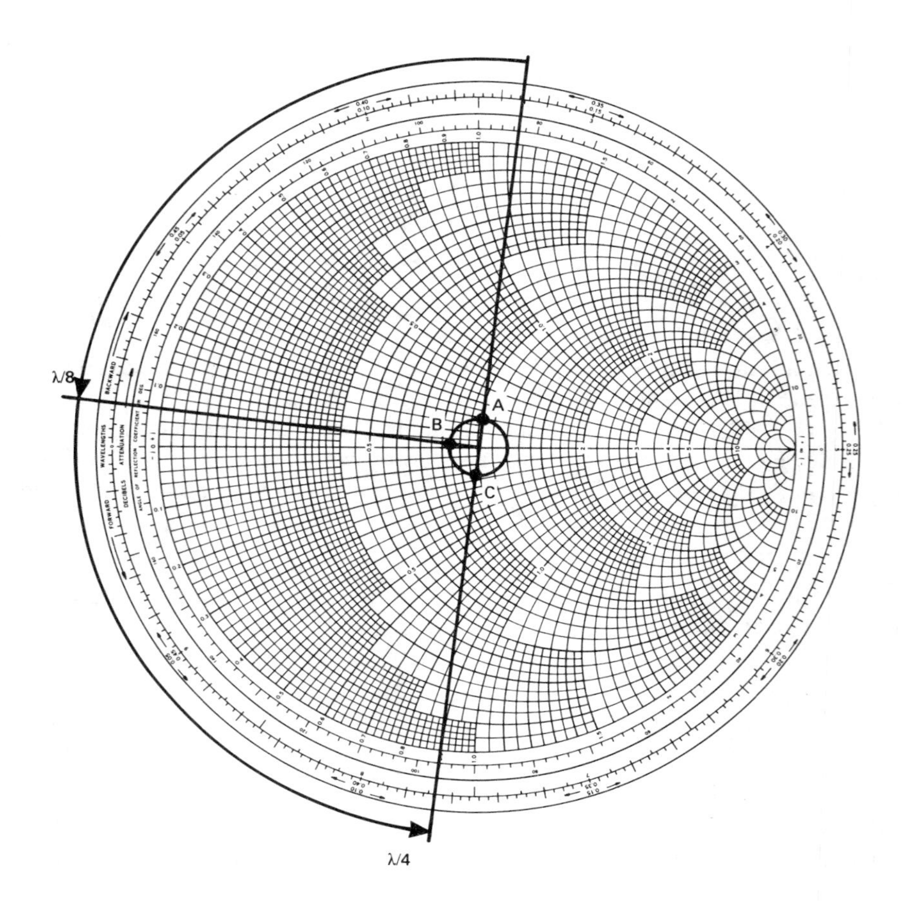

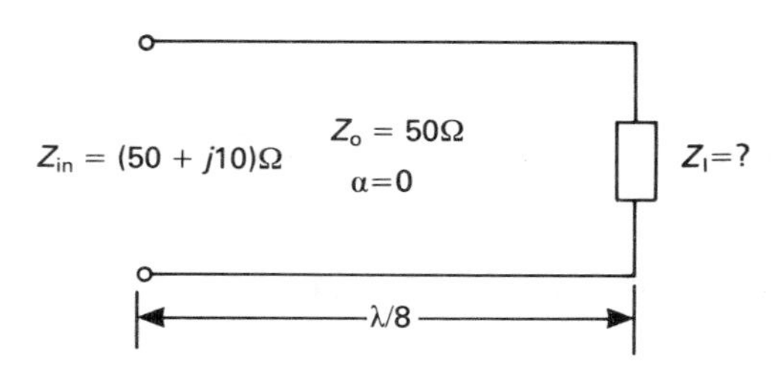

Fig. 4.17 — Solution for example 7.

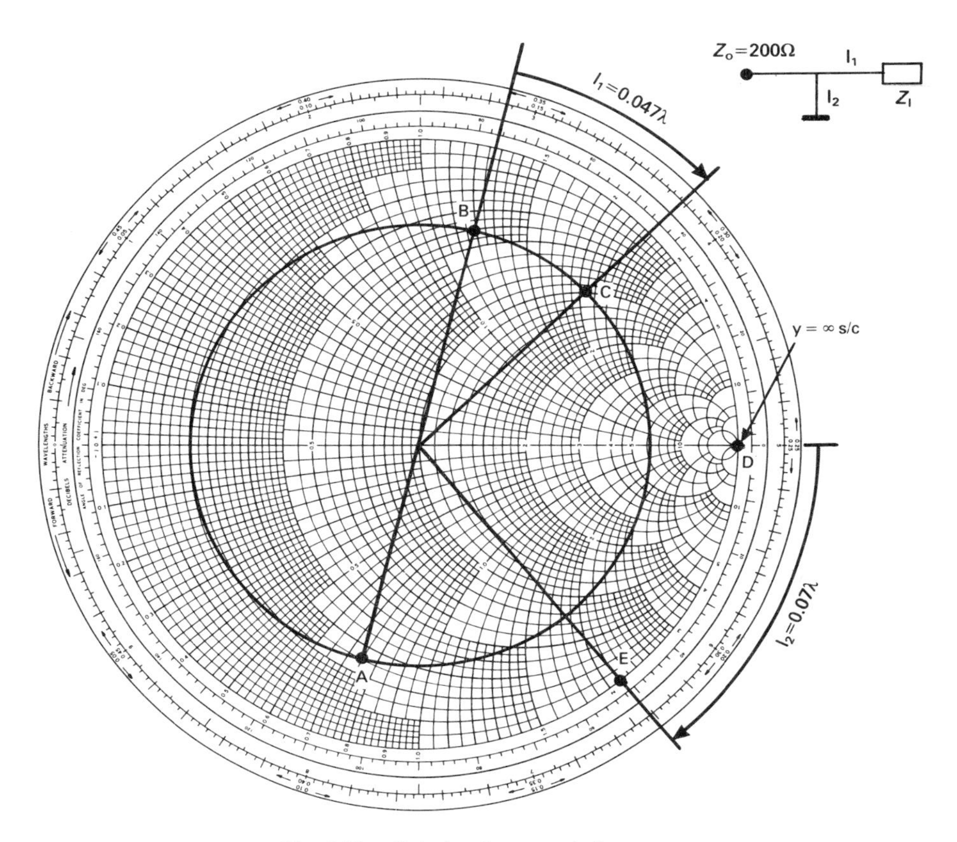

Fig. 4.18 — Solution for example 8.

3. Move from 'B' towards generator on constant VSWR circle until it intersects unit resistance circle at 'C'. Distance BC represents the length $l_1 = 0.047\lambda$.
4. A short circuit stub is desired for matching. Hence plot short circuit as 'D' since $y = \infty$ at this point.
5. To cancel the imaginary part of the load at the stub position l_1, the imaginary value $+j2.1$ at 'C' must be cancelled by the conjugate of $-j2.1$. This value is found at 'E'.
6. The short circuit represents a load as far as the stub is concerned. Thus if we travel from 'D' towards the generator until 'E', we obtain the length l_2 required to cancel the imaginary part. Whence $l_2 = 0.07\lambda$.

SPECIAL CASES OF TRANSMISSION LINE TERMINATIONS

Although it is possible to terminate a transmission line in an infinite number of ways, there are three cases of special interest. At the receiving end the line can be terminated by a short circuit, an open circuit or it can be terminated in its characteristic impedance as shown in Fig. 4.6.

Short circuited line

For a short circuited line $Z_r = 0$. Substituting into the equation,

$$Z_s = Z_o \left[\frac{Z_r + jZ_o \tan \beta l}{Z_o + jZ_o \tan \beta l} \right] \tag{4.67}$$

$$Z_s = jZ_o \tan \beta l \tag{4.68}$$

Obviously, for $l = 0$, the input or source impedance seen into the line is $Z_s = 0$. Now let us see what happens if we choose line lengths which are particular fractions of a wavelength.

For $l = \lambda/4$ we obtain

$$Z_s = jZ_o \tan \beta l = jZ_o \tan 90° = \infty\ \Omega \tag{4.69}$$

A quarter wave short circuited line has thus the interesting property of infinite input impedance. This is a factor which is being exploited for biasing purposes of active and passive devices (see Chapter 5).

For $l = \lambda/8$

$$Z_s = jZ_o \tan \beta l = jZ_o \tag{4.70}$$

Again, this is an interesting practical case in that the short circuited $\lambda/8$ line can be used to create a load of characteristic impedance Z_o.

Open circuited line

Substituting $Z_s = \infty$ into eqn. 4.67, we obtain for the input impedance

$$Z_s = -jZ_o \cot \beta l = -jZ_o \frac{1}{\tan \beta l} \tag{4.71}$$

The open circuited line thus behaves in a reciprocal way to that of the short circuited line.

Matched line

If

$$Z_r = Z_o \text{ we have } Z_s = Z_o\ . \tag{4.72}$$

The physical interpretation of this is that the line length between the source and load impedance can be any length without upsetting the matching.

QUARTER WAVE TRANSFORMER

A logical extension of the above is to use eqn. 4.67 for $l = \lambda/4$. Hence with

$$\tan \beta l = \frac{2\pi}{\lambda} = \tan\frac{\pi}{2} = \infty$$

$$Z_s = Z_o\left[\frac{Z_r/\tan \beta l + jZ_o}{Z_o/\tan \beta l + jZ_r}\right] = Z_o\left[\frac{0 + jZ_o}{0 + jZ_r}\right] = \frac{Z_o^2}{Z_r} \qquad (4.73)$$

or

$$Z_o = \sqrt{Z_s Z_r} = Z_t \qquad (4.74)$$

The symbol Z_t is perhaps more appropriate than Z_o in the context of an impedance transformer. Z_t is thus the geometric mean of the impedances to be matched.

Example 9

A coaxial line is to be used to match a 50 Ω signal source to a load of 112.5 Ω. Hence calculate the impedance of the quarter wave transformer section.

Solution

$$Z_t = \sqrt{Z_s Z_r} = \sqrt{50 \times 112.5} = 75\,\Omega$$

Example 10

An aerial of 75 Ω impedance is to be matched to a microwave amplifier of 50 Ω input impedance. What is the impedance of the quarter wave transformer?

Solution

$$Z_t = \sqrt{Z_s Z_r} = \sqrt{75 \times 50} = 61.24\,\Omega$$

An alternative to stub matching is thus the use of an impedance transformer matching network. The schematic of an impedance transformer matching network is shown in Fig. 4.19(a). In order to illustrate its design consider a transistor amplifier with a centre frequency of 10 GHz. From the transistor data sheet we obtain for the input reflection coefficient 0.44, 165°.

The first step is to enter the reflection coefficient onto the Smith chart as z. Then draw a line from z through the centre of the chart to y in the usual way. Point y represents the normalized admittance of the transistor which is [Fig. 4.19(b)].

$$y = g + jb = (2.4 - j0.65)S$$

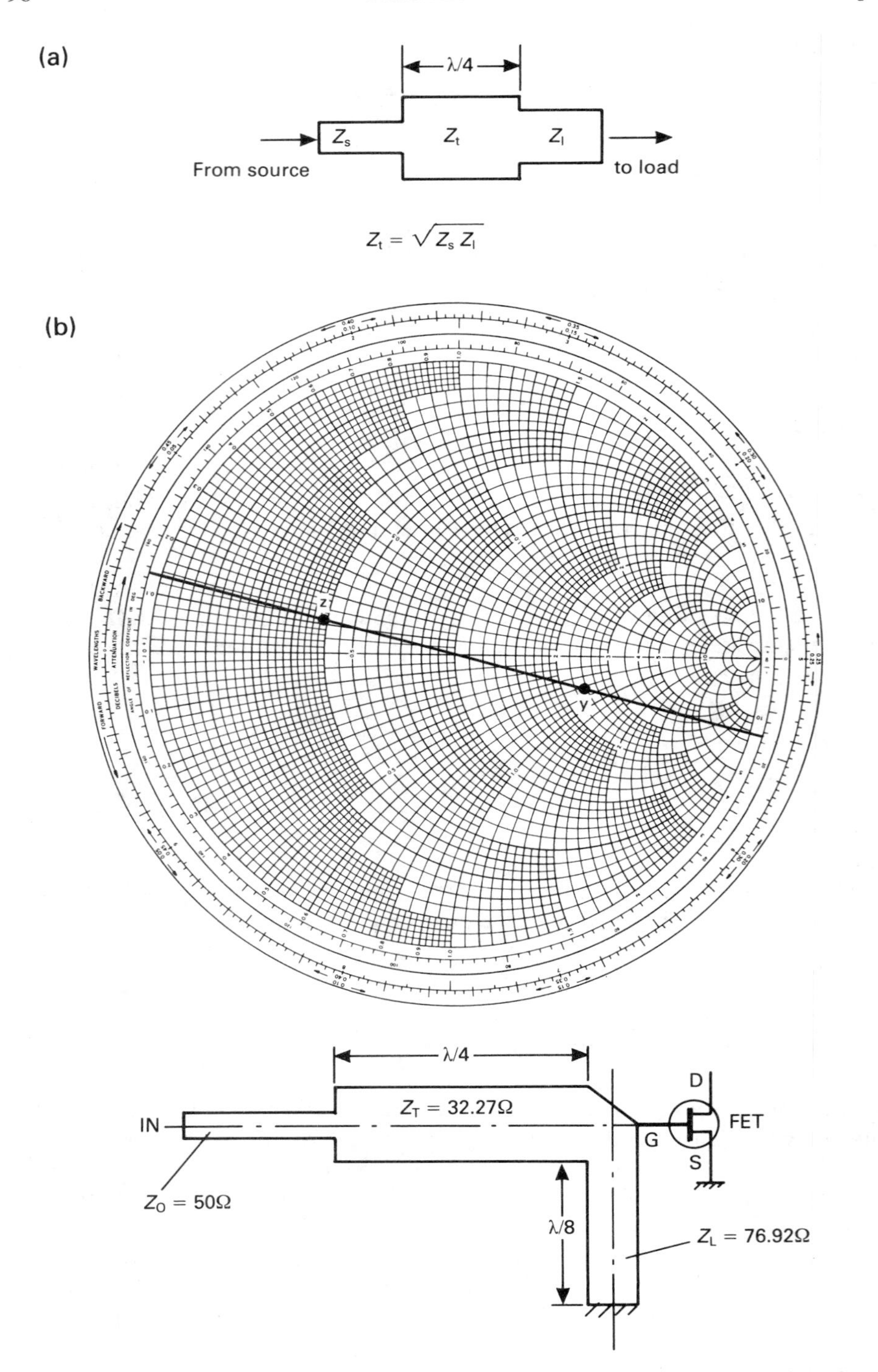

Fig. 4.19 — (a) Quarter wave impedance transformer. (b) Use of a quarter wave impedance transformer.

Upon de-normalization one obtains

$$y = G + jB = (0.048 - j0.013)S$$

Since an impedance transformer can only be used for resistive terminations, the input impedance of the transistor must be made real. This can be done by means of a stub one eigth of a wavelength long and of impedance Z_o. This was shown earlier on.

Hence

$$jX_s = jZ_o \tan \beta l = jZ_o \qquad \text{for} \qquad l = \lambda/8$$

The necessary value of Z_o required to cancel the susceptance of $-j0.013S$ is

$$jZ_o = \frac{1}{jB} = \frac{1}{-j0.013} = j76.92\,\Omega$$

The stub has thus an impedance of $76.92\,\Omega$ and is $\lambda/8$ long. All that remains to be done is to match the $50\,\Omega$ source impedance by the $\lambda/4$ transformer section to the now real load of $0.048S$ of the transistor which is $Z_{in} = 1/G$. Thus

$$Z_{in} = \frac{1}{0.048} = 20.83\,\Omega$$

$$Z_t = \sqrt{Z_s Z_{in}} = \sqrt{50 \times 20.83} = 32.27\,\Omega$$

The final layout of the transistor input matching network is then as shown in Fig. 4.19(b). Naturally, sharp bends are mitred to provide for better performance.

Example 11

(a) Find the input impedance of a short circuited $50\,\Omega$ ideal transmission line of length $\lambda/4$. (b) Repeat for an open circuited line. Comment on the results obtained.

Solution

For an ideal line $\alpha = 0$. From eqn. (4.68)

$$Z_s = jZ_o \tan \beta l = jZ_o \tan \frac{2\pi}{\lambda} l = j50 \tan \pi 2 = j\infty\ \Omega$$

This shows that Z_s for a short circuited $\lambda/4$ line is independent of Z_o.

(b)

$$Z_s = \frac{-jZ_o}{\tan \beta l} = \frac{-j50}{\tan \pi 2} = -j0\,\Omega$$

This shows that Z_s for an open circuit $\lambda/4$ line is independent of Z_o. The short and open circuit lie at opposite ends of the Smith chart, i.e. 0.25 wavelength apart. An open circuited line can thus be created by adding $\lambda/4$ to a short circuited line and vice versa. Thus:

$$Z_s = jZ_o \tan\left[2\pi\left(\frac{l + 0.25\lambda}{\lambda}\right)\right] = j50\tan\left[2\pi\left(\frac{1}{4}+\frac{1}{4}\right)\right]$$

$$Z_s = j50\tan\pi/2 = j0\,\Omega$$

This feature can be exploited for amplifier stub matching purposes.

Using the computer program ZIN, the input impedance for different line terminations or loads and line length can be calculated and plotted. Some examples are given in Fig. 4.6(b to d).

```
RUN
Sending End Impedance Zs
Sending End Impedance Zs
0........End
1........Calculation.
2........Graph.
3........Instructions.
4........Examples.
Enter your choice


INPUT line parameters.
INPUT line parameters.
INPUT Z0  (OHM).............50
INPUT F   (GHz).............1
INPUT ATT.(dB/cm)...........0
INPUT Er ...................1
INPUT LENGTH (cm)...........7.5
INPUT LOAD RESIST.(Ohm).....50
INPUT LOAD REACTANCE (Ohm)..0
Are The values correct (Y/N) ?


          RESULTS
          RESULTS

Line length..........7.5cm
Frequency of.........1GHz
Load of..............50 + j0
Attenuation of.......0dB/cm
Permittivity of......1

Calculated sending end impedance Zs :
50 + j0 Ohms.


Press SPACE to continue
```

```
L.
   10 REM This programme is called ZIN
   20 REM This programme calculates the impedance as seen into a transmission
        line
   30 MODE 7
   40 *FX5,1
   50 *FX6,0
   60 *FX8,4
   70 H=0:end=1
   80 REPEAT
   90 VDU 23;8202;0;0;0;
  100 CLS
  110 REPEAT
  120 PROCmenu
  130 UNTIL REPLY$>="0" AND REPLY$<="4"
  140 IF REPLY$="0" THEN GOTO 160
  150 ON VAL(REPLY$) GOSUB 340,2460,2710,2770
  160 IF REPLY$="0" THEN PROCend
  170 UNTIL end=0
  180 MODE 7
  190 END
  200 REM:
  210 DEFPROCmenu
  220 CLS
  230 PRINT TAB(5,5) CHR$(141)"Sending End Impedance Zs"
  240 PRINT TAB(5,6) CHR$(141)"Sending End Impedance Zs"
  250 PRINT TAB(10,12) "0........End"
  260 PRINT TAB(10,14) "1........Calculation."
  270 PRINT TAB(10,16) "2........Graph."
  280 PRINT TAB(10,18) "3........Instructions."
  290 PRINT TAB(10,20) "4........Examples."
  300 PROCchoice
  310 ENDPROC
  320 :
  330 :
  340 REPEAT
  350 PROCinput
  360 PROChappy
  370 UNTIL H=1
  380 PROCcalc
  390 PROCresults
  400 RETURN
  410 :
  420 DEFPROCcalc
  430 REM * input or sending end impedance Zs *
  440 REM * for a uniform transmission line *
  450 REM * without or with loss *
  460 REM -L GIVES MOVEMENT TOWARDS LOAD , i.e.FORWARD
  470 REM
  480 REM +L GIVES MOVEMENT TOWARDS GENERATOR , i.e. BACKWARD
  490 REM F=FREQUENCY [GHz]
  500 REM L=LINE LENGTH [cm]
  510 REM A2=ATTENUATION [dB/cm]
  520 REM Z0=CHRARCTERISTIC IMPEDANCE [Ohm]
  530 REM E=RELATIVE PERMITTIVITY
  540 REM HOME
  550LET R1=R
  560LET X1=X
  570W1=30 / (F*SQR(E))
  580R2=(R1*R1-Z0*Z0+X1*X1)/((R1+Z0)*(R1+Z0)+X1*X1)
  590LET X2=2*X1*Z0/((R1+Z0)*(R1+Z0)+X1*X1)
  600IF X2 <>0 THEN 640
```

```
610IF R2>=0 THEN 630
620G=-3.1415927 GOTO 740
630G=1E-20:GOTO 740
640IF R2<=0 THEN 670
650IF X2=0 THEN 600
660LET G=ATN(X2/R2):GOTO 740
670IF R2<>0 THEN 710
680IF X2>=0 THEN  700
690LET G=-1.5707963:GOTO 740
700LET G=1.5707963:GOTO 740
710IF R2>=0 THEN 740
720IF X2 =0 THEN 600
730G=3.1415927+ATN(X2/R2)
740LET T1=G
750LET M1=SQR(R2*R2+X2*X2)
760LET A2=A2/8.686
770LET T2=T1-(4*3.1415927*L)/W1
780LET M2=M1*EXP(-(2*A2*L))
790LET D=1-2*M2*COS(T2)+M2*M2
800IF D=0 THEN LET D=1E-20
810LET R1=Z0*(1-M2*M2)/D
820LET X1=Z0*2*M2*SIN(T2)/D
830 ENDPROC
840 :
850 DEFPROCresults
860 CLS
870 q=INT(A2*1000*8.686+0.5)/1000
880 PRINT CHR$(141);CHR$(135);CHR$(157);"           RESULTS"
890 PRINT CHR$(141);CHR$(135);CHR$(157);"           RESULTS"
900 PRINT
910 PRINT TAB(1,6)"Line length..........";L;"cm"
920 PRINT TAB(1,7)"Frequency of.........";F;"GHz"
930 PRINT TAB(1,12)"Load of..............";R;" + j";X
940 PRINT TAB(1,8)"Attenuation of.......";A2;"dB/cm"
950 PRINT TAB(1,9)"Permittivity of......";E;" "
960 PRINT
970 PRINT TAB(1,15)"Calculated sending end impedance Zs :"
980 PRINT TAB(1,18)"";INT(1000*R1/1000+0.5);" + j";INT(1000*X1/1000+0.5);" Ohm
s."
990PRINT
1000PRINT
1010 PROCspace(1,10,24)
1020 ENDPROC
1030 :
1040 DEFPROCchoice
1050 PRINT TAB(10,24)Enter your choice ";:REPLY$=GET$
1060 REPLY$=LEFT$(REPLY$,1)
1070 ENDPROC
1080:
1090 DEFPROCinput
1100 CLS
1110 PRINT TAB(5,2) CHR$(141);"INPUT line parameters."
1120 PRINT TAB(5,3) CHR$(141);"INPUT line parameters."
1130PRINT TAB(0,5)"INPUT Z0  (OHM).............";:
1140 REPEAT:INPUT TAB(25,5) Z0:PROCcheck(Z0,1):UNTIL H=1
1150PRINT TAB(0,7)"INPUT F   (GHz)............";:
1160 REPEAT:INPUT TAB(25,7) F;:PROCcheck(F,0.0000001):UNTIL H=1
1170PRINT TAB(0,9)"INPUT ATT.(dB/cm)...........";:
1180 REPEAT:INPUT TAB(25,9) A2:PROCcheck(A2,0):UNTIL H=1
1190PRINT TAB(0,11)"INPUT Er ...................";:
1200 REPEAT:INPUT TAB(25,11) E:PROCcheck(E,0.000000001):UNTIL H=1
1210 PRINT TAB(0,13)"INPUT LENGTH (cm)...........";:
```

```
1220 REPEAT:INPUT TAB(25,13) L:PROCcheck(L,-10E5):UNTIL H=1
1230 PRINT TAB(0,15)"INPUT LOAD RESIST.(Ohm).....";:
1240 INPUT TAB(25,15) R
1250PRINT TAB(0,17)"INPUT LOAD REACTANCE (Ohm)..";:
1260 INPUT TAB(25,18) X
1270 ENDPROC
1280 :
1290 DEFPROChappy
1300 H=0
1310 *FX15,1
1320 PRINT TAB(2,24)Are The values correct (Y/N) ?";
1330 G$=GET$
1340 IF G$="Y" OR G$="y" THEN H=1
1350 ENDPROC
1360 :
1370 DEFPROCcheck(I%,J%)
1380 H=0
1390 IF J%<=I% THEN H=1
1400 IF I%<J% THEN H=0
1410 ENDPROC
1420 :
1430 DEFPROCaxis
1440 RESTORE
1450 VDU 29,100;100;
1460 MOVE 0,0
1470 DRAW 1150,0
1480 MOVE 0,0
1490 DRAW 0,800
1500 GCOL 0,1
1510 PRINT TAB(5,31) "Wave lengths along line (lambda)."
1520 FOR var1=1 TO 18
1530 READ A$:PRINT TAB(0,var1+5) A$
1540 NEXT
1550 MOVE 0,0
1560 FOR var3=0 TO lamda STEP lamda/4
1570 VDU 5
1580 MOVE (var3*scale),0:READ A$:PRINT A$
1590 MOVE (var3*scale),20:DRAW (var3*scale),30
1600 VDU 4
1610 NEXT
1620 VDU 5
1630 FOR var3=0 TO maxr1 STEP (maxr1/2)
1640 imp=INT(var3)
1650 MOVE -350,(var3*scaley)+35:PRINT imp
1660 MOVE 0,(var3*scaley)+35:DRAW -10,(var3*scaley)+35
1670 NEXT
1680 VDU4
1690 ENDPROC
1700 :
1710 DEFPROCspace(D%,B%,C%)
1720 *FX15,0
1730 IF D%=1 THEN PRINT TAB(B%,C%) "Press SPACE to continue ";
1740 REPEAT:UNTIL INKEY(-99)
1750 ENDPROC
1760 :
1770 DEFPROCscaley
1780 maxr1=0
1790 var4=0
1800 PROCbored
1810 FOR var4=0 TO lamda STEP step
1820 L=var4
1830 PROCcalc
```

```
1840 IF R1>maxr1 THEN maxr1=R1
1850 PRINT TAB(18,8) CHR$(141);CHR$(133);lamda-var4;"      "
1860 PRINT TAB(18,9) CHR$(141);CHR$(133);lamda-var4;"      "
1870 NEXT
1880 scaley=(800/maxr1)
1890 ENDPROC
1900 :
1910 DEFPROCfkeys
1920 *KEY0  50|M10|M0|M1|M10|M50|M50|M
1930 *KEY1  50|M10|M0|M1|M1|M50|M-50|M
1940 ENDPROC
1950 :
1960 DEFPROCbored
1970 CLS
1971 *LOAD"SCRNDMP"
1972 CALL&900
1980 PRINT TAB(0,2) CHR$(132);CHR$(157);CHR$(129);"   Graph values being calcul
ated.    "
1990 PRINT TAB(5,4) " Countdown to PLOT."
2000 PRINT TAB(18,8) CHR$(141);CHR$(133);var4
2010 PRINT TAB(18,9) CHR$(141);CHR$(133);var4
2020 PRINT TAB(10,12)"INPUT VALUES"
2030 PRINT TAB(0,14)"Characteristic impedance   (Ohm)     ";Z0
2040 PRINT TAB(0,15)"Input frequency            (GHz)     ";F
2050 PRINT TAB(0,16)"Input attenuation          (dB/cm)   ";A2
2060 PRINT TAB(0,17)"Permittivity                         ";E
2070 PRINT TAB(0,18)"Line length                (cm)      ";L
2080 PRINT TAB(0,21)"Load               (Ohm)   ";R;"+ j";X
2090 ENDPROC
2100 :
2110 DEFPROCend
2120 CLS
2130 PRINT TAB(5,10) "The program will now terminate."
2140 FOR X=0 TO 5000:NEXT
2150 end=0
2160 ENDPROC
2170 :
2180 DEFPROCinstructions
2190 CLS
2200 PRINT 'CHR$(141);CHR$(132);CHR$(157);"        Instructions "
2210 PRINT CHR$(141);CHR$(132);CHR$(157);"        Instructions "
2220 PRINT TAB(5,6)"When aloadis placed at the end of a transmission line,the l
ine produces a transformation effect whereby theload impedanceat the end of the
line will  appear different when viewed from the    sending end";
2230 PRINT "of the line.";
2240 PRINT '"The equation to work out the sending end impedance in a cable of l
ength  L, is  given below;"
2250 PRINT'CHR$(141)"Zs   Zt cosh(ul) + Zo sinh(ul)"
2260 PRINTCHR$(141)"Zs   Zt cosh(ul) + Zo sinh(ul)"
2270 PRINT CHR$(141)"__ = ______________________________"
2280 PRINT CHR$(141)"__ = ______________________________"
2290 PRINT CHR$(141)"Zo   Zt sinh(ul) + Zo cosh(ul)"
2300 PRINT CHR$(141)"Zo   Zt sinh(ul) + Zo cosh(ul)"
2310 PROCspace(1,10,24)
2320 CLS:PRINT'CHR$(141);CHR$(132);CHR$(157);"        Instructions "
2330 PRINT CHR$(141);CHR$(132);CHR$(157);"        Instructions "
2340 PRINT TAB(5)'"INPUTS"
2350 PRINT '"Characteristic Impedance.Z0This is   the impeadance of the transmi
ssion line.It is measured in Ohms."
2360 PRINT "Frequency.FThe frequency to be sent  down the line.It is entered in
terms of Gigahertz,(GHz)."
2370 PRINT "Attenuation.AThe amount that the     signal is attenuated along the
```

```
line for every centimeter travelled.Measured in dB/cm."
2380 PRINT "Line Length.LThe length of the transmission line in centimeters."
2390 PRINT "Effective permittivity.EThe substrate permittivity between the tran
smission line and the ground plane."
2400 PRINT "Input Resistance.RThe real part of   the input resistance measured
in Ohms."
2410 PRINT "Input Reactance.XThe reactance of the input,in Ohm"
2420 PROCspace(1,10,24)
2430 ENDPROC
2440 :
2450 :
2460 REM GRAPH
2470 REPEAT
2480 PROCinput
2490 PROChappy
2500 UNTIL H=1
2510 lamda=(3E10/(F*1E9))
2520 scale=(1150/lamda)
2530 step=(lamda/200)
2540 PROCscaley
2550 MODE 1
2560 VDU 23;8202;0;0;0;
2570 PROCaxis
2580 VDU 29,100;100;
2590  GCOL 0,2
2600 FOR var=0 TO lamda STEP step
2610 L=var
2620 PROCcalc
2630 Z%=INT(1000*R1/1000+0.5)
2640 IF var=0 THEN MOVE (var*scale),(Z%*5)+35
2650 DRAW (var*scale),(Z%*scaley)+35
2660 NEXT var
2670 PROCspace(1,10,0)
2680 MODE 7
2690 RETURN
2700 :
2710 REM INSTRUCTIONS
2720 CLS
2730 PROCinstructions
2740 RETURN
2750 DATA S,e,n,d,i,n,g, , , , , , ,E,n,d, ,Zs
2760 DATA 0,1/4,1/2,3/4,1
2770 REM EXAMPLES
2780 CLS
2790 *KEY 0
2800 *KEY 1
2810 PRINT CHR$(141);CHR$(129);CHR$(157);CHR$(135);"          Examples
"
2820 PRINT CHR$(141);CHR$(129);CHR$(157);CHR$(135);"          Examples
"
2830 PRINT TAB(5,6)"F0..........Example Set 1"
2840 PRINT TAB(5,8)"F1..........Example Set 2"
2850 PRINT TAB(5,10)"SPACE.......Main Menu."
2860 PRINT TAB(5,14)"After you have chosen an option at the main menu screen,on
e of the listed  REDfunction keys can be used.these are located at the top of th
e keyboard.";
2870 PRINT "You will be required to press the Y key at  the end of the input sc
reen when using  the examples."
2880 PROCspace(1,10,24)
2890 PROCfkeys
2900 RETURN
```

L.

```
 10 REM This programme is called SMITH
 20 REM This programme draws a                        Smith chart and allows
        the display of resistance,                    reactance and impedance values
 30 REM ***    MAIN ROUTINE   ***
 40 MODE 1
 50   PROCTITLE
 60   PROCSETUP
 70   PROCCIRCLE_GEN
 80   PROCRES_LOCI
 90   PROCCIRCLE_GEN
100   PROCINDUCT
110   PROCBI_LIN
120   PROCPOS_RC
130   PROCCAPAC
140   PROCBI_LIN
150   PROCNEG_RC
160   PROCZPLOT
170   PROCVSWR
180   PROCRERUN
190 CLS
200 PRINT TAB(1,3)"To run the program again :"
210 PRINT TAB(1,5)"Press 'SHIFT' and 'BREAK' together."
220 END
230 REM ***     Setup Procedure     ***
240 REM         ---------------
250 DEF PROCSETUP
260 CLS
270 N=60: H=410: K=700: R=400: G=1
280 FLAG=0: U=1E20: D=2*PI/N
290 MOVE K-R,H:PLOT 21,K+R,H
300 COLOUR 129
310 PRINT TAB(1,1)"**SMITH CHART--IMPEDANCE COORDINATES**"
320 COLOUR 128:COLOUR 1
330 PRINT TAB(7,15)"^"
340 PRINT TAB(2,16)"GEN. |"
350 PRINT TAB(2,23)"LOAD |"
360 PRINT TAB(7,24)"v"
370 PRINT TAB(4,20)"(S/C)"
380 PRINT TAB(35,20)"(O/C)"
390 COLOUR 2
400 PRINT TAB(35,12)"+J"
410 PRINT TAB(3,19)"0/0.5~"
420 PRINT TAB(35,26)"-J"
430 PRINT TAB(35,19)"0.25~"
440 PRINT TAB(0,30)"(~ WAVELENGTH)"
450 COLOUR 3
460 MOVE K-12,H+10:VDU5:PRINT"+":VDU4
470 ENDPROC
480 REM ***  Clear top of screen  ***
490 REM      -------------------
500 DEF PROCCLEAR
510 PRINT TAB(0,2)"                                        "
520 PRINT TAB(0,3)"                                        "
530 PRINT TAB(0,4)"                                        "
540 PRINT TAB(0,5)"                                        "
550 ENDPROC
560 REM ***   Plotting Z points   ***
570 REM       -----------------
580 DEF PROCZPLOT
590 PROCCLEAR
600 PRINT TAB(1,3)"Do you wish to plot an IMPEDANCE POINT  (Y OR N) ?"
```

```
 610 LET TEST$ =GET$
 620 IF TEST$ = "N" THEN 1160
 630 IF TEST$ = "Y" THEN 650
 640 GOTO 610
 650 PROCCLEAR
 660 PRINT TAB(1,3)"Enter PHYSICAL RESISTANCE value (OHMS)"
 670 PRINT TAB(1,4)"(e.g. 100)"
 680 INPUT A
 690 COLOUR 2
 700 PRINT TAB(1,5)"  VERIFY (Y OR N)"
 710 COLOUR 3
 720 LET G$=GET$
 730 IF G$ = "Y" THEN 760
 740 IF G$ = "N" THEN 650
 750 GOTO 720
 760 IF A=0 THEN A=1E-10
 770 PROCCLEAR
 780 PRINT TAB(1,3)"Enter PHYSICAL REACTANCE value (OHMS)"
 790 PRINT TAB(1,4)"(e.g. -50 or 40)"
 800 INPUT B
 810 COLOUR 2
 820 PRINT TAB(1,5)"  VERIFY (Y OR N)"
 830 COLOUR 3
 840 LET G$=GET$
 850 IF G$ = "Y" THEN 880
 860 IF G$ = "N" THEN 770
 870 GOTO 840
 880 IF B=0 THEN B=1E-10
 890 PROCCLEAR
 900 PRINT TAB(1,3)"Enter PHYSICAL NORMALISATION RESISTANCE"
 910 PRINT TAB(1,4)"(OHMS) (e.g 50 for 50 ohm system)"
 920 INPUT ZO
 930 COLOUR 2
 940 PRINT TAB(1,5)"  VERIFY (Y OR N)"
 950 COLOUR 3
 960 LET G$=GET$
 970 IF G$ = "Y" THEN 1030
 980 IF G$ = "N" THEN 890
 990 GOTO 960
1000 IF ZO=0 THEN 890
1010 REM *** MAP Z ONTO CHART ***
1020 REM     ----------------
1030 A = A/ZO:B=B/ZO
1040 REAL = (A+1)*(A-1)+B*B
1050 IM = B*((A+1)-(A-1))
1060 N=(A+1)*(A+1)+B*B
1070 REAL = REAL/N*R
1080 IM = IM/N*R
1090 MOVE REAL+K-10,IM+H+10:VDU5:PRINT"*":VDU4
1100 PROCCLEAR
1110 PRINT TAB(1,3)"All impedance points done ? (Y OR N)"
1120 LET TEST$ = GET$
1130 IF TEST$ ="Y" THEN 1160
1140 IF TEST$ ="N" THEN 650
1150 GOTO 1120
1160 ENDPROC
1170 REM ***   I/P Normalised XL   ***
1180 REM       -----------------
1190 DEF PROCINDUCT
1200 PROCCLEAR
1210 PRINT TAB(1,3)"Enter NORMALISED INDUCTIVE reactances"
1220 INPUT B:A=0
1230 COLOUR 2
```

```
1240 PRINT TAB(1,5)"  VERIFY (Y OR N)"
1250 COLOUR 3
1260 LET G$=GET$
1270 IF G$="Y" THEN 1300
1280 IF G$="N" THEN 1200
1290 GOTO 1260
1300 IF B=0 THEN B=0.0001
1310 MOVE K+R,H
1320 ENDPROC
1330 REM ***   I/P Normalised XC   ***
1340 REM       -----------------
1350 DEF PROCCAPAC
1360 PROCCLEAR
1370 PRINT TAB(1,3)"Enter NORMALISED CAPACITIVE reactances"
1380 INPUT B:A=0:FLAG=1
1390 COLOUR 2
1400 PRINT TAB(1,5)"  VERIFY (Y OR N)"
1410 COLOUR 3
1420 LET G$=GET$
1430 IF G$="Y" THEN 1460
1440 IF G$="N" THEN 1360
1450 GOTO 1420
1460 IF B=0 THEN B=0.0001
1470 MOVE K+R,H
1480 ENDPROC
1490 REM ***  Bi-Linear Transform  ***
1500 REM      -------------------
1510 DEF PROCBI_LIN
1520 REAL = (A+1)*(A-1)+B*B
1530 IM = B*((A+1)-(A-1))
1540 N = (A+1)*(A+1)+B*B
1550 REAL = REAL/N*R
1560 IM =IM/N*R
1570 A=B
1580 X=SQR((R-REAL)^2+IM^2)
1590 Y=R/A
1600 Z=R/A
1610 AS=ACS((Z^2+Y^2-X^2)/2/Z/Y)
1620 ENDPROC
1630 REM * Positive Reactance Circle *
1640 REM   -------------------------
1650 DEF PROCPOS_RC
1660 FOR M = PI TO PI+AS STEP D/8
1670   Y = H+R/A+R/A * COS(M)
1680   X=K+R+R/A*SIN(M)
1690   PLOT 21,X,Y
1700 NEXT M
1710 PROCCLEAR
1720 PRINT TAB(1,3)"All INDUCTIVE loci done ? (Y OR N)"
1730 LET TEST$ = GET$
1740 IF TEST$ = "Y" THEN 1770
1750 IF TEST$ = "N" THEN 100
1760 GOTO 1730
1770 ENDPROC
1780 REM * Negative Reactance Circle *
1790 REM   -------------------------
1800 DEF PROCNEG_RC
1810 FOR M = 2*PI TO 2*PI-AS STEP -D/8
1820   Y=H-R/A+ABS(R/A)*COS(M)
1830   X=K+R+ABS(R/A)*SIN(M)
1840   PLOT 21,X,Y
1850 NEXT M
```

```
1860 PROCCLEAR
1870 PRINT TAB(1,3)"All CAPACITIVE loci done ? (Y OR N)"
1880 LET TEST$ = GET$
1890 IF TEST$ = "Y" THEN 1920
1900 IF TEST$ = "N" THEN 130
1910 GOTO 1880
1920 ENDPROC
1930 REM **  'R' Circle Generation  **
1940 REM     --------------------
1950 DEF PROCCIRCLE_GEN
1960 MOVE K+R/U,H+G*R
1970 FOR M=0 TO 2*PI+D STEP D
1980 Y=H+G*R*COS(M)
1990 X=K+G*R*SIN(M)+R/U
2000  PLOT 5,X,Y
2010 NEXT M
2020 IF U =1E20 THEN 2090
2030 PROCCLEAR
2040 PRINT TAB(1,3)"All RESISTANCE loci done ? (Y OR N)"
2050 LET TEST$ = GET$
2060 IF TEST$ = "Y" THEN 2090
2070 IF TEST$ = "N" THEN 80
2080 GOTO 2050
2090 ENDPROC
2100 REM ***  I/P 'R' Loci Values  ***
2110 REM      -------------------
2120 DEF PROCRES_LOCI
2130 PROCCLEAR
2140 PRINT TAB(1,3)"Enter NORMALISED RESISTANCE LOCI value"
2150 INPUT RLO
2160 IF RLO <= 0 THEN 2130
2170 RLO = 1/RLO
2180 U=RLO+1
2190 G=RLO/U
2200 ENDPROC
2210 REM ***      Start Again ?      ***
2220 REM          -------------
2230 DEF PROCRERUN
2240 PROCCLEAR
2250 COLOUR 2
2260 PRINT TAB(1,3)"Type 'N' for a new chart or 'Q' to quit"
2270 COLOUR 3
2280 LET G$ = GET$
2290 IF G$="Q" THEN 2320
2300 IF G$="N" THEN 60
2310 GOTO 2280
2320 ENDPROC
2330 REM ***        Title Page       ***
2340 REM            ----------
2350 DEF PROCTITLE
2360 COLOUR 129:COLOUR 2:CLS
2370 PRINT TAB(5,8)"*****************************"
2380 PRINT TAB(5,9)"*                           *"
2390 PRINT TAB(5,10)"*     Generation Of The     *"
2400 PRINT TAB(5,11)"*        Smith Chart        *"
2410 PRINT TAB(5,12)"*                           *"
2420 PRINT TAB(5,13)"*****************************"
2430 PRINT TAB(7,27)"Press any key to continue"
2440 LET G$=GET$
2450 COLOUR 128:COLOUR 3
2460 ENDPROC
2470 REM ***     VSWR Circles     ***
2480 REM         ------------
```

```
2490 DEF PROCVSWR
2500 PROCCLEAR
2510 PRINT TAB(1,3)"WOULD YOU LIKE TO PLOT A VSWR CIRCLE?"
2520 PRINT TAB(1,4)"(Y OR N)"
2530 LET G$=GET$
2540 IF G$="Y" THEN 2570
2550 IF G$="N" THEN 2730
2560 GOTO 2530
2570 PROCCLEAR
2580 PRINT TAB(1,3)"ENTER THE VSWR VALUE"
2590 INPUT V
2600 IF V <= 0 THEN 2570
2610 V=1/V:U=V+1:G=V/U
2620 X=700+(400/U)-(G*400)
2630 R=X-700
2640 G=1:N=60:K=700
2650 U=1E20:H=410:D=2*PI/N
2660 MOVE K+R/U,H+G*R
2670 FOR M=0 TO 2*PI+D STEP D
2680   Y=H+G*R*COS(M)
2690   X=K+G*R*SIN(M)+R/U
2700   PLOT 21,X,Y
2710 NEXT M
2720 GOTO 2500
2730 ENDPROC
2740 END
```

REFERENCES

[1] Ramo, S. and Whinnery, J. R., *Fields and waves in communication electronics*, John Wiley, 1984.

[2] Matthaei, G., Young, L., and Jones, E. M. T., *Microwave filters, impedance matching networks and coupling structures*, Artech House Inc. 1985.

5

Theoretical and practical aspects of narrow-band microwave amplifier design

This chapter covers the most relevant theoretical and practical aspects of narrow-band microwave amplifier design. First of all the scattering parameters are discussed; this is followed by an explanation and example of how to transfer these onto a Smith chart where they can be used for the solution of practical problems. We then turn to amplifier concepts, design considerations and limitations. Practical amplifier design examples both, in CAD and longhand, are followed by the not unimportant bias considerations and circuit examples.

SCATTERING MATRIX AND PARAMETERS

From lossless transmission line theory we know that such a line is completely characterized by its characteristic impedance Z_o and its electrical line length θ, namely:

$$Z_o = \sqrt{\frac{L}{C}} \quad \text{and } \theta = \frac{2\pi l}{\lambda} = \frac{2\pi l f}{v}$$

where v is the signal velocity in the transmission line and where l is its physical length. If this line is terminated at its ports as shown in Fig. 5.1, then the appropriate voltage, current and power relationships can be derived. From the incident (i) and reflected

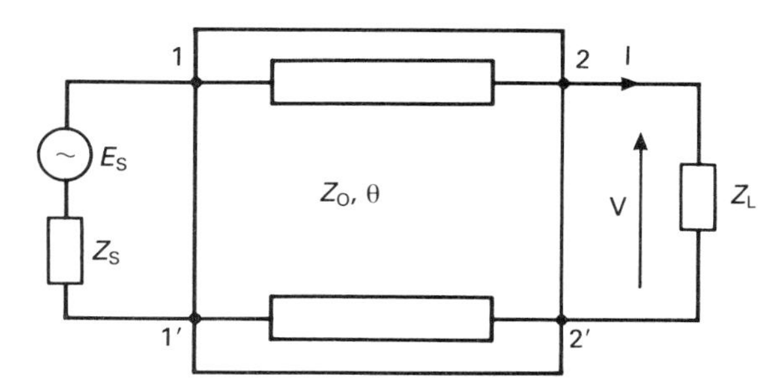

Fig. 5.1 — Terminated transmission line.

waves (r) we can set up the following relationships:

$$V = V_i + V_r \tag{5.1}$$

$$I = \frac{V_i - V_r}{Z_o} \tag{5.2}$$

or

$$V_i = 1/2\ (V + I\ Z_o) \tag{5.3}$$

$$V_r = 1/2\ (V - I\ Z_o) \tag{5.4}$$

By definition of the reflection coefficient Γ

$$\Gamma = V_r/V_i \tag{5.5}$$

Using eqns (5.3) to (5.5) we obtain for the load reflection coefficient Γ_l

$$\Gamma_l = \frac{V - I\ Z_o}{V + I\ Z_o} = \frac{(V/I) - Z_o}{(V/I) + Z_o} = \frac{Z_l - Z_o}{Z_l + Z_o} \tag{5.6}$$

Defining V_i and V_r into new variables gives:

$$a = \frac{V_i}{\sqrt{Z_o}} = \frac{V + I\ Z_o}{2\sqrt{Z_o}} \tag{5.7}$$

$$b = \frac{V_r}{\sqrt{Z_o}} = \frac{V - I\ Z_o}{2\sqrt{Z_o}} \tag{5.8}$$

and hence

$$\Gamma = b/a \tag{5.9}$$

The square of the magnitude of a and b represents the power. The power incident on the load impedance Z_l is thus

$$P_l = |a|^2 \tag{5.10}$$

and the power reflected from Z_l, which is thus the power incident on the source impedance Z_s is

$$P_s = |b|^2 \tag{5.11}$$

For a generalized network having port 1 and port 2 as shown in Fig. 5.2, the incident and reflected waves may be defined more specifically from eqns (5.7) and (5.8) as:

$$a_1 = \frac{V_1 + I_1 Z_o}{2\sqrt{Z_o}} \tag{5.12}$$

$$b_1 = \frac{V_1 - I_1 Z_o}{2\sqrt{Z_o}} \tag{5.13}$$

$$a_2 = \frac{V_2 + I_2 Z_o}{2\sqrt{Z_o}} \tag{5.14}$$

$$b_2 = \frac{V_2 - I_2 Z_o}{2\sqrt{Z_o}} \tag{5.15}$$

Historically, network analysis has been carried out by using, amongst others, z- and y-parameters. With the availability and wider use of transistors in the early 1960s h-parameters were extensively used. These parameters rely on the measurement of voltages and currents. As a result of the increasing use of radio frequencies, physicists and engineers were striving to develop devices and circuits for higher and higher frequencies. Beyond 1 GHz, however, device and circuit parasitics assume an important role and it is not possible to conduct precise voltage and current measurements [1,2,3]. That is where the concept of S-parameters proves useful.

S-parameters are transmission and reflection coefficients and are expressed in terms of power as far as the input and output of a two-port is concerned. In order to explain this in more detail consider the two-port shown in Fig. 5.2. Here a is the

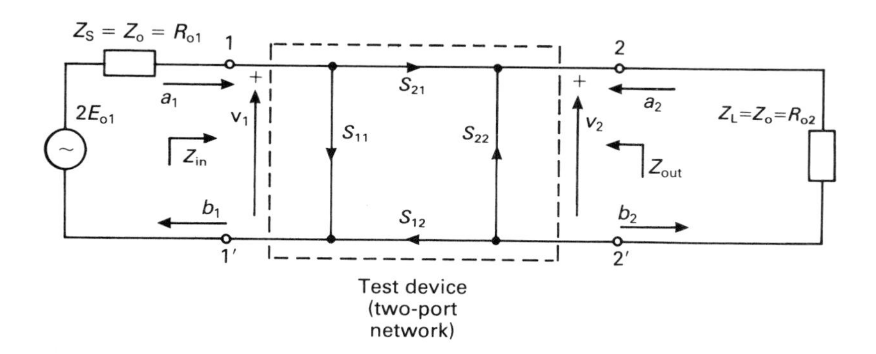

Fig. 5.2 — Two-port and associated parameters.

signal entering a port and b is the signal leaving a port. In particular a_1 is entering port 1 and leaves as b_2 at port 2. Any signal a_2 entering the output port, i.e. by virtue of reflection, leaves port 1 as b_1. Real circuits have imperfections which in this case means that they exhibit attenuation and reflection, however small. Thus, part of the incident signal a_1 is transmitted as S_{21} and part is reflected as S_{11}. Consequently S_{11} is the input reflection coefficient and S_{21} is the transmission coefficient. Using the same reasoning for port 2, any incident signal a_2 is decomposed into two components, the reflected component S_{22} and the transmitted component S_{12}. It is evident that S_{12} transmits in the reverse direction and if the circuit or device were unidirectional, S_{12} would be zero. In summary, S_{11} and S_{22} are the input and output reflection coefficients of the two-port. The transmission coefficients (either gain or attenuation) are denoted by S_{21} and S_{12}.

The foregoing relationship may be expressed mathematically in the following form

$$b_1 = S_{11}\,a_1 + S_{12}\,a_2 \tag{5.16}$$

$$b_2 = S_{21}\,a_1 + S_{22}\,a_2 \tag{5.17}$$

In matrix form eqns (5.16) and (5.17) become

$$\begin{bmatrix} b_1 \\ b_2 \end{bmatrix} = \begin{bmatrix} S_{11} & S_{12} \\ S_{21} & S_{22} \end{bmatrix} \begin{bmatrix} a_1 \\ a_2 \end{bmatrix} = [S] \begin{bmatrix} a_1 \\ a_2 \end{bmatrix} \tag{5.18}$$

where the scattering matrix is

$$[S] = \begin{bmatrix} S_{11} & S_{12} \\ S_{21} & S_{22} \end{bmatrix} \tag{5.19}$$

Radio frequency and microwave circuits typically use characteristic impedances of 50 Ω. Thus, if the two-port is driven by a signal from a 50 Ω source, and if the output is terminated into 50 Ω, then there will be no reflection from the output, since the latter is matched. Thus a_2 is zero. Equations (5.1) and (5.2) therefore reduce to

$$b_1 = S_{11}\,a_1 \quad \text{or} \quad S_{11} = \left.\frac{b_1}{a_1}\right|_{a_2=0} \tag{5.20}$$

$$b_2 = S_{21}\,a_1 \quad \text{or} \quad S_{21} = \left.\frac{b_2}{a_1}\right|_{a_2=0} \tag{5.21}$$

Reversing the arrangement, i.e. driving port 2 and terminating port 1, gives

$$b_1 = S_{12}\,a_2 \quad \text{or} \quad S_{12} = \left.\frac{b_1}{a_2}\right|_{a_1=0} \tag{5.22}$$

$$b_2 = S_{22}\, a_2 \qquad \text{or} \qquad S_{22} = \left.\frac{b_2}{a_2}\right|_{a_1=0} \tag{5.23}$$

From eqns (5.20) and (5.23) it can be seen that S_{11} and S_{22} are power ratios, or more specifically ratios of reflected to incident power. That is of course the same as the reflection coefficient (as discussed in Ch. 4), denoted by the symbol Γ. Since the input and output port of Fig. 5.2 are denoted by the suffix 1 and 2, the input and the output reflection coefficient may be written as Γ_1 and Γ_2. Obviously, if one were to call the input the 'source' and the output the 'load', then the respective reflection coefficients are Γ_s and Γ_l.

By setting either $a_1 = 0$ or $a_2 = 0$, the physical meaning of the scattering parameters can be explained. Equation (5.5) states the relationship of the reflected and incident wave and can be expressed in S-parameter form as

$$S_{11} = \left.\frac{b}{a_1}\right|_{a_2=0}$$

Using eqns (5.11) and (5.12)

$$S_{11} = \frac{(V_1 - I_1\, Z_o)/2\sqrt{Z_o}}{(V_1 + I_1\, Z_o)/2\sqrt{Z_o}} = \frac{V_1 - I_1\, Z_o}{V_1 + I_1\, Z_o} \tag{5.24}$$

Noting that $Z_{in} = V_1/I_1$ and $Z_{out} = V_2/I_2$

$$S_{11} = \frac{Z_{in} - Z_o}{Z_{in} + Z_o} = \Gamma_s \tag{5.25}$$

This is of course the reflection coefficient of the input. Similarly, we obtain for the output reflection coefficient

$$S_{22} = \Gamma_l = \frac{Z_{out} - Z_o}{Z_{out} + Z_o} \tag{5.26}$$

Equations (5.25) and (5.26) can be solved for Z_{in} and Z_{out}, thus

$$Z_{in} = Z_o\, \frac{1 + S_{11}}{1 - S_{11}} \tag{5.27}$$

$$Z_{out} = Z_o\, \frac{1 + S_{22}}{1 - S_{22}} \tag{5.28}$$

It is this relationship between impedance and reflection coefficient which forms the foundation for Smith chart calculations. The two reflection coefficients S_{11} and S_{22} can be plotted on the Smith chart and directly converted into impedances, as will be shown. Solutions for matching problems are thus easily obtained.

In order to obtain S_{12} and S_{21} we proceed as follows, using eqn. (5.20):

$$S_{21} = \frac{b_2}{a_1}\bigg|_{a_2=0}$$

If we now drive the network of Fig. 5.2 with a voltage E_{01}, then half of this voltage drops across Z_o of the signal source and half of it is available across the input of the two-port network in the form of V_1. This implies a matched condition. Hence, for $a_2 = 0$ the input a_1 can be expressed as

$$a_1 = \frac{E_{01}}{\sqrt{Z_o}}, \text{ or from eqn. (5.14)}$$

$$a_2 = 0 = \frac{V_2 + I_2\, Z_o}{2\sqrt{Z_o}} \tag{5.29}$$

Substitution into eqn. (5.15) yields

$$b_2 = \frac{V_2 - 2\, Z_o}{2\sqrt{Z_o}} = \frac{V_2}{\sqrt{Z_o}}$$

Hence

$$S_{21} = \frac{b_2}{a_1} = \frac{V_2}{E_{01}} \tag{5.30}$$

which is the ratio of output to input voltage, i.e. the forward transmission coefficient. Similarly, we obtain for the reverse transmission coefficient

$$S_{12} = \frac{V_1}{E_{02}} \tag{5.31}$$

Through the use of matched conditions advantages may be obtained in that the power reflection coefficients (S_{11}, S_{22}) reduce to voltage reflection coefficients which may be measured with commercially available voltage measuring equipment. This applies equally to S_{12} and S_{21} in that they also reduce to voltage ratios which can be measured with a vector voltmeter.

S-PARAMETER TRANSFER

In the previous sections *S*-parameters were derived and their physical meaning explained. Here it is shown how these *S*-parameters may be transferred onto the Smith chart, the ultimate purpose for this being the use in the design of matching networks. Specifically this could be the input and output matching network for a single amplifier stage. When designing an amplifier one has to bear in mind its prime function. For the purpose of illustration let us consider a small signal amplifier using a FET working at a centre frequency of 10 GHz. Transistor data are usually presented in the form of a data sheet as shown in Fig. 5.3. There are frequently two sets of data, one for low noise operation (3 V, 10 mA) and one for high gain operation (3 V, 30 mA). This information is transferred onto a Smith chart for the parameters S_{11}, S_{22} and on a polar diagram for S_{12}, S_{21}. This gives a visual appreciation of their behaviour with frequency and allows matching network design. This will be shown in a later section.

Referring to Fig. 5.3, the procedure of transfer is as follows:

(a) Select frequency, e.g. 2 GHz
(b) Select *S*-parameter to be transferred, either S_{11} or S_{22}. If S_{11} is selected, then $S_{11} = 0.91,\ -60°$
(c) Draw line from $r = 1.0$ at angle of $-60°$ as shown in Fig. 5.4(a).
(d) If S_{11} was selected, multiply the radius of the Smith chart (84 mm on the full sized original)by the magnitude of S_{11}. This will give 84 mm $\times$ 0.91 = 76.4 mm. Mark off as '2 GHz'.
(e) Repeat for all other frequencies, i.e. 3 to 12 GHz.
(f) Angles on lower half of Smith chart are negative; on upper half positive
(g) Note that there are two different scales for S_{12} and S_{21} on the polar diagram, Fig. 5.4(b), when transferring these parameters

Transfer of the transmission *S*-parameters S_{12} and S_{21} [point (g) above] is self explanatory. Ensure that the axis of the polar diagram and the transmission curves are clearly marked as S_{12} and S_{21}. Finally, observe that the smaller the distribution scatter of the *S*-parameters, the better the transistor. If the transistor were not frequency dependent, then this would manifest itself in a single dot for each of the four sets of *S*-parameters on the Smith chart and polar diagram.

AMPLIFIERS

Amplifiers, and in particular microwave amplifiers, can be put into two categories: amplifiers using bipolar transistors or FETs and narrow/wideband amplifiers. The frequency range over which they amplify may cover the range from about 1 GHz to about 30 GHz. Clearly, the design approach varies depending on the main objective to be achieved. Here we restrict ourselves to small signal narrow-band designs with the objective of either achieving maximum gain or minimum noise figure. Narrow-band refers to amplifiers whose bandwidth is typically about 10% of its centre frequency. Diverse literature exists [4,5,6,7] dealing with other features of these amplifiers.

The source and load impedances, as usual for microwave circuits, are 50 Ω resistive, whilst the active devices used for amplification do have a port impedance

NE720, LOW COST GENERAL PURPOSE GaAs MESFET

NE72084 Typical common source scattering parameters

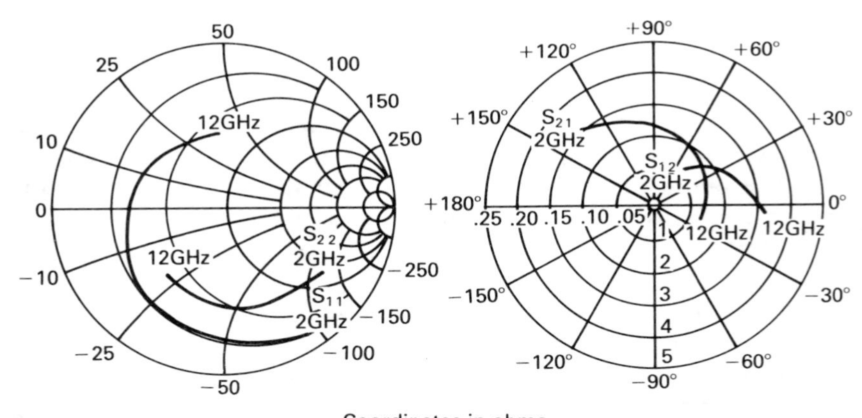

Coordinates in ohms
Frequency in GHz
(V_{DS} = 3 V, I_{DS} = 10 mA)

S—MAGN AND ANGLES

V_{DS} = 3 V, I_{DS} = 10 mA

Frequency (MHz)	S_{11}		S_{21}		S_{12}		S_{22}	
2000	0.91	−60	3.02	131	0.07	54	0.70	−36
3000	0.82	−81	2.65	108	0.09	39	0.66	−49
4000	0.76	−102	2.36	92	0.11	30	0.64	−63
5000	0.72	−122	2.17	76	0.12	21	0.62	−73
6000	0.66	−139	1.97	58	0.12	13	0.61	−80
7000	0.61	−156	1.81	45	0.12	9	0.61	−90
8000	0.56	−176	1.69	34	0.12	5	0.60	−97
9000	0.53	167	1.62	18	0.13	2	0.59	−105
10000	0.50	142	1.59	4	0.13	0	0.57	−111
11000	0.48	114	1.49	−9	0.14	−5	0.55	−116
12000	0.48	94	1.37	−20	0.15	−9	0.50	−128

V_{DS} = 3 V, I_{DS} = 30 mA

Frequency (MHz)	S_{11}		S_{21}		S_{12}		S_{22}	
2000	0.88	−65	4.04	125	0.06	55	0.61	−35
3000	0.78	−94	3.58	101	0.08	39	0.55	−51
4000	0.69	−121	3.11	81	0.09	30	0.50	−66
5000	0.62	−148	2.74	62	0.10	23	0.45	−80
6000	0.59	−173	2.46	45	0.10	19	0.44	−93
7000	0.57	166	2.17	29	0.11	15	0.42	−110
8000	0.56	148	2.01	14	0.11	12	0.42	−122
9000	0.56	128	1.88	−0	0.12	9	0.42	−138
10000	0.57	107	1.78	−15	0.14	6	0.43	−152
11000	0.58	87	1.64	−33	0.15	−6	0.43	−174
12000	0.60	72	1.47	−46	0.16	−13	0.41	166

Fig. 5.3 — Table of typical microwave transistor *S*-parameters.

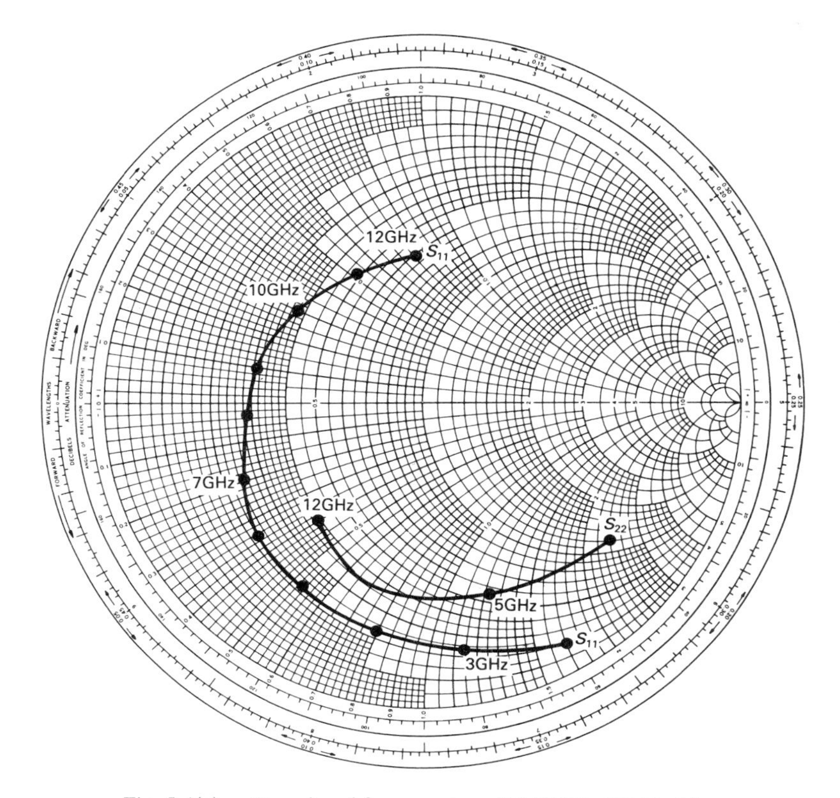

Fig. 5.4(a) — Transfer of S-parameters, NE 72084 of Table 5.3.

which may vary considerably from 50 Ω. This means that they represent different reflection coefficients to both, the source and load impedances of the amplifier. To overcome this problem matching needs to be employed. The following outlines the essential aspects underlying the theory of amplifier design for maximum gain.

MAXIMUM GAIN DESIGN

The block diagram for a single and two-stage amplifier is shown in Fig. 5.5. As was explained in a previous section, S-parameters are used to design the amplifiers. The source reflection coefficient Γ_s and the load reflection coefficient Γ_l must match the input reflection coefficient Γ_{sm} and the output reflection coefficient Γ_{lm} of the active device. Here the suffix m relates to a matched condition. This means that the active device, either bipolar transistor or FET is unconditionally stable, i.e. the stability factor k is larger than unity, as will be shown later.

Depending on the matching used or otherwise, the gain of an amplifier may be defined in various ways. Replacing Fig. 5.5 by the general block diagram in Fig. 5.6, we can define the power gain G in a general manner as the ratio

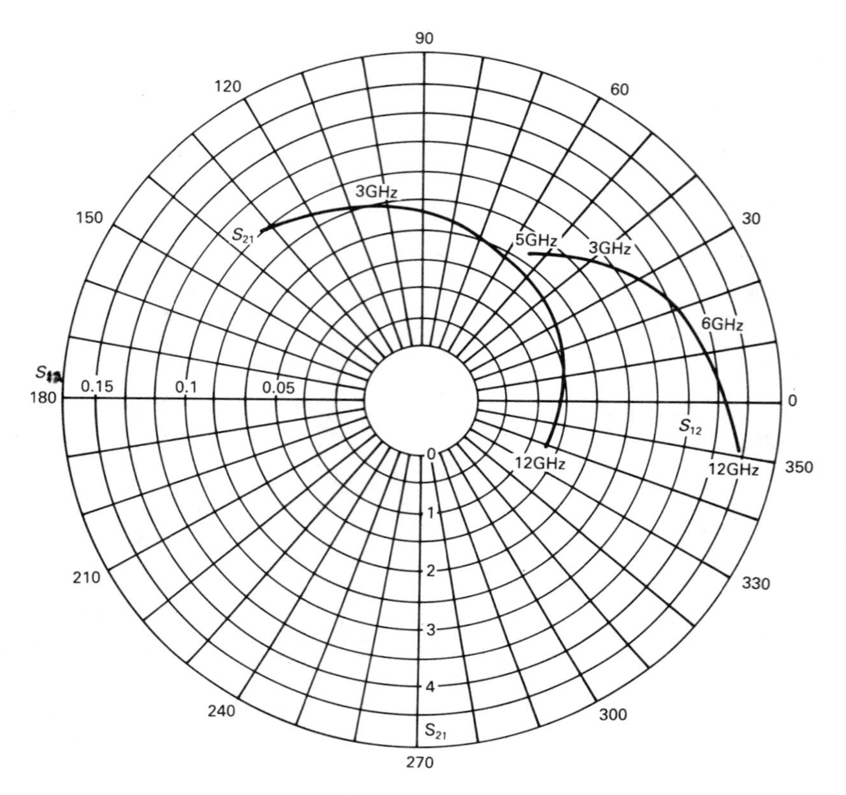

Fig. 5.4(b) — Transfer of S-parameters, NE 72084 of Table 5.3.

$$G = \frac{\text{power delivered to the load}}{\text{power fed into network}} = \frac{P_2}{P_1} = \frac{|a_2|^2 - |b_2|^2}{|a_1|^2 - |b_1|^2} \tag{5.32}$$

The second way of defining gain is to use the optimum power transfer between source and network (or device) input impedance. This means that the input power P_1 becomes the maximum power available, $P_{1\,\text{av}}$, from the source. Hence

$$G_\text{T} = \frac{P_2}{P_1} = \frac{P_2}{P_{1\,\text{av}}} = G_\text{u} \tag{5.33}$$

The gain as expressed in eqn. (5.33) is also referred to as transducer gain G_T or unilateral gain G_u. There exists a special case for the unilateral gain when the output of the network or active device is terminated by a matched load, i.e. when $\Gamma_1 = \Gamma_\text{o}{}^*$. Under this special condition the unilateral gain is known as available gain G_av.

The general analysis of two-port networks is well established [6,7] and only the most relevant equations are stated here and explained. From the above gain definitions and with the aid of Fig. 5.6 we obtain for the transducer power gain

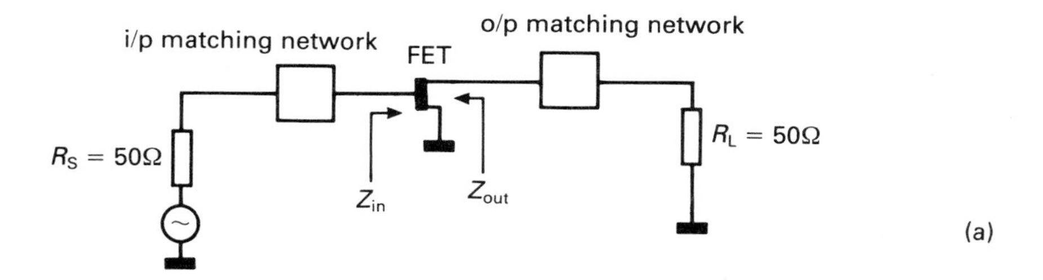

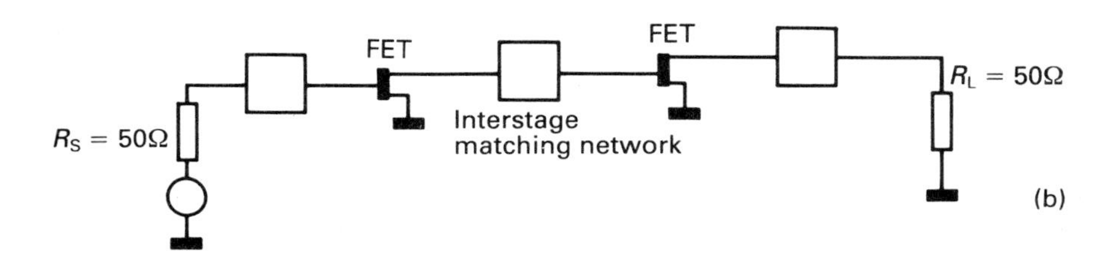

Fig. 5.5 — (a) single stage and (b) two-stage narrow band amplifier. Biasing not shown.

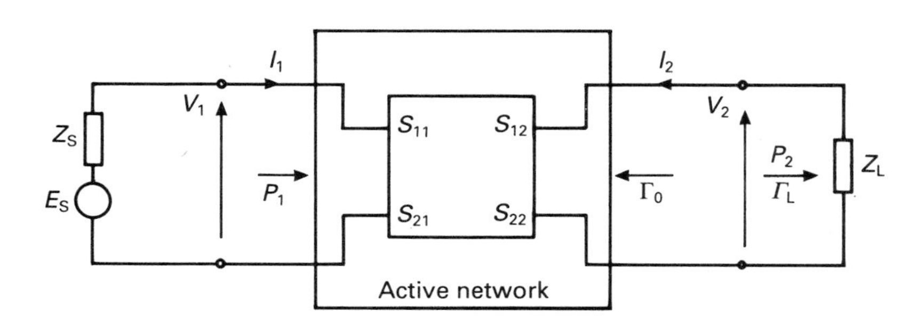

Fig. 5.6 — Generalised presentation of Fig. 5.5(a).

$$G = \frac{P_2}{P_{1\,av}} = \frac{\{Re\ Z_l\}|-I_2|^2}{|E_s|^2/4\{Re\ Z_s\}} \quad \text{or} \tag{5.34}$$

$$G = |S_{21}|^2 \frac{(1-|\Gamma_s|)(1-|\Gamma_l|)}{|(1\text{-}\Gamma_s\ S_{11})(1-\Gamma_l\ S_{22}) - \Gamma_s\Gamma_l S_{12} S_{21}|^2} \tag{5.35}$$

If one assumes the circuit to be unilateral, i.e. $S_{12} = 0$, then eqn. (5.35) simplifies to

$$G_u = |S_{21}|^2 \frac{1 - |\Gamma_s|^2}{|1 - \Gamma_s S_{11}|^2} \times \frac{1 - |\Gamma_l|^2}{|1 - \Gamma_l S_{22}|^2} \tag{5.36}$$

If the assumed network is in reality a microwave amplifier, then S_{21} is the forward power gain as read off the table in Fig. 5.3. S_{21} is thus a fixed parameter and depends only on the bias conditions of the bipolar transistor or GaAs FET. The other remaining parameters are not only interrelated but are also related, viz:

$$\Gamma_s = \frac{Z_s - Z_o}{Z_s + Z_o} \tag{5.37}$$

and

$$\Gamma_l = \frac{Z_l - Z_o}{Z_l + Z_o} \tag{5.38}$$

Equation 5.38 may be further simplified by choosing $\Gamma_s = S_{11}{}^*$ and $\Gamma_l = S_{22}{}^*$. Thus

$$G_{u,\max} = |S_{21}|^2 \frac{1}{1 - |S_{11}|^2} \times \frac{1}{1 - |S_{22}|^2} \tag{5.39}$$

is the maximum gain which may be achieved with the circuit under unilateral conditions. This very special case shows that the gain depends solely on the transistor S-parameters. For maximum gain conditions for a non-unilateral circuit, expression (5.39) becomes more complicated.

Two impedances can be matched conjugately. The same applies to reflection coefficients, e.g. if the source reflection coefficient is the conjugate of the load reflection coefficient. Thus the following conditions hold true for Fig. 5.6:

$$\Gamma_l = \Gamma_s{}^* = S_{11} + \frac{S_{12} S_{21} \Gamma_l}{1 - \Gamma_l S_{22}} \tag{5.40}$$

$$\Gamma_o = \Gamma_l{}^* = S_{22} + \frac{S_{12} S_{21} \Gamma_s}{1 - |_s S_{11}} \tag{5.41}$$

For simultaneous matching of source and load reflection coefficient one obtains the special cases

$$\Gamma_{sm} = C_1{}^*[B_1 + \sqrt{B_1{}^2 - 4|C_1|^2}]/(2|C_1|^2) \tag{5.42}$$

$$\Gamma_{lm} = C_2{}^*[B_2 + \sqrt{B_2{}^2 - 4|C_2|^2}]/(2|C_2|^2) \tag{5.43}$$

where

$$B_1 = 1 + |S_{11}| - |S_{22}| - |D|^2 \tag{5.44}$$

$$B_2 = 1 - |S_{11}| + |S_{22}| - |D|^2 \tag{5.45}$$

$$C_1 = S_{11} - D\ S_{22}{}^* \tag{5.46}$$

$$C_2 = S_{22} - D\ S_{11}{}^* \tag{5.47}$$

$$D = S_{11}\ S_{22} - S_{12}\ S_{21} \tag{5.48}$$

The plus sign in eqns (5.42) and (5.43) is used when B_1 or B_2 are negative and the negative sign is used when B_1 and B_2 are positive. From eqns (5.37) and (5.38)

$$\Gamma_{sm} = \frac{Z_{sm} - Z_o}{Z_{sm} + Z_o} \tag{5.49}$$

and

$$\Gamma_{lm} = \frac{Z_{lm} - Z_o}{Z_{lm} + Z_o} \tag{5.50}$$

The real parts of the matched source and load impedances Z_{sm} and Z_{lm} must always be positive to ensure stability, and for this reason $|\Gamma_{sm}| < 1$ and $|\Gamma_{lm}| < 1$. The resistance of a transistor against oscillation or its stability can be computed from the following expression

$$k = \frac{1 - |S_{11}|^2 - |S_{22}|^2 + |D|}{2\ |S_{12}\ S_{21}|} > 1 \tag{5.51}$$

with D as defined in eqn. (5.48). This is a necessary and sufficient condition when simultaneously matching a two-port;in practical terms this could be an amplifier. Substituting eqns (5.42) to (5.51) into eqn. (5.35) gives the maximum transducer gain G_{max}:

$$G_{max} = \left|\frac{S_{21}}{S_{12}}\right|\ (k + \sqrt{k^2 - 1}) \tag{5.52}$$

if $|S_{12} S_{21}| = 0$ and $B_1 < 0$ and

$$G_{max} = \left|\frac{S_{21}}{S_{12}}\right| (k - \sqrt{k^2 - 1}) \tag{5.53}$$

TRANSISTOR STABILITY CALCULATION

The following is an example as to how the stability factor k may be calculated in longhand. The S-parameters are assumed as follows:

$$\begin{aligned} S_{11} &= 0.445,\ 160.1^\circ \\ S_{12} &= 0.170,\ 133.5^\circ \\ S_{21} &= 2.978,\ 114.7^\circ \\ S_{22} &= 0.679,\ \ 78.9^\circ \end{aligned}$$

The first step is to convert the S-parameters into rectangular form, whence

$$\begin{aligned} S_{11} &= -0.420 + j\,0.15 \\ S_{12} &= -0.120 + j\,0.12 \\ S_{21} &= -1.245 + j\,2.706 \\ S_{22} &= \ \ 0.131 + j\,0.666 \end{aligned}$$

The individual terms are then calculated and substituted into the following two equations:

$$k = \frac{1 + |D| - |S_{11}|^2 - |S_{22}|^2}{2|S_{12}\, S_{21}|}$$

where

$$D = S_{11}\, S_{22} - S_{12}\, S_{21}\ .$$

Thus

$$S_{11}\, S_{22} = (-0.42 + j0.15)\ (0.131 + j0.66) = -0.155 - j0.26$$

$$S_{12}\, S_{21} = (-0.12 + j0.12)\ (-1.245 + j2.707) = -\ 0.1753 - j0.4741$$

$$|S_{12}\, S_{21}| = \sqrt{0.1753^2 + 0.4741^2} = 0.5055$$

$$D = (-0.155 - j0.26) - (-0.1753 - \ j0.4741) = -0.02 + j0.214$$

$$|D|^2 = 0.046$$

$$k = \frac{1 + 0.046 - 0.198 - 0.461}{1.011} = 0.383$$

This shows that the transistor is unstable since $k < 1$.

The following is a printout of the above stability calculation using the computer program KFACTOR, which uses only a fraction of the time required for the longhand calculation.

```
RUN
      ( k-factor calculation )


      Please type ´ 1 ´ to access
      programme

                   OR

      Type  ´ 0 ´(ZERO) to end the
      programme?1



* This programme calculates        *
* the k-factor of transistors      *
*                                  *
* Obtain S11,S12,S21 and S22 from  *
* FET or bipolar transistor data   *
* sheet                            *
*                                  *
* input magnitude & angle (degree) *
*         e.g. as 0.909,-54        *
*                                  *
* If you have made a mistake whilst*
* inputting any of the four S-     *
* parameters,then do not worry .   *
* You have a chance for correction *


     PRESS ANY KEY TO CONTINUE
     =========================


   INPUT:MAGNITUDE , ANGLE (DEGREES)
   ---------------------------------


       INPUT S11?0.445,160.1

       INPUT S12?0.170,133.5

       INPUT S21?2.878,114.7

       INPUT S22?0.679,78.9


       BEFORE CARRYING ON PLEASE

       CHECK THE DATA AGAIN.IF

       IT IS  CORRECT TYPE  Y

       OTHERWISE        TYPE  N
```

```
     data given in POLAR FORM
     -------------------------

          S11    S12    S21    S22

 MOD.     0.445  0.170  2.978  0.679

 ANGLE    160.100133.500114.70078.900

    changing to RECTANGULAR FORM
    ----------------------------

          S11    S12    S21    S22

 REAL     -0.418 -0.117 -1.244 0.131

 IMAG.    0.151  0.123  2.706  0.666


 PRESS ANY KEY TO CONTINUE


S11S22 =-0.156 (REAL)   -0.259 (IMAG)

S21S12 =-0.188 (REAL)   -0.470 (IMAG)

DELTA =0.214(MOD.) 81.276 (ANGLE)

          K =0.382

 The transistor is stable if k is larger than '1'
```

```
L.
   10 REM This programme is called KFACTOR
   20 REM This programme calculates the transistor stability factor k
   30 @%=131850
   40 CLS
   50 MODE 4
   60 VDU 19,1,0,0,0,0
   70 VDU 19,2,2,0,0,0
   80 PRINT"       ( k-factor calculation )      "
   90 PRINT
  100 PRINT
  110 PRINT"       Please type ' 1 ' to access
               programme"
  120 PRINT
  130 PRINT"                      OR              "
  140 PRINT
  150 INPUT"       Type  ' 0 '(ZERO) to end the
               programme",KC
  160 IF KC=1 GOTO 200
  170 IF KC=0 GOTO 60
  180 IF KC<>0 OR KC<>1 OR KC<>2 OR KC<>3 OR KC<>4 OR THEN GOTO 60
  190 IF KC=A$ THEN GOTO 60
  200 CLS
  210 MODE 4
  220 VDU 19,1,0,0,0,0
  230 VDU 19,2,2,0,0,0
  240 PRINT""'""
  250 PRINT"* This programme calculates        *"
  260 PRINT"* the k-factor of transistors      *"
  270 PRINT"*                                  *"
  280 PRINT"* Obtain S11,S12,S21 and S22 from  *"
  290 PRINT"* FET or bipolar transistor data   *"
  300 PRINT"* sheet                            *"
  310 PRINT"*                                  *"
```

```
320 PRINT"* input magnitude & angle (degree) *"
330 PRINT"*          e.g. as 0.909,-54         *"
340 PRINT"*                                    *"
350 PRINT"* If you have made a mistake whilst*"
360 PRINT"* inputting any of the four S-       *"
370 PRINT"* parameters,then do not worry .     *"
380 PRINT"* You have a chance for correction *"
390 PRINT
400 PRINT""'""
410 PRINT"      PRESS ANY KEY TO CONTINUE"
420 PRINT"      ========================="
430 A$=GET$
440 IF A$="0"GOTO 450
450 CLS
460 VDU 19,1,0,0,0,0
470 VDU 19,2,2,0,0,0
480 PRINT""''""
490 PRINT"   INPUT:MAGNITUDE , ANGLE (DEGREES)"
500 PRINT"   --------------------------------"
510 PRINT""'""
520 INPUT"        INPUT S11",M11,A11
530 PRINT
540 INPUT"        INPUT S12",M12,A12
550 PRINT
560 INPUT"        INPUT S21",M21,A21
570 PRINT
580 INPUT"        INPUT S22",M22,A22
590 PRINT""''""
600 PRINT"        BEFORE CARRYING ON PLEASE"
610 PRINT
620 PRINT"        CHECK THE DATA AGAIN.IF  "
630 PRINT
640 PRINT"        IT  IS  CORRECT TYPE  Y"
650 PRINT
660 PRINT"        OTHERWISE         TYPE  N"
670 B$=GET$:IF B$="" GOTO 670 ELSE IF B$= "Y" GOTO 1040 ELSE GOTO 680
680 CLS
690 PRINT""''""
700 PRINT"               (MOD.)  (ANGLE)"
710 PRINT
720 PRINT"     1)S11     ";M11;"   ";A11
730 PRINT
740 PRINT"     2)S12     ";M12;"   ";A12
750 PRINT
760 PRINT"     3)S21     ";M21;"   ";A21
770 PRINT
780 PRINT"     4)S22     ";M22;"   ";A22
790 PRINT""''""
800 PRINT" INPUT LINE NUMBER TO CHANGE OR "
810 PRINT
820 INPUT" 'O' (ZERO) TO CONTINUE. ",NO
830 IF NO=0 GOTO 1040
840 IF NO=1 GOTO 880
850 IF NO=2 GOTO 920
860 IF NO=3 GOTO 960
870 IF NO=4 GOTO 1000
880 CLS
890 PRINT""''""
900 INPUT"INPUT S11",M11,A11
910 GOTO 680
920 CLS
930 PRINT""''''""
940 INPUT"INPUT S12",M12,A12
950 GOTO 680
960 CLS
970 PRINT""''""
980 INPUT"INPUT S21",M21,A21
990 GOTO 680
1000 CLS
1010 PRINT""''''""
1020 INPUT"INPUT S22",M22,A22
1030 GOTO 680
```

```
1040 R11=M11*COS(A11*PI/180)
1050 I11=M11*SIN(A11*PI/180)
1060 R12=M12*COS(A12*PI/180)
1070 I12=M12*SIN(A12*PI/180)
1080 R21=M21*COS(A21*PI/180)
1090 I21=M21*SIN(A21*PI/180)
1100 R22=M22*COS(A22*PI/180)
1110 I22=M22*SIN(A22*PI/180)
1120 CLS
1130 PRINT
1140 PRINT"     data given in POLAR FORM"
1150 PRINT"     -------------------------"
1160 PRINT
1170 PRINT TAB(9);"S11";TAB(16);"S12";TAB(23);"S21";TAB(30);"S22"
1180 PRINT
1190 PRINT TAB(1);"MOD.";TAB(9);M11;TAB(16);M12;TAB(23);M21;TAB(30);M22
1200 PRINT
1210 PRINT TAB(1);"ANGLE";TAB(9);A11;TAB(16);A12;TAB(23);A21;TAB(30);A22
1220 PRINT
1230 PRINT
1240 PRINT"    changing to RECTANGULAR FORM"
1250 PRINT"    ----------------------------"
1260 PRINT
1270 PRINT TAB(9);"S11";TAB(16);"S12";TAB(23);"S21";TAB(30);"S22"
1280 PRINT
1290 PRINT TAB(1);"REAL";TAB(9);R11;TAB(16);R12;TAB(23);R21;TAB(30);R22
1300 PRINT
1310 PRINT TAB(1);"IMAG.";TAB(9);I11;TAB(16);I12;TAB(23);I21;TAB(30);I22
1320 PRINT""'''""
1330 PRINT" PRESS ANY KEY TO CONTINUE"
1340 A$=GET$
1350 IF A$="0"GOTO 1360
1360 CLS
1370 A=R11*R22
1380 B=(I11*I22)*(-1)
1390 C=I11*R22
1400 D11=I22*R11
1410 X=A+B
1420 Y=C+D11
1430 PRINT
1440 PRINT"S11S22 =";X;" (REAL)   ";Y;" (IMAG)"
1450 PRINT
1460 A1=R21*R12
1470 B11=(I21*I12)*(-1)
1480 C1=R21*I12
1490 D1=I21*R12
1500 X11=A1+B11
1510 Y11=C1+D1
1520 PRINT"S21S12 =";X11;" (REAL)   ";Y11" (IMAG)"
1530 Z=X-X11
1540 Z1=Y-Y11
1550 MT=(Z^2+Z1^2)^.5
1560 AT1=ATN(Z1/Z)
1570 AT=AT1*(180/PI)
1580 PRINT
1590 PRINT"DELTA =";MT;"(MOD.) ";AT;" (ANGLE)"
1600 K1=1+(MT)^2-(M11)^2-(M22)^2
1610 K2=2*M12*M21
1620 K=K1/K2
1630 PRINT
1640 PRINT"            K =";K
1650 PRINT
1660 PRINT
1670 PRINT" The transistor is stable if k is larger than '1'"
1680 B1=1+M11^2-M22^2-MT^2
1690 B2=1+M22^2-M11^2-MT^2
1700 PRINT
1710 E=R11-Z*R22-Z1*I22
1720 F=I11+Z*I22-Z1*R22
1730 MC1=(E^2+F^2)^.5
1740 AC=ATN(F/E)
1750 AC1=AC*180/PI
1760 IF K<1 GOTO 1780
```

```
1770 PRINT
1780 E1=R22-Z*R11-Z1*I11
1790 F1=I22+Z*I11-Z1*R11
1800 MC2=(E1^2+F1^2)^.5
1810 AC3=ATN(F1/E1)
1820 AC2=AC3*180/PI
1830 IF K<1 GOTO 1850
1840 PRINT
1850 IF K<1 GOTO 2080
1860 MAO=(M21/M12)*(K-(K^2-1)^.5)
1870 MAG=10*LOG(MAO)
1880 P=(B1-(B1^2-4*MC1^2)^.5)/(2*MC1^2)
1890 P1=P*E
1900 P2=-P*F
1910 MRSM=(P1^2+P2^2)^.5
1920 ARS=ATN(P2/P1)
1930 ARSM=ARS*180/PI
1940 PRINT
1950 P3=(B2-(B2^2-4*MC2^2)^.5)/(2*MC2^2)
1960 P4=P3*E1
1970 P5=P3*(-F1)
1980 MRLM=(P4^2+P5^2)^.5
1990 ARL=ATN(P5/P4)
2000 ARLM=ARL*180/PI
2010 PRINT
2020 PRINT
2030 PRINT""'""
2040 PRINT"   DO YOU WISH TO RUN THE"
2050 INPUT"   PROGRAM AGAIN (Y/N)",A$
2060 IF A$="Y" GOTO 40
2070 GOTO 60
2080 MSO=M21/M12
2090 MSG=LOG(MSO)*10
2100 PRINT
2110 PRINT"         ";   ;"      "
2120 END
```

PRACTICAL AMPLIFIER DESIGN

In previous sections and chapters we have discussed, amongst others, transmission lines, substrate materials, matching, the stability factor and the Smith chart. Now we bring together all these strands in order to design a single stage microwave amplifier. The technique which we adopt here is to use microstrip with a 50 Ω line impedance throughout as shown in Fig. 5.7. For the moment we concern ourselves solely with the input and output matching networks, consisting of the microstrip length L_1, L_2, L_3 and L_4.L_2 and L_3 are referred to as the input and output line respectively. The matching stubs L_1 and L_4 may be either open or short circuited. The gaps in the input and output feedlines are for capacitors to stop a d.c. path into the source and load. The gap between the input and output lines accommodates the microwave transistor and is typically around 2 mm.

Biasing of the transistor takes place via two quarter wave transmission lines (*g*) which have solder pads (e,f) at one end. Opposite these pads is another which is shimmed to ground. A capacitor of typically between 100 and 1000 pF is soldered onto the pads, thus creating an effective RF short circuit at one end of the bias line. If the bias line is now made a quarter-wave long, then we obtain an infinite impedance at that end of the line which is attached to the transistor, e.g. the gate/drain or base/collector as the case may be. The width of the bias line should be made as narrow as

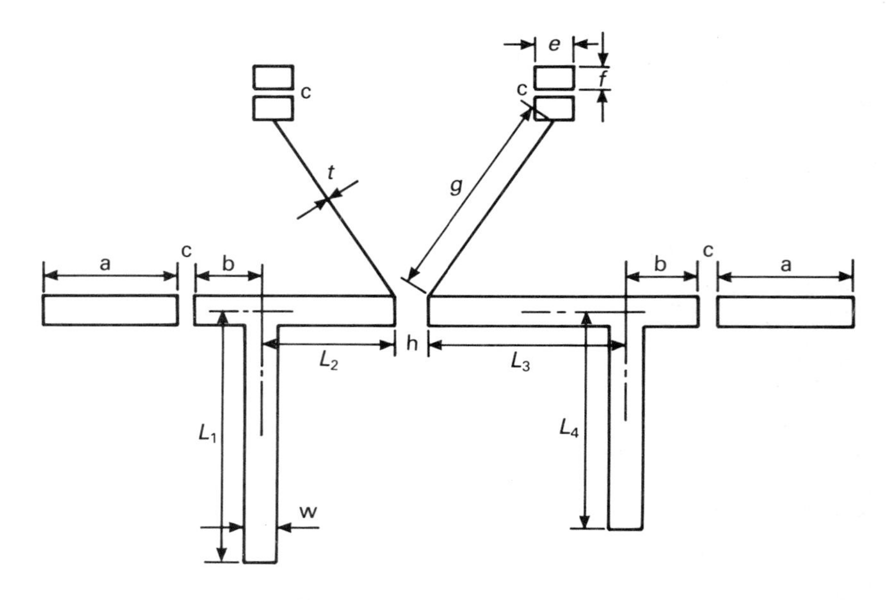

Fig. 5.7 — Sample layout of an amplifier using 50 Ω microstrip throughout. C = capacitor gap.

possible, since this will enhance the bias line impedance and assist the creation of a high impedance transistor feedpoint.

A short circuited matching stub may be created by drilling a hole at the end of the stub through the substrate and electrically join the stub end and groundplane. The shorter the connection the better. Now consider the following problem.

Problem

An amplifier working at a centre frequency of 6 GHz is to be driven by a 50 Ω source. (a) Calculate k, G_{max}, Γ_{sm}, Γ_{lm}. Design a single short circuit stub matching network to optimize gain. The substrate permittivity is 10.2. Utilize the S-parameters given below. (b) Hence obtain by means of the Smith chart

(1) the electrical length of the input line and stub;
(2) the physical length of the input line and stub.

$S_{11} = 0.614, \ -167°$ $\qquad S_{12} = 0.045, 65°$
$S_{21} = 2.187, 32.4°$ $\qquad S_{22} = 0.716, \ -83°$

Solution:

(a) The stability factor may be calculated as shown before either long hand or through the use of program KFACTOR. The stability factor is about 1.2, which means the transistor is unconditionally stable.
The maximum is

$$G_{max} = \left|\frac{S_{21}}{S_{12}}\right| \ (k - \sqrt{k^2 - 1}) = 47.54 \ (1.1296 - 0.525) = 28.72$$

$$G_{max}\ [\text{dB}] = 10 \log 28.72 = 14.58$$

The reflection coefficients are obtained in the following manner:

$$B_1 = 1 + |S_{11}|^2 - |S_{22}|^2 - |\Delta|^2 = 1 + 0.376 - 0.512 - 0.117$$

$$B_1 = 0.747 \qquad B_1^2 = 0.558$$

$$B_2 = 1 - |S_{11}|^2 + |S_{22}|^2 - |\Delta|^2 = 1 - 0.376 + 0.512 - 0.117$$

$$B_2 = 1.019 \qquad B_2^2 = 1.0383$$

$$C_1 = S_{11} - \Delta\ S_{22}^*$$

$$= (-0.559 - j0.1339) - [(-0.1345 + j0.3144)(0.0873 + j0.7107)]$$

$$= -0.363 - j0.0658$$

$$|C_1| = 0.3688 \qquad |C_1|^2 = 0.136$$

$$\phi = \text{arc tan}\ (0.0658/0.363) = 10.25°$$

or

$$\phi = -169.75° \text{ since } \tan\phi = \tan[-(180° - \phi)]$$

and $\tan(-\phi) = -\tan\phi$

$$C_2 = S_{22} - \Delta\ S_{11}^*$$

$$= (0.0873 - j0.7107) - [(-0.1345 + j0.3144)(-0.5992 + j0.1339)]$$

$$= -0.0488 - j0.5044$$

$$|C_2| = (0.0488^2 + 0.5044^2)^{0.5} = 0.507$$

$$|C_2|^2 = 0.2567$$

$$\phi = \text{arc tan}\ \frac{-0.5044}{0.0488}\ \text{arc tan}\ (-10.33) = -84.47°$$

$$\Gamma_{sm} = \frac{C_1^*[B_1 + \sqrt{B_1^2 - 4|C_1|^2}]}{2|C_1|^2}$$

Use negative sign before square root, since B is positive.

$$\Gamma_{sm} = \frac{(-0.363 + j0.0658)[0.747 - \sqrt{0.558 - 4 \times 0.136}]}{0.272} =$$

$$= -0.8388 + j0.152$$

$$|\Gamma_{sm}| = \sqrt{0.7035 + 0.0231} = 0.8524$$

$$\phi = \text{arc tan}\left(-\frac{0.152}{0.8388}\right) = -10.25°$$

$$\Gamma_{lm} = \frac{C_2[B_2 - \sqrt{B_2{}^2 - 4|C_2|^2}]}{2|C_2|^2}$$

$$= \frac{0.0488 + j0.5044\ [1.019 - \sqrt{1.019^2 - 4 \times 0.2567}]}{0.5134}$$

$$= 0.0866 + j0.8955$$

$$|\Gamma_{lm}| = 0.9 \qquad \text{and} \qquad \phi = 84°$$

(b) Using the Smith chart of Fig 5.8:

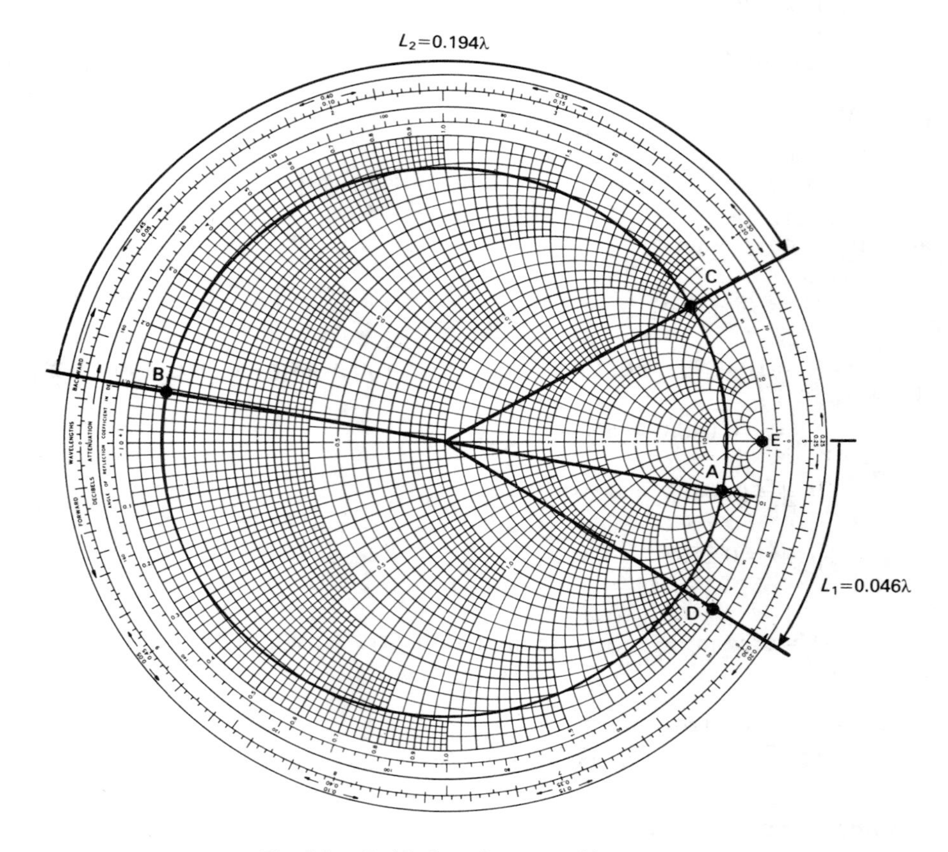

Fig. 5.8 — Smith chart, input matching.

— plot Γ_{sm} at A giving $z = (5.2 - j6.5)\ \Omega$;
— go to admittance mode and plot y at B;
— draw a circle from B TOWARDS the generator until it intersects with the unity circle at C;
— angle BC $= L_2 = 0.194\ \lambda$;
— at C we have $y = (1 + j3.4)\ S$. Thus we need $-j3.4$ to cancel the susceptance of $+j3.4$;
— capacitive susceptance of $-j3.4$ can be found at D;
— since we use a short circuit stub, the short circuit lies at E;
— go from E TOWARDS the generator until D. ED $= L_1 = 0.046\ \lambda$. ED represents the angle through which we have to go to obtain $-j3.4\ S$;
— the electrical lengths are thus $0.194\ \lambda$ and $0.046\ \lambda$ respectively;
— $\lambda = c/f$. For 6 GHz this gives a wavelength of 50 mm. The electrical length are thus $L_2 = 9.7$ mm and $L_1 = 2.3$ mm;
— Owing to the substrate permittivity of 10.2, the physical line lengths are reduced by $(\varepsilon_r)^{0.5} = 3.19$, giving $L_2 = 3$ mm and $L_1 = 0.72$ mm.

Very short line length can be increased by adding integral numbers of $\lambda/2$.

The following is an example of calculating a complete single transistor amplifier stage. The flowchart is given in Fig. 5.9. After loading the program AMP the display as shown on page 134 appears on the video display unit (VDU) requesting the inputting of the S-parameters. Since we want to design the amplifier from conception we choose menu number 1 first in order to check for transistor stability, since this will determine the type of matching approach.

Following this, the S-parameters of the transistor to be used have to be entered. Assuming a centre, frequency of 6 GHz, these parameters can be obtained from the manufacturer's data. In this example we use the same transistor as before and hence the same S-parameters:

$$S_{11} = 0.614,\ -167^\circ \qquad S_{12} = 0.045,\ 65^\circ$$

$$S_{21} = 2.187,\ 32.4^\circ \qquad S_{22} = 0.716,\ -83^\circ$$

From this we obtain $k = 1.16$, which means that the transistor is unconditionally stable since $k > 1$. It is also seen that the maximum available gain is 14.4 dB. The calculations may be repeated or, by pressing N, we can return to the menu.

By pressing menu number 2 one enters the matching program. In the previous program the source and load reflection coefficients $\Gamma_{sm} = R_{SM}$ and $\Gamma_{lm} = R_{LM}$ were calculated [7]. Hence inputting these one obtains various matching possibilities. For the purpose of illustration choose an output line length L_3 of $0.33\ \lambda$ and an open circuit stub of $0.29\ \lambda$. For the source one may select an input line length of $L_2 = 0.192\ \lambda$ and a short circuit stub of length $L_1 = 0.047\ \lambda$. It should be recalled that these lengths are electrical lengths which need to be converted into a physical length depending on the substrate parameters. This is done by choosing menu number 3. Assume a substrate height of 1.27 mm. The physical values of the amplifier are thus obtained by this program. As can be seen, a physical wavelength of 19.227 mm is

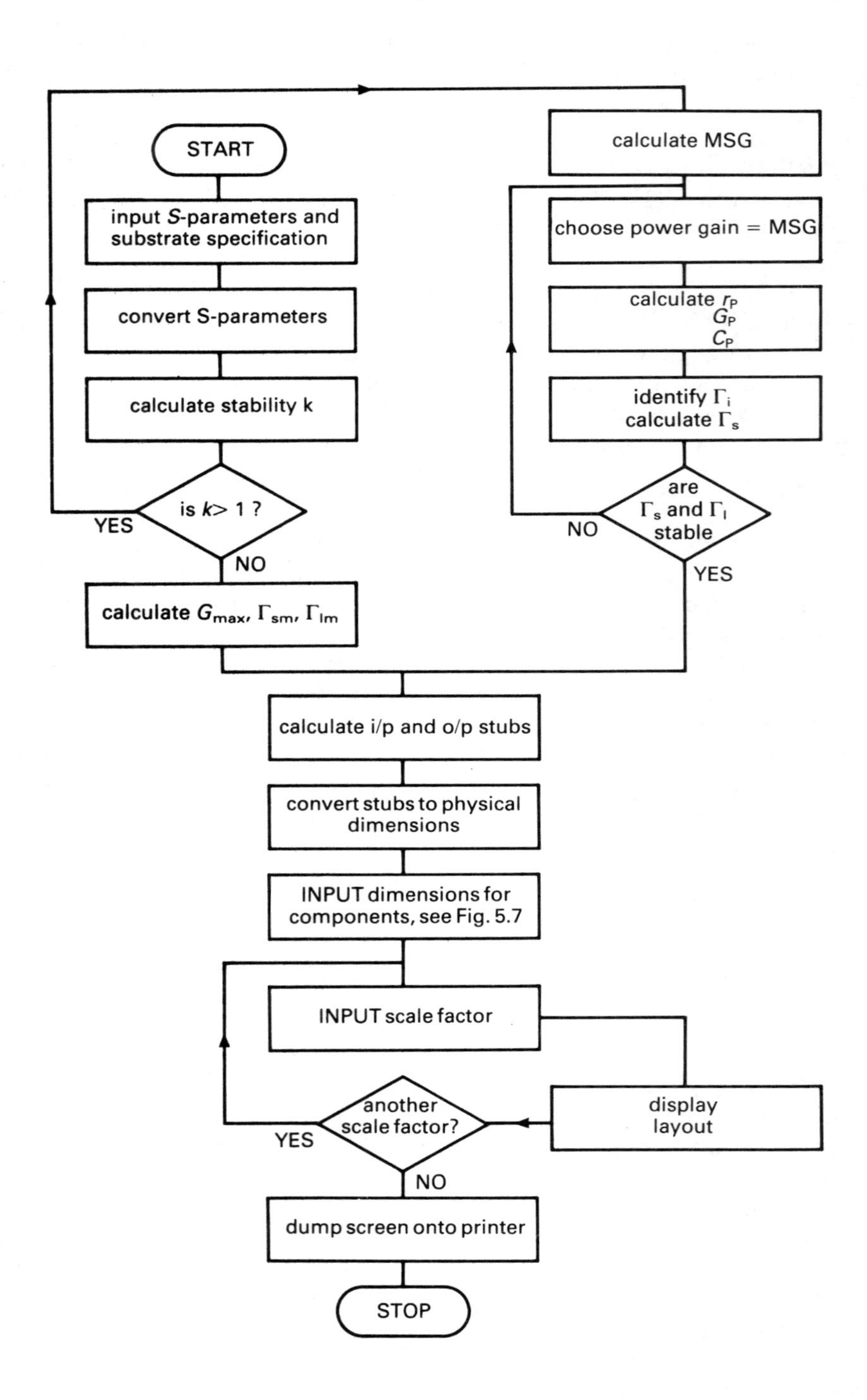

Fig. 5.9 — Flowchart of amplifier program.

obtained and a microstrip width of 1.193 mm. The matching circuits need then be multiplied by this wavelength. This results in an output line length L_3 of $0.33 \times 19.227 = 6.34$ mm. Similarly, $L_4 = 5.57$ mm, $L_2 = 3.6$ mm, $L_1 = 0.90$ mm.

If these dimensions are either too large or small then they may be modified either by choosing a different matching configuration as per menu number 2, or by choosing different substrate parameters which need then be entered into the program of menu number 3 or by adding half wavelengths.

Having calculated the matching network or networks, the physical layout may now be obtained through the use of menu number 4 which plots the layout of the amplifier on a printer.

```
**********************************
*                                *
* LEEDS POLYTECHNIC              *
* ----------------------------   *
*                                *
* DIVISION OF INFORMATION AND    *
*                                *
* ENGINEERING SYSTEMS            *
*                                *
* CALVERLEY STREET               *
*                                *
* LEEDS , LS1 3HE   ,UK          *
*                                *
*    Microwave software          *
*                                *
*    amplifier design.           *
*                                *
*                                *
*  1)stability calculations      *
*                                *
*  2)matching calculations       *
*                                *
*  3)microstrip calculations     *
*                                *
*  4)amplifier layout            *
*                                *
*  For a hard copy of 4) use     *
*  programme AMPLOT              *
**********************************

PLEASE TYPE YOUR CHOICE NUMBER 1,2,3 or 4

TYPE '0'(ZERO) TO END THE PROGRAM?1


* Program No. 1 calculates the      *
* stability factor                  *
*                                   *
* Values of four parameters S11 ,   *
* S12,S21,S22 should be known       *
* before running the program.       *
* Input magnitude & angle (degree)  *
* respectively,e.g. as 0.123,-45.2  *
*                                   *
* If you made a mistake,you are     *
* given another chance to correct   *


       PRESS ANY KEY TO CONTINUE
       =========================
```

```
INPUT:MAGNITUDE , ANGLE (DEGREES)
---------------------------------

    INPUT S11?0.614,-167
    INPUT S12?0.045,65
    INPUT S21?2.187,32.4
    INPUT S22?0.716,-83



    BEFORE WE CARRY ON PLEASE
    CHECK THE DATA AGAIN,IF
    THEY ARE CORRECT TYPE  Y
    OTHERWISE          TYPE  N

  DATA GIVEN IN POLAR FORM
  -------------------------

        S11     S12     S21     S22
MOD.    0.614   0.045   2.187   0.716
ANGLE   -167.000
                65.000  32.400  -83.000

  CHANGING TO RECTANGULAR FORM
  ----------------------------

        S11     S12     S21     S22
REAL    -0.598  0.019   1.847   0.087
IMAG.   -0.138  0.041   1.172   -0.711


PRESS ANY KEY TO CONTINUE PRINTING



S11S22 =-0.150 (REAL)   0.413 (IMAG)
S21S12 =-0.013 (REAL)   0.098 (IMAG)
DELTA =0.344(MOD.) -66.425 (ANGLE)
          K =1.163
          B1=0.746
          B2=1.017
```

```
C1  =0.368 (MAG.)  10.608 (ANG.)

C2  =0.505 (MAG.)  -84.495 (ANG.)

RSM =0.854 (MAG.) -10.608 (ANG.)

RLM =0.891 (MAG.) 84.495 (ANG.)

MAG =14.421 (DB)


    DO YOU WISH TO RUN THE
    PROGRAM AGAIN (Y/N)?


             This part of the programme calculates the
             matching sections for the input and output
             of the active amplifier device

     PRESS ANY KEY TO CONTINUE

INPUT the LOAD REFLECTION COEFF. MAGNITUDE

AND ANGLE (DEGREE)?0.891,84.495


MATCHING IS POSSIBLE AT

                      0.330 WAVELENGTHS

USING AN O/C STUB OF LENGTH

                      0.290 WAVELENGTHS

OR A S/C STUB OF LENGTH

                      0.040 WAVELENGTHS

MATCHING IS ALSO POSSIBLE AT

                      0.405 WAVELENGTHS

USING AN O/C STUB OF LENGTH

                      0.210 WAVELENGTHS

OR A S/C STUB OF LENGTH

                      0.460 WAVELENGTHS


    PRESS ANY KEY TO CONTINUE
```

```
INPUT the SOURCE REFLECTION COEFF.,MAGNITUDE
AND ANGLE (DEGREE)?0.854,-10.608


MATCHING IS POSSIBLE AT
                      0.192 WAVELENGTHS
USING AN O/C STUB OF LENGTH
                      0.297 WAVELENGTHS
OR A S/C STUB OF LENGTH
                      0.047 WAVELENGTHS
MATCHING IS ALSO POSSIBLE AT
                      0.279 WAVELENGTHS
USING AN O/C STUB OF LENGTH
                      0.203 WAVELENGTHS
OR A S/C STUB OF LENGTH
                      0.453 WAVELENGTHS

    PRESS ANY KEY TO CONTINUE
```

```
This part of the programme calculates the microstrip
width and effective permittivity.
It also calculates the effective wavelength

   PRESS ANY KEY TO CONTINUE

  INPUT ZO?50
  INPUT ER?10.2
  INPUT the HEIGHT OF SUBSTRATE (MM)?1.27
  INPUT the FREQUENCY (GHZ)?6

 RATIO OF W/H =0.939
 GIVEN H =1.270,THUS W =1.193 (MM)
 FREESPACE WAVELENGTH =50.000 (MM)
 EEFF =     6.762
 EFF WAVELENGTH =19.227 (MM)
 EFF HALFWAVE =9.614 (MM)
 EFF QUARTERWAVE =4.807 (MM)
```

AMPLIFIER ARTWORK

The amplifier has now been designed and the key physical dimensions are available. From the foregoing sections these are the input and output lines, the input and output stubs, and the two bias lines. This is shown in Fig. 5.7. The input and output matching sections are separated by the FET width h which can be obtained from the data sheet and is typically 2 mm. In order to make absolutely sure that there is no radio frequency at the end of the quarter wave bias lines, we connect a capacitor to ground. In order to be able to solder these capacitors to the bias line and furthermore stopping the thin bias line being stripped from the substrate, solder pads are provided. Pad size may be selected via dimensions e and f. The gap c between the pads is governed by the chip capacitor used and is typically 1 mm. The remainder of the circuit dimensions are self-explanatory and can be chosen to suit the required layout.

All that needs to be done now is to provide the input and output connections for the amplifier sockets. For this purpose a very short length of transmission line a and b is connected to the input and output stub with a gap c for the d.c. blocking capacitor.

Clearly, if frequent designs are made it is imperative to provide for computer drawn artwork. Fig. 5.10(a) is a copy of an example of such artwork, as drawn on a printer. A plotter will give better results but is more expensive. Measured results are shown in Fig. 5.10(b) and (c).

BIASING CIRCUITS

Microwave amplifiers employ active devices in order to obtain amplification. These are predominantly bipolar transistors and GaAs FETs. In order to obtain the desired performance in terms of noise figure or gain the transistors must be biased. The following sections outline briefly the basic aspects of bias networks for FETs and bipolar transistors.

Low power bias circuits

When designing a bias supply for a transistor, one has to understand the active device reasonably well to avoid pitfalls. Unlike a MOSFET whose gate is insulated by an oxide layer, the GaAs FET has no gate insulation, i.e. there is a direct path for current flow. Thus, the gate voltage of GaAs FET (n-channel) must always be negative or at most zero, or the narrow gate will short circuit owing to excessive gate current. A typical family of curves for a GaAs FET is shown in Fig. 5.11 which demonstrates that the gate voltage V_{gs} controls the FET channel current, i.e. the current I_d flowing from the drain to the source. The transfer function of a n-channel GaAs FET is shown in Fig. 5.12. The importance of this transfer function lies in the fact that it is being used to set the operating point and hence bias conditions. Since the transconductance of the FET is given by $g_m = dI_d/dV_{gs}$ the operating point in Fig. 5.12 is directly related to g_m.

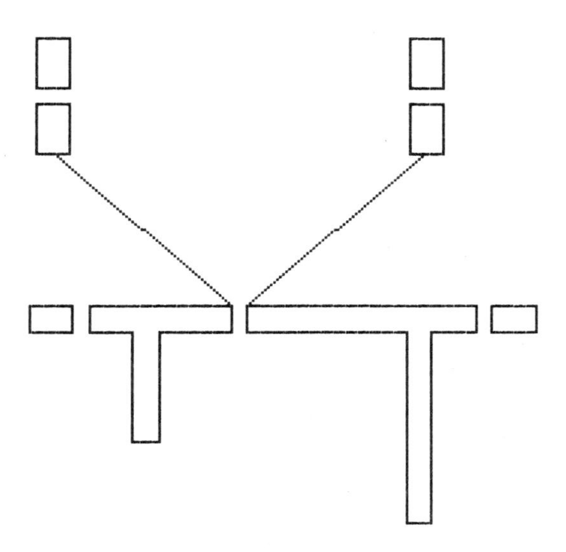

Fig. 5.10(a) — Artwork of amplifier as printed out on an EPSON L800 printer.

There are two fundamental ways of designing a bias circuit as shown in Fig. 5.13, using a dual or a single power supply. Consider a dual power supply first as shown in Fig. 5.13(a). Referring to Fig. 5.12, the transconductance of a FET is given by the well known equation

$$\mathrm{I_d} = I_{\mathrm{dss}}\left(1 - \frac{V_{\mathrm{gs}}}{V_{\mathrm{p}}}\right)^2 \tag{5.54}$$

where I_d = drain current
I_dss = drain-source saturation current
V_gs = gate-source voltage
V_p = pinch off voltage

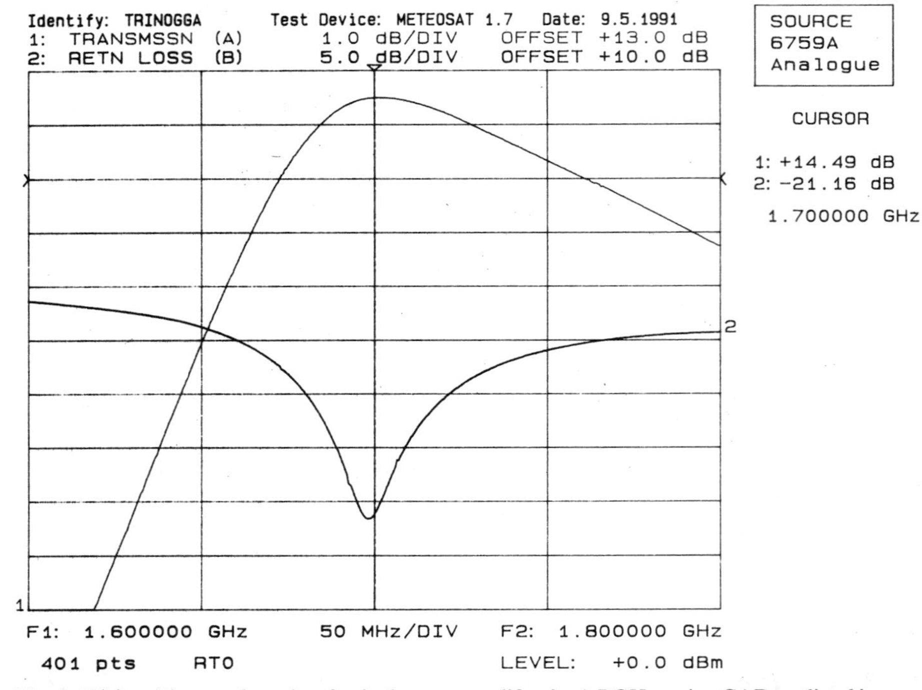

Fig. 5.10(b) — Measured results of a single stage amplifier for 1.7 GHz, using CAD outlined in this chapter.

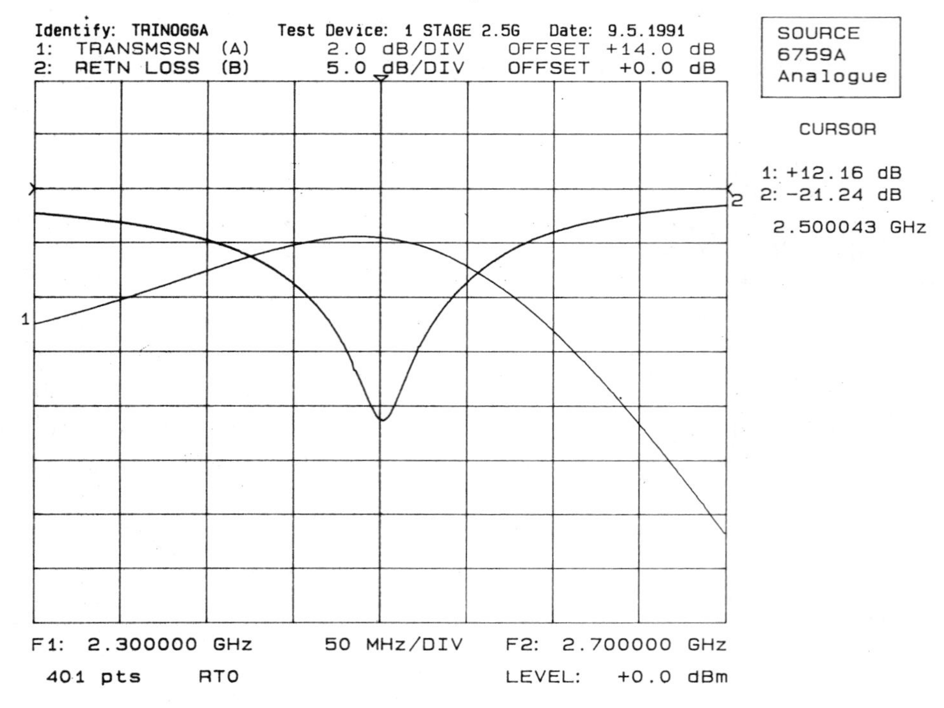

Fig. 5.10(c) — Measured results of a single stage 2.5 GHz amplifier, using CAD outlined in this chapter.

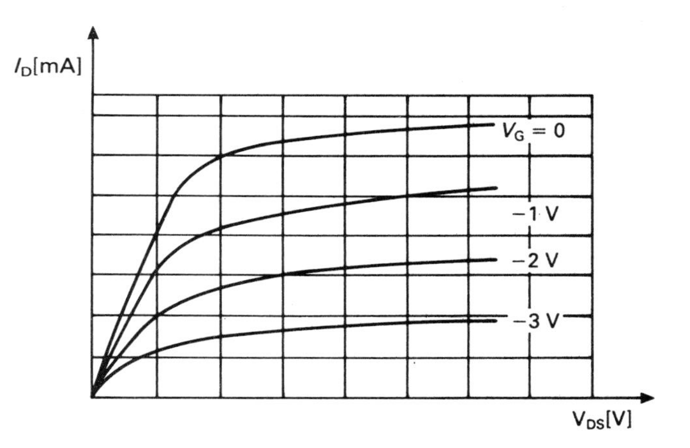

Fig. 5.11 — Typical GaAs FET d.c. characteristics.

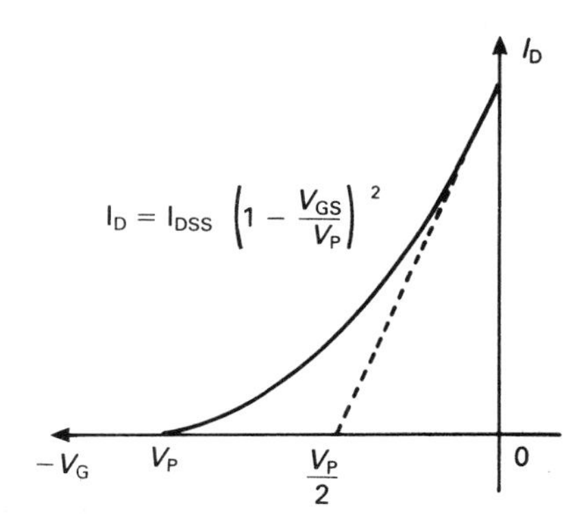

Fig. 5.12 — Square-law characteristic.

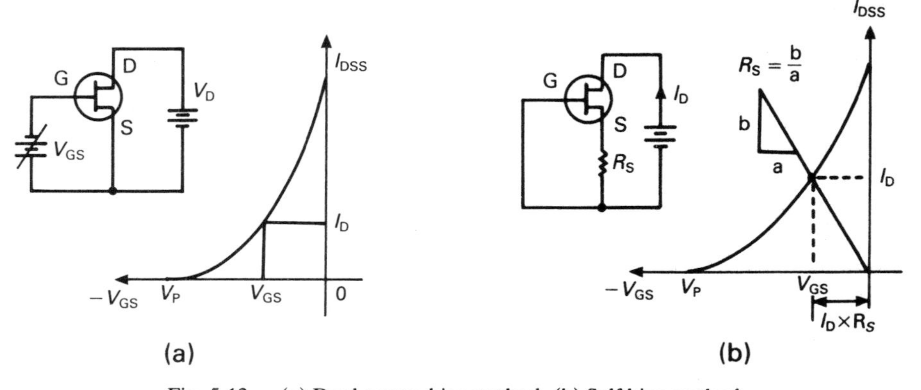

Fig. 5.13 — (a) Dual source bias method. (b) Self bias method

Upon differentiation one obtains for the transconductance g_m

$$g_m = \frac{d\,I_d}{d\,V_{gs}} = -\frac{2\,I_{dss}}{V_p}\left(1 - \frac{V_{gs}}{V_p}\right) \tag{5.55}$$

As indicated earlier, the gate-source bias voltage must be larger than the pinch-off voltage V_p to allow for any current flow and smaller than zero to stop the drain current from exceeding its permissible maximum. Hence

$$V_p < V_{gs} < 0 \tag{5.56}$$

From eqn. (5.54) the gate voltage for any bias may be obtained as

$$V_{gs} = V_p\left(1\sqrt{\frac{I_d}{I_{dss}}}\right) \tag{5.57}$$

Sometimes it may be more desirable to use a single power supply source to bias a transistor circuit. This leads to the self-bias or auto-bias of the transistor and is now being explained with reference to Fig. 5.13(b). The drain current flows through the FET source resistance R_s and creates a voltage across the gate-source terminals

$$V_{ds} = -\,I_d\,R_s \tag{5.58}$$

Combining eqns (5.57) and (5.58) permits the calculation of R_s. Whence

$$R = \frac{V_{gs}}{I_d} = -\frac{V_p}{I_d}\left(1\sqrt{\frac{I_d}{I_{dss}}}\right) \tag{5.59}$$

There is, however, a drawback associated with self-bias in that the source leads of the transistor cannot be directly connected to ground as is desirable for high frequency applications. Any component in the source lead exhibits loss or inductance, resulting in reduced amplifier performance.

Two examples of a schematic layout of a dual and single supply are shown in Fig. 5.14. The capacitors C_1 are input/output d.c. blocking capacitors. The inductors L provide the necessary path for the direct current, but block any microwave frequency from entering the supply.

As a general rule, bias supplies for FETs should be designed in such a way that the gate biases off whilst the drain voltage is rising smoothly to its nominal value, and then the gate voltage is gradually ramped on. This reduces the risk of FET overloading and destruction.

A practical bias supply which can be used to drive up to two amplifier stages [5] is shown in Fig. 5.15. The circuit requires a dual power supply and is suitable for biasing small signal amplifiers requiring drain currents up to about 50 mA at a drain-source

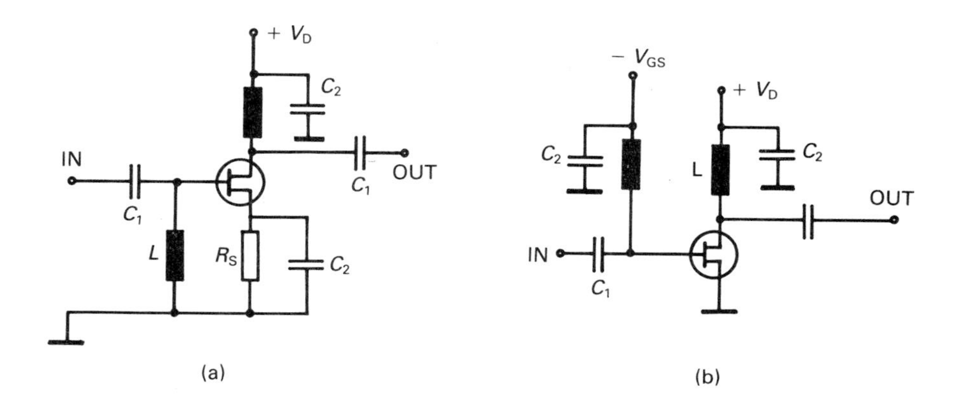

Fig. 5.14 — (a) Single power supply; (b) dual power supply.

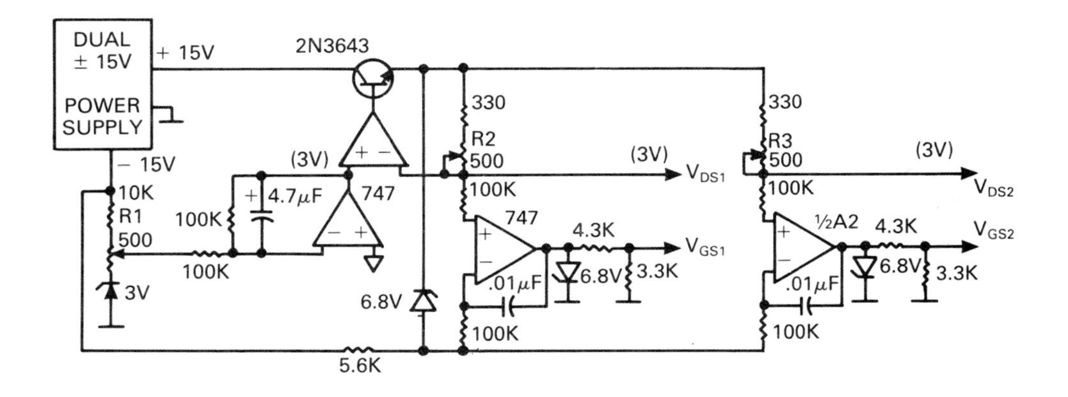

Fig. 5.15 — FET biasing circuit for two stages.

voltage of 3 V. The drain currents can be set by means of R_2/R_3, whilst V_{DS} is set with the help of R_1. Assume for example a typical two-stage amplifier with the high gain. Referring to Fig. 5.15, V_{DS1}/V_{GS1} can be set to the required bias for low noise operation such as $V_{DS} = 3$ V and $I_{DS1} = 10$ mA; in the maximum gain case $V_{DS2} = 3$ V and $I_{DS} = 30$ mA. The drain current can be measured in either case by temporarily connecting a milliampere meter into the drain circuit.

Another bias supply powered from a single power supply is shown in Fig. 5.16. As in the previous case, it can be used to drive up to two transistor stages. If it is required to drive only one stage, components R_3 may be omitted.

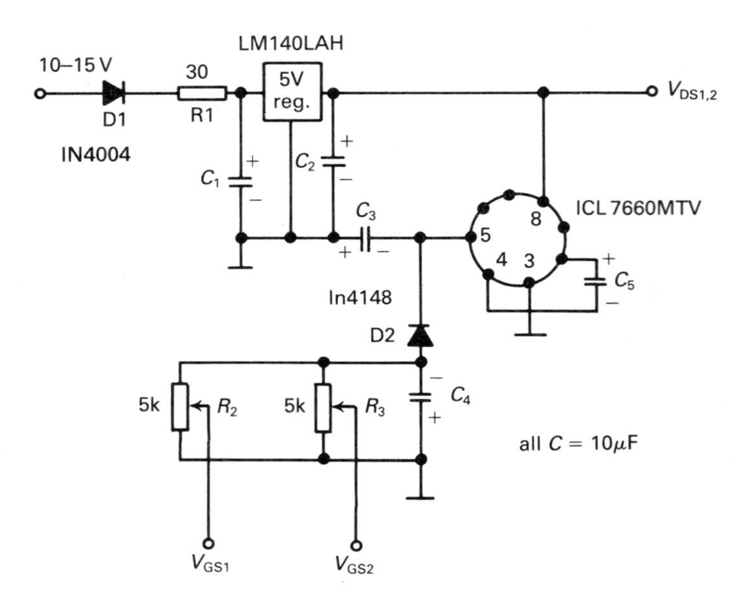

Fig. 5.16 — Alternative FET biasing circuit for two stages.

The bias circuit for a bipolar microwave transistor is shown in Fig. 5.17. The bias current is determined by the choice of resistors R_2 and R_3. The collector current is limited by R_4. It is good practice to decouple the bias circuit by means of capacitors C.

High power bias circuits

On occasions low power microwave signals need boosting by means of medium power driver stages or even high power output stages. It is thus obvious that bias supplies are needed which can supply the necessary higher voltages and currents. A typical driver stage at 10 GHz for example requires a drain current of 170 mA. Table 5.1 is intended to give an appreciation of the gate currents involved in the bias of some high power GaAs FETs.

Table 5.1 — Maximum allowable gate currents for some NEC power FETs

NE9000	0.5 mA
NE9002	2.6 mA
NE9004	5 mA
NE9008	8 mA
NE8684	10 mA
NE8008	15 mA
NE3716	20 mA

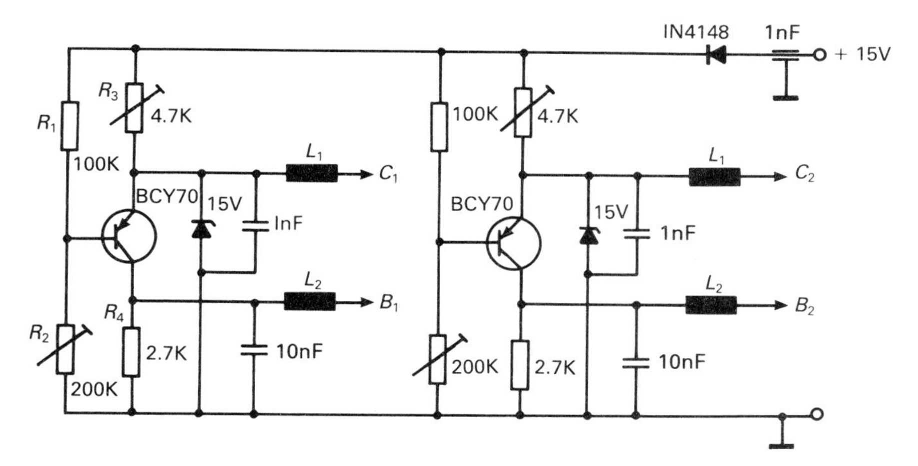

Fig. 5.17 — Two-stage biasing supply for bipolar transistors. B = Base, C = collector.

An example of a gate bias circuit incorporating a series feedback resistor R is shown in Fig. 5.18. As the gate current increases the voltage drop across R increases with subsequent drop in gate bias. This will bias the FET on. The operating drain current can be set by means of V_R. It is advantageous to incorporate a low-pass filter configuration in the gate bias circuit. This will reduce the chance of transistor oscillation and indeed will prevent any microwave power from entering the bias supply.

As indicated earlier, it is sometimes not desirable to self-bias a transistor, especially if it is being used at very high frequencies. Furthermore, it is not desirable from a thermal point of view in the case of power amplifiers or oscillators since any resistor in the source path will reduce heatsinking. Also, it is important to consider the way in which bias is applied. It is good practice to bias the gate off, then apply the drain voltage and finally bias the gate of the FET on. Fig. 5.19 shows the schematic of three approaches to biasing power GaAs FETs. Of these circuits that of (c) is perhaps the most useful. On switching power on, the bias circuit is supplied with voltages + V and − V. Transistor T_2 will be biased on, and in turn T_1. The speed with which V_D will attain its nominal value is governed by the time constant CR of the circuit. When the power supply is switched off, T_2 is biased off and C discharges via D_1 into the power supply side of the circuit, thus preventing an excessive V_D and hence I_d in the FET. A more practical bias supply for a microwave amplifier driver stage is shown in Fig. 5.20.

PROBLEMS

Question 1

Calculate the stability factor k for the following S-parameters in long-hand:

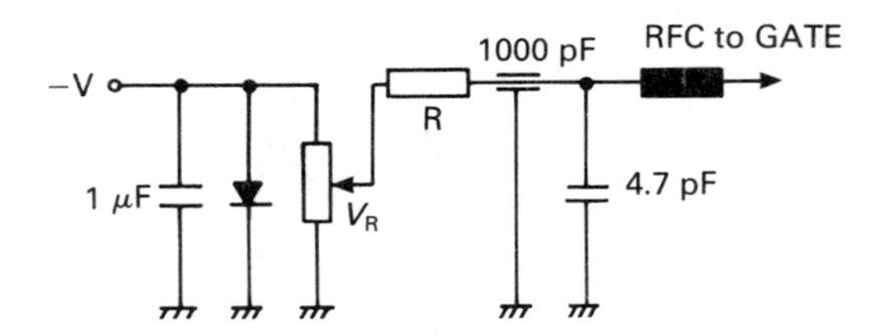

Fig. 5.18 — FET gate biasing circuit.

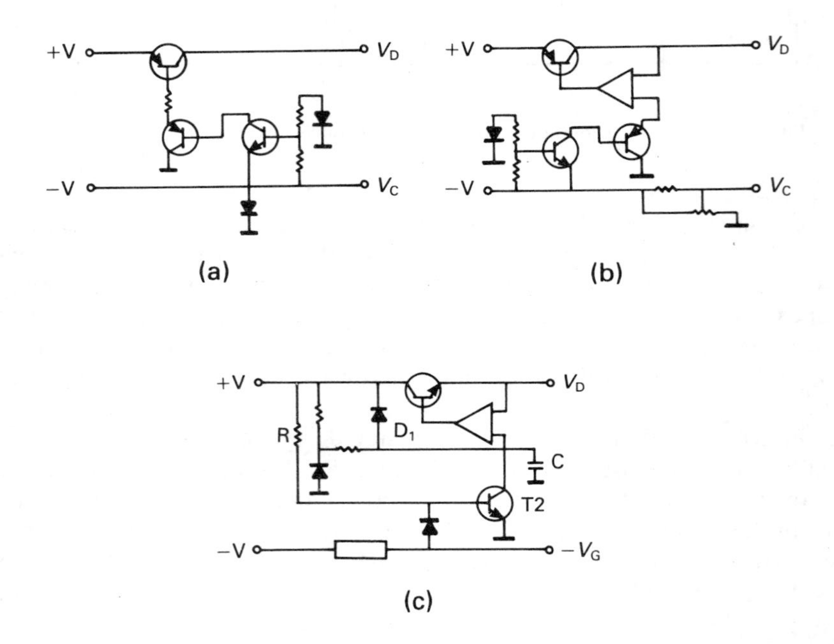

Fig. 5.19 — Power FET biasing circuits, schematic.

$S_{11}=0.6,$	$170°$	$S_{11}=0.75,$	$111°$
$S_{12}=0.1,$	$-22°$	$S_{12}=0.15,$	$15°$
$S_{21}=1.7,$	$10°$	$S_{21}=2.50,$	$80°$
$S_{22}=0.5,$	$-150°$	$S_{22}=0.55,$	$-80°$
(Answer: $k=1.242$)		(Answer: $k=0.481$)	

Question 2

By means of program KFACTOR calculate the stability factor of the transistor documented in the table of Fig. 5.3 at various frequencies. Verify the results in question 1.

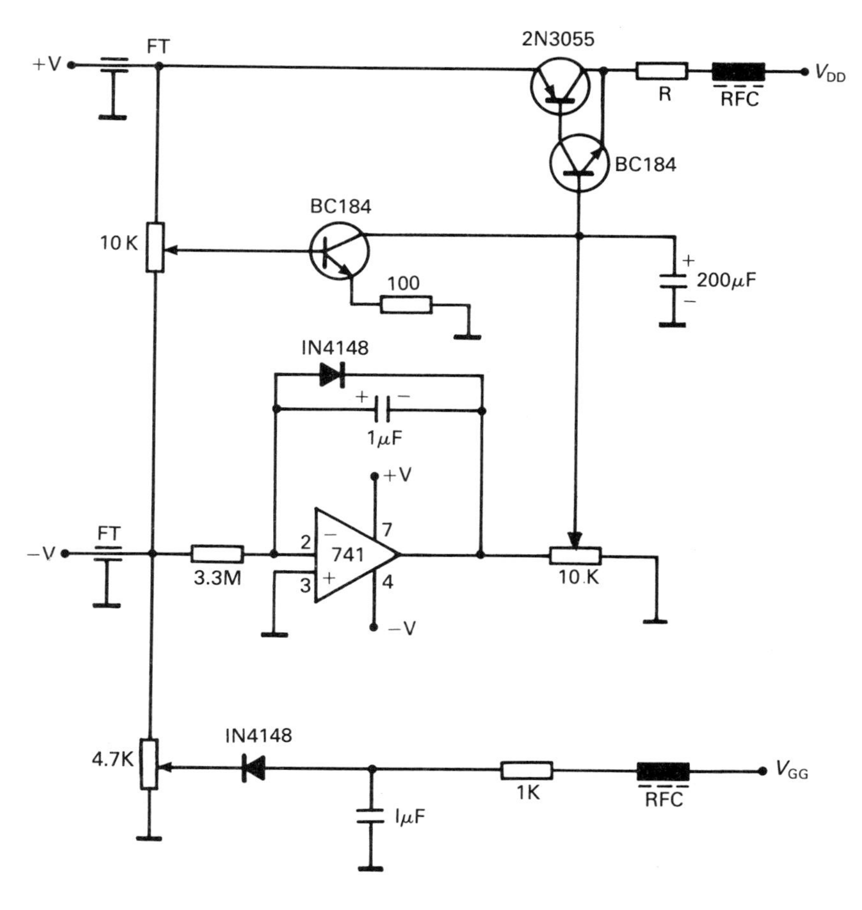

Fig. 5.20 — Biasing circuit for amplifier driving stage. Choose R for current limiting < 170 mA.

L.

```
 10 REM This programme is called AMPLOT
 20 REM This programme plots the amplifier layout based on the calculation of
        programme AMP
 30 @%=131850
 40 CLS
 50 MODE 4
 60 VDU 19,1,0,0,0,0
 70 VDU 19,2,2,0,0,0
 80 PRINT
 90 PRINT"************************"
100 PRINT"*                      *"
110 PRINT"*     Amplifier plot   *"
120 PRINT"*                      *"
130 PRINT"************************"
```

```
140 PRINT
150 PRINT
160 PRINT
170 INPUT"TYPE '1' to START programme",KC
180  IF KC=1 GOTO 240
190 IF KC=0 GOTO 60
200 IF KC<>0 OR KC<>1 OR KC<>2 OR KC<>3 OR KC<>4 OR THEN GOTO 60
210 IF KC=A$ THEN GOTO 60
220 A$=GET$
230 IF A$="0" GOTO 4310
240 CLS
250 MODE 4
260 VDU19,1,0,0,0,0
270 VDU19,2,2,0,0,0
280 REM"AMPLIFIER DESIGN"
290 PRINT
300 INPUT"     INPUT length (L2)",L12
310 L2S=(L12*1280/225.5)
320 PRINT
330 INPUT"     INPUT width (W)",W
340 WVS=(W*1024/163.5)
350 WHS=(W*1280/225.5)
360 PRINT
370 INPUT"     INPUT length (L3)",L13
380 L3S=(L13*1280/225.5)
390 PRINT
400 INPUT"     INPUT length (L1)",L11
410 L1S=(L11*1024/163.5)
420 PRINT
430 INPUT"     INPUT length (L4)",L14
440 L4S=(L14*1024/163.5)
450 PRINT
460 INPUT"     INPUT bias line (g)",g1
470 gS=(g1*1200/202)
480 PRINT
490 INPUT"    INPUT length (b) between stub and C",Z1
500 ZS=(Z1*1280/225.5)
510 PRINT
520
530 INPUT"     INPUT feedline length (a)  ",A1
540 AS=(A1*1280/225.5)
550 PRINT
560 INPUT"     INPUT transistor gap (h)  ",H1
570 HS=(H1*1280/225.5)
580 PRINT
590 INPUT"    INPUT capacitor gaps (c) ",C1
600 CS=(C1*1280/225.5)
610 C2S=(C1*1024/163.5)
620 PRINT
630 INPUT"     INPUT bias pad dimension (e,f) ",M1,N1
640 MS=(M1*1280/225.5)
650 NS=(N1*1024/163.5)
660 GOSUB 1190
670 CLS
680 VDU19,1,3,0,0,0
690 MODE 4
700 MOVE 633,512
710 DRAW (633-L2-(WH/2)-Z),512
720 DRAW (633-L2-(WH/2)-Z),(512-WV)
730 DRAW (633-L2-(WH/2)),(512-WV)
740 MOVE 633,(512-WV)
750 DRAW (633-L2+(WH/2)),(512-WV)
```

```
 760 DRAW (633-L2+(WH/2)),((512-WV)-L1+(WV/2))
 770 DRAW (633-L2-(WH/2)),((512-WV)-L1+(WV/2))
 780 DRAW (633-L2-(WH/2)),(512-WV)
 790 MOVE 633,512:DRAW 633,(512-WV)
 800 O=(633+H)
 810 MOVE O,512
 820 DRAW(O+L3+Z+(WH/2)),512
 830 DRAW(O+L3+Z+(WH/2)),(512-WV)
 840 DRAW(O+L3+(WH/2)),(512-WV)
 850 DRAW(O+L3+(WH/2)),((512-WV)-L4+(WV/2))
 860 DRAW(O+L3-(WH/2)),((512-WV)-L4+(WV/2))
 870 DRAW(O+L3-(WH/2)),(512-WV)
 880 DRAW O,(512-WV)
 890 DRAW O,512
 900 MOVE (C+O+L3+Z+(WH/2)),512
 910 DRAW (C+O+L3+Z+(WH/2)+A),512
 920 DRAW (C+O+L3+Z+(WH/2)+A),(512-WV)
 930 DRAW (C+O+L3+Z+(WH/2)),(512-WV)
 940 DRAW (C+O+L3+Z+(WH/2)),512
 950 MOVE (633-C-Z-L2-(WH/2)),512
 960 DRAW (633-C-Z-L2-(WH/2)-A),512
 970 DRAW (633-C-Z-L2-(WH/2)-A),(512-WV)
 980 DRAW (633-C-Z-L2-(WH/2)),(512-WV)
 990 DRAW (633-C-Z-L2-(WH/2)),512
1000 VDU 29,633;512;
1010 MOVE 0,0
1020 X1=g*COS(RAD(135)):Y1=g*SIN(RAD(135))
1030 DRAW X1,Y1
1040 VDU 29,0;512;
1050 MOVE 0,0
1060 X2=g*COS(RAD(45)):Y2=g*SIN(RAD(45))
1070 DRAW X2,Y2
1080 OX=(g*COS(RAD(135))):OY=(g*SIN(RAD(135)))
1090 VDU 29 ,(OX+633);(OY+512);
1100 MOVE 0,0:DRAW M/2,0:DRAW M/2,N:DRAW -M/2,N:DRAW -M/2,0:DRAW 0,0
1110 MOVE M/2,(N+C2):DRAW M/2,((2*N)+C2):DRAW -M/2,((2*N)+C2):DRAW -M/2,(N+C2):
DRAW M/2,(N+C2)
1120 OX1=(g*COS(RAD(45))):OY1=(g*SIN(RAD(45)))
1130 VDU 29 ,(OX1+O);(OY1+512);
1140 MOVE 0,0:DRAW M/2,0:DRAW M/2,N:DRAW -M/2,N:DRAW -M/2,0:DRAW 0,0
1150 MOVE M/2,(N+C2):DRAW M/2,((2*N)+C2):DRAW -M/2,((2*N)+C2):DRAW -M/2,(N+C2):
DRAW M/2,(N+C2)
1160 INPUT TAB(0,29),"WOULD YOU LIKE TO TRY ANOTHER SCALE(Y/N)",B$
1170 IF B$="Y" GOTO 660
1180 GOTO 1350
1190 INPUT "ENTER SCALE FACTOR",SF
1200 L2=L2S*SF
1210 WV=WVS*SF
1220 WH=WHS*SF
1230 L3=L3S*SF
1240 L1=L1S*SF
1250 L4=L4S*SF
1260 g=gS*SF
1270 Z=ZS*SF
1280 A=AS*SF
1290H=HS*SF
1300C=CS*SF
1310 C2=C2S*SF
1320 M=MS*SF
1330 N=NS*SF
1340 RETURN
1350 END
```

```
L.

 10 REM This programme is called AMP
 20 REM This programme allows a complete amplifier to be designed
 30 @%=131850
 40 CLS
 50 MODE 128
 60 VDU 19,1,0,0,0,0
 70 VDU 19,2,2,0,0,0
 80 PRINT
 90 PRINT"                  **********************************"
100 PRINT"                  *                                *"
110 PRINT"                  * LEEDS POLYTECHNIC              *"
120 PRINT"                  * ----------------------------   *"
130 PRINT"                  *                                *"
140 PRINT"                  * DIVISION OF INFORMATION AND    *"
150 PRINT"                  *                                *"
160 PRINT"                  * ENGINEERING SYSTEMS            *"
170 PRINT"                  *                                *"
180 PRINT"                  * CALVERLEY STREET               *"
190 PRINT"                  *                                *"
200 PRINT"                  * LEEDS , LS1 3HE  ,UK           *"
210 PRINT"                  *                                *"
220 PRINT"                  *    Microwave software          *"
230 PRINT"                  *                                *"
240 PRINT"                  *    amplifier design.           *"
250 PRINT"                  *                                *"
260 PRINT"                  *                                *"
270 PRINT"                  *  1)stability calculations      *"
280 PRINT"                  *                                *"
290 PRINT"                  *  2)matching calculations       *"
300 PRINT"                  *                                *"
310 PRINT"                  *  3)microstrip calculations     *"
320 PRINT"                  *                                *"
330 PRINT"                  *  4)amplifier layout            *"
340 PRINT"                  *                                *"
350 PRINT"                  *  For a hard copy of 4) use     *"
360 PRINT"                  *  programme AMPLOT              *"
370 PRINT"                  **********************************"
380 PRINT
390 PRINT
400 PRINT"                  PLEASE TYPE YOUR CHOICE NUMBER 1,2,3 or 4"
410 PRINT
420 INPUT"                  TYPE '0'(ZERO) TO END THE PROGRAM",KC
430 IF KC=1 GOTO 500
440 IF KC=2 GOTO 3300
450 IF KC=3 GOTO 4180
460  IF KC=4 GOTO 4870
470 IF KC=0 GOTO 60
480 IF KC<>0 OR KC<>1 OR KC<>2 OR KC<>3 OR KC<>4 OR THEN GOTO 60
490 IF KC=A$ THEN GOTO 60
500 CLS
510 MODE 128
520 VDU 19,1,0,0,0,0
530 VDU 18,2,2,0,0,0
540 PRINT""''''"".
550 PRINT"                  * Program No. 1 calculates the     *"
560 PRINT"                  * stability factor                 *"
570 PRINT"                  *                                  *"
580 PRINT"                  * Values of four parameters S11 ,  *"
590 PRINT"                  * S12,S21,S22 should be known      *"
600 PRINT"                  * before running the program.      *"
610 PRINT"                  * Input magnitude & angle (degree) *"
620 PRINT"                  * respectively,e.g. as 0.123,-45.2 *"
625 PRINT"                  *                                  *"
630 PRINT"                  * If you made a mistake,you are    *"
640 PRINT"                  * given another chance to correct  *"
650 PRINT""''""
660 PRINT"                         PRESS ANY KEY TO CONTINUE"
670 PRINT"                         ========================"
680 A$=GET$
690 IF A$="0"GOTO 700
700 CLS
710 VDU 19,1,0,0,0,0
720 VDU 18,2,2,0,0,0
730 PRINT""''""
740 PRINT"   INPUT:MAGNITUDE , ANGLE (DEGREES)"
750 PRINT"   --------------------------------"
```

```
 760 PRINT""''""
 770 INPUT"        INPUT S11",M11,A11
 780 PRINT
 790 INPUT"        INPUT S12",M12,A12
 800 PRINT
 810 INPUT"        INPUT S21",M21,A21
 820 PRINT
 830 INPUT"        INPUT S22",M22,A22
 840 PRINT""'''''""
 850 PRINT"        BEFORE WE CARRY ON PLEASE"
 860 PRINT
 870 PRINT"        CHECK THE DATA AGAIN,IF"
 880 PRINT
 890 PRINT"        THEY ARE CORRECT TYPE  Y"
 900 PRINT
 910 PRINT"        OTHERWISE        TYPE  N"
 920 B$=GET$:IF B$="" GOTO 920 ELSE IF B$= "Y" GOTO 1290 ELSE GOTO 930
 930 CLS
 940 PRINT""'''''""
 950 PRINT"              (MOD.)  (ANGLE)"
 960 PRINT
 970 PRINT"      1)S11     ";M11;"  ";A11
 980 PRINT
 990 PRINT"      2)S12     ";M12;"  ";A12
1000 PRINT
1010 PRINT"      3)S21     ";M21;"  ";A21
1020 PRINT
1030 PRINT"      4)S22     ";M22;"  ";A22
1040 PRINT""'''""
1050 PRINT" INPUT LINE NUMBER TO CHANGE OR "
1060 PRINT
1070 INPUT" '0' (ZERO) TO CONTINUE. ",NO
1080 IF NO=0 GOTO 1290
1090 IF NO=1 GOTO 1130
1100 IF NO=2 GOTO 1170
1110 IF NO=3 GOTO 1210
1120 IF NO=4 GOTO 1250
1130 CLS
1140 PRINT""'''""
1150 INPUT"INPUT S11",M11,A11
1160 GOTO 930
1170 CLS
1180 PRINT""'''""
1190 INPUT"INPUT S12",M12,A12
1200 GOTO 930
1210 CLS
1220 PRINT""'''""
1230 INPUT"INPUT S21",M21,A21
1240 GOTO 930
1250 CLS
1260 PRINT""'''""
1270 INPUT"INPUT S22",M22,A22
1280 GOTO 930
1290 R11=M11*COS(A11*PI/180)
1300 I11=M11*SIN(A11*PI/180)
1310 R12=M12*COS(A12*PI/180)
1320 I12=M12*SIN(A12*PI/180)
1330 R21=M21*COS(A21*PI/180)
1340 I21=M21*SIN(A21*PI/180)
1350 R22=M22*COS(A22*PI/180)
1360 I22=M22*SIN(A22*PI/180)
1370 CLS
1380 PRINT
1390 PRINT"     DATA GIVEN IN POLAR FORM"
1400 PRINT"     ------------------------"
1410 PRINT
1420 PRINT TAB(9);"S11";TAB(16);"S12";TAB(23);"S21";TAB(30);"S22"
1430 PRINT
1440 PRINT TAB(1);"MOD.";TAB(9);M11;TAB(16);M12;TAB(23);M21;TAB(30);M22
1450 PRINT
1460 PRINT TAB(1);"ANGLE";TAB(9);A11;TAB(16);A12;TAB(23);A21;TAB(30);A22
1470 PRINT
1480 PRINT
1490 PRINT"    CHANGING TO RECTANGULAR FORM"
1500 PRINT"    ----------------------------"
1510 PRINT
1520 PRINT TAB(9);"S11";TAB(16);"S12";TAB(23);"S21";TAB(30);"S22"
1530 PRINT
```

```
1540 PRINT TAB(1);"REAL";TAB(9);R11;TAB(16);R12;TAB(23);R21;TAB(30);R22
1550 PRINT
1560 PRINT TAB(1);"IMAG.";TAB(9);I11;TAB(16);I12;TAB(23);I21;TAB(30);I22
1570 PRINT""''''""
1580 PRINT" PRESS ANY KEY TO CONTINUE PRINTING "
1590 A$=GET$
1600 IF A$="0"GOTO 1610
1610 CLS
1620 A=R11*R22
1630 B=(I11*I22)*(-1)
1640 C=I11*R22
1650 D11=I22*R11 ,
1660 X=A+B
1670 Y=C+D11
1680 PRINT
1690 PRINT"S11S22 =";X;" (REAL)    ";Y;" (IMAG)"
1700 PRINT
1710 A1=R21*R12
1720 B11=(I21*I12)*(-1)
1730 C1=R21*I12
1740 D1=I21*R12
1750 X11=A1+B11
1760 Y11=C1+D1
1770 PRINT"S21S12 =";X11;" (REAL)    ";Y11" (IMAG)"
1780 Z=X-X11
1790 Z1=Y-Y11
1800 MT=(Z^2+Z1^2)^.5
1810 AT1=ATN(Z1/Z)
1820 AT=AT1*(180/PI)
1830 PRINT
1840 PRINT"DELTA =";MT;"(MOD.) ";AT;" (ANGLE)"
1850 K1=1+(MT)^2-(M11)^2-(M22)^2
1860 K2=2*M12*M21
1870 K=K1/K2
1880 PRINT
1890 PRINT"              K =";K
1900 B1=1+M11^2-M22^2-MT^2
1910 B2=1+M22^2-M11^2-MT^2
1920 PRINT
1930 PRINT"              B1=";B1
1940 PRINT
1950 PRINT"              B2=";B2
1960 E=R11-Z*R22-Z1*I22
1970 F=I11+Z*I22-Z1*R22
1980 MC1=(E^2+F^2)^.5
1990 AC=ATN(F/E)
2000 AC1=AC*180/PI
2010 IF K<1 GOTO 2040
2020 PRINT
2030 PRINT"C1  =";MC1;" (MAG.)  ";AC1" (ANG.)"
2040 E1=R22-Z*R11-Z1*I11
2050 F1=I22+Z*I11-Z1*R11
2060 MC2=(E1^2+F1^2)^.5
2070 AC3=ATN(F1/E1)
2080 AC2=AC3*180/PI
2090 IF K<1 GOTO 2120
2100 PRINT
2110 PRINT"C2  =";MC2;" (MAG.)  ";AC2;" (ANG.)"
2120 IF K<1 GOTO 2380
2130 MAO=(M21/M12)*(K-(K^2-1)^.5)
2140 MAG=10*LOG(MAO)
2150 P=(B1-(B1^2-4*MC1^2)^.5)/(2*MC1^2)
2160 P1=P*E
2170 P2=-P*F
2180 MRSM=(P1^2+P2^2)^.5
2190 ARS=ATN(P2/P1)
2200 ARSM=ARS*180/PI
2210 PRINT
2220 PRINT"RSM =";MRSM;" (MAG.) ";ARSM;" (ANG.)"
2230 P3=(B2-(B2^2-4*MC2^2)^.5)/(2*MC2^2)
2240 P4=P3*E1
2250 P5=P3*(-F1)
2260 MRLM=(P4^2+P5^2)^.5
2270 ARL=ATN(P5/P4)
2280 ARLM=ARL*180/PI
2290 PRINT
2300 PRINT"RLM =";MRLM;" (MAG.) ";ARLM;" (ANG.)"
2310 PRINT
```

```
2320 PRINT"MAG =";MAG;" (DB)"
2330 PRINT""''''""
2340 PRINT"   DO YOU WISH TO RUN THE"
2350 INPUT"   PROGRAM AGAIN (Y/N)",A$
2360 IF A$="Y" GOTO 40
2370 GOTO 60
2380 MSO=M21/M12
2390 MSG=LOG(MSO)*10
2400 PRINT
2410 PRINT" MSG =";MSG;" (DB)"
2420 G1=MT^2-M11^2
2430 G2=G1^2
2440 G=G2^.5
2450 E2=(R22*Z+I22*Z1-R11)/G1
2460 F2=(I11+I22*Z-R22*Z1)/G1
2470 MCS=(E2^2+F2^2)^.5
2480 AC1=ATN(F2/E2)
2490 AB=(AC1)*180/PI
2500 PRINT
2510 PRINT" Cs =";MCS;" (MAG.) ";AB;" (ANG.)"
2520 MRS=(M12*M21)/G
2530 PRINT
2540 PRINT" RS =";MRS;" (MAG.) "
2550 H1=MT^2-M22^2
2560 H2=H1^2
2570 H3=H2^.5
2580 E3=(R11*Z+I11*Z1-R22)/H1
2590 F3=(I22+I11*Z-R11*Z1)/H1
2600 MCL=(E3^2+F3^2)^.5
2610 AL=ATN(F3/E3)
2620 ACL=AL*180/PI
2630 PRINT
2640 PRINT" CL =";MCL;" (MAG.) ";ACL;" (ANG.)"
2650 MRL=(M21*M12)/H3
2660 PRINT
2670 PRINT" RL =";MRL;" (MAG.)"
2680PRINT
2690 PRINT"    PRESS ANY KEY TO CONTINUE"
2700 A$=GET$
2710 IF A$="0" GOTO2720
2720 CLS
2730 PRINT""''''""
2740 INPUT "  INPUT the POWER GAIN  (dB)",G
2750 PRINT""''""
2760 G1=G/10
2770 GP=(10^G1)/(M21^2)
2780 V=1+GP*(M22^2-MT^2)
2790 E4=((R22-Z*R11-Z1*I11)*GP)/V
2800 F4=((-I22-Z*I11+Z1*R11)*GP)/V
2810 MCP=(E4^2+F4^2)^.5
2820 ACP1=ATN(F4/E4)
2830 ACP=ACP1*180/PI
2840 PRINT
2850 PRINT"  CP =";MCP;" (MAG.) ";ACP;" (ANG.)"
2860 V3=V^2
2870 V2=V3^.5
2880 RP=(1-(2*K)*(M12*M21)*GP+(M12*M21)^2*GP^2)^.5/V2
2890 PRINT
2900 PRINT"  Rp =";RP
2910 W4=MCP-RP
2920 PRINT
2930 IGL=W4*(SIN(ACP1))
2940 RGL=W4*(COS(ACP1))
2950 W1=(IGL^2+RGL^2)^.5
2960 ACW1=(ATN(IGL/RGL))*(180/PI)
2970 ACW=ACW1-180
2980 PRINT"  GL =";W1;" (MAG.) ";ACW;"(ANG.)"
2990 E5=R11-RGL*Z+IGL*Z1
3000 F5=I11-RGL*Z1-IGL*Z
3010 MN=(E5^2+F5^2)^.5
3020 AN=(ATN(F5/E5))*180/PI
3030 E6=1-RGL*R22+IGL*I22
3040 F6=-RGL*I22-IGL*R22
3050 MD=(E6^2+F6^2)^.5
3060 AD=(ATN(F6/E6))*180/PI
3070 MGS=MN/MD
3080 AGS=-(AN-AD)
3090 PRINT
```

```
3100 PRINT"  GS =";MGS;" (MAG.) ";AGS;" (ANG.)"
3110 AU=AGS*PI/180
3120 E7=MGS*COS(AU)
3130 F7=MGS*SIN(AU)
3140 XO=E7-E2
3150 YO=F7-F2
3160 RSO=(MRS+.02)
3170 PRINT
3180 IF XO^2+YO^2<RSO^2 GOTO 3220
3190 PRINT""''""
3200 PRINT"  SELECTED GAMMA IS ACCEPTABLE"
3210 GOTO 3240
3220 PRINT"",,,,""
3230 PRINT"  SELECTED GAMMA IS UNSTABLE."
3240 PRINT
3250 PRINT "  WOULD YOU LIKE TO TRY ANOTHER"
3260 PRINT
3270 INPUT"  GAIN VALUE?(Y/N)",B$
3280 IF B$="Y" GOTO 2720
3290 GOTO 40
3300 CLS
3310 VDU 19,1,0,0,0,0
3320 VDU 19,2,2,0,0,0
3330 PRINT""''''''""
3340 PRINT"                    This part of the programme calculates the
                           matching sections for the input and output
                           of the active amplifier device "
3350 PRINT""'''''""
3360 PRINT"      PRESS ANY KEY TO CONTINUE"
3370 A$=GET$
3380 IF A$="O" GOTO 3400
3390 FOR I=1 TO 2
3400 CLS
3410 VDU 19,1,0,0,0,0
3420 VDU 19,2,2,0,0,0
3430 IF I=2 GOTO 3470
3440 PRINT""'''''""
3450 PRINT "INPUT the LOAD REFLECTION COEFF. MAGNITUDE"
3460 GOTO 3500
3470 CLS
3480 PRINT""'''''""
3490 PRINT"INPUT the SOURCE REFLECTION COEFF.,MAGNITUDE"
3500 PRINT
3510 INPUT"AND ANGLE (DEGREE)",M3,A3
3520 A4=-A3*PI/180
3530 P7=M3*COS(A4)
3540 Q7=M3*SIN(A4)
3550 A7=(1-Q7^2-P7^2)/((1+P7)^2+Q7^2)
3560 B7=(-2*Q7)/((1+P7)^2+Q7^2)
3570 BAD=((A7^3)-2*A7^2+A7*(1+B7^2))^.5
3580 DEN=A7^2+B7^2-A7
3590 IF ABS(DEN)<1E-6 THEN GOTO 4060
3600 PANBX1=(B7+BAD)/DEN
3610 PANBX2=(B7-BAD)/DEN
3620 X2=ATN(PANBX1)/(2*PI)
3630 X1=ATN(PANBX2)/(2*PI)
3640 Q8=((1-2*A7+A7^2+B7^2)/(A7))^.5
3650 Q9=-Q8
3660 X5=ATN(1/Q8)/(2*PI)
3670 X6=ATN(1/Q9)/(2*PI)
3680 X3=ATN(-Q8)/(2*PI)
3690 X4=ATN(-Q9)/(2*PI)
3700 IF X1<0 THEN X1=X1+.5
3710 IF X2<0 THEN X2=X2+.5
3720 IF X3<0 THEN X3=X3+.5
3730 IF X4<0 THEN X4=X4+.5
3740 IF X5<0 THEN X5=X5+.5
3750 IF X6<0 THEN X6=X6+.5
3760 CLS
3770 PRINT""''""
3780 PRINT"MATCHING IS POSSIBLE AT"
3790 PRINT
3800 PRINT"                          ";X1;" WAVELENGTHS"
3810 PRINT
3820 PRINT"USING AN O/C STUB OF LENGTH "
3830 PRINT
3840 PRINT"                          ";X3;" WAVELENGTHS"
3850 PRINT
```

```
3860 PRINT"OR A S/C STUB OF LENGTH "
3870 PRINT
3880 PRINT"                         ";X5;" WAVELENGTHS"
3890 PRINT
3900 PRINT"MATCHING IS ALSO POSSIBLE AT "
3910 PRINT
3920 PRINT"                         ";X2;" WAVELENGTHS"
3930 PRINT
3940 PRINT"USING AN O/C STUB OF LENGTH"
3950 PRINT
3960 PRINT"                         ";X4;" WAVELENGTHS"
3970 PRINT
3980 PRINT"OR A S/C STUB OF LENGTH "
3990 PRINT
4000 PRINT"                         ";X6;" WAVELENGTHS"
4010 PRINT""''""
4020 PRINT"    PRESS ANY KEY TO CONTINUE"
4030 A$=GET$
4040 IF A$="O" GOTO 4050
4050 GOTO 4100
4060 PRINT"YOU CANNOT MATCH PURE SUSCEPTANCES!"
4070 PRINT"    PRESS ANY KEY TO CONTINUE"
4080 A$=GET$
4090 IF A$="O" GOTO 4100
4100 NEXT I
4110 CLS
4120 PRINT""''''''''""
4130 PRINT"  WOULD YOU LIKE TO TRY THE "
4140 PRINT
4150 INPUT"  PROGRAM AGAIN (Y/N)",A$
4160 IF A$="Y" GOTO 40
4170 GOTO 40
4180 CLS
4190 VDU 19,1,0,0,0,0
4200 VDU 19,2,2,0,0,0
4210 PRINT""''''""
4220 PRINT" This part of the programme calculates the microstrip"
4230 PRINT
4240 PRINT"width and effective permittivity."
4250 PRINT
4260 PRINT"It also calculates the effective wavelength"
4270 PRINT""''''''''''""
4280 PRINT"    PRESS ANY KEY TO CONTINUE"
4290 A$=GET$
4300 IF A$="O" GOTO 4310
4310 CLS
4320 VDU 19,1,0,0,0,0
4330 VDU 19,2,2,0,0,0
4340 PRINT""'''''""
4350 INPUT "    INPUT ZO",ZO
4360 PRINT
4370 INPUT "    INPUT ER",ER
4380 PRINT
4390 INPUT "    INPUT the HEIGHT OF SUBSTRATE (MM)",HI
4400 PRINT
4410 INPUT "    INPUT the FREQUENCY (GHZ)",F
4420 H=(ZO*SQR(2*(ER+1))/119.9)+((ER-1)/(2*(ER+1)))*(LN(PI/2)+(1/ER)*LN(4/PI))
4430 IF ZO<=44-2*ER GOTO 4460
4440 RATIO=1/(EXP(H)/8-1/(4*EXP(H)))
4450 GOTO 4480
4460 D=59.95*(PI)^2/ZO/SQR(ER)
4470 RATIO=2/PI*(D-1-LN(2*D-1))+(ER-1)/PI/ER*(LN(D-1)+0.293-0.517/ER)
4480 IF ZO<=63-2*ER GOTO 4520
4490 RD=1/(1-((1/2/H)*(ER-1)/(ER+1)*(LN(PI/2)+(1/ER)*(LN(4/PI)))))^2
4500 EF=(ER+1)/2*RD
4510 GOTO 4530
4520 EF=ER/(0.96+ER*(.109-.004*ER)*(LOG(10+ZO)-1))
4530 W=RATIO*HI
4540 LO=300/F
4550 LEFF=LO/SQR(EF)
4560 LQ=LEFF/4
4570 LH=LEFF/2
4580 CLS
4590 PRINT""''''""
4600 PRINT"  RATIO OF W/H =";RATIO
4610 PRINT
4620 PRINT"  GIVEN H =";HI;",THUS W =";W;" (MM)"
4630 PRINT
```

```
4640 PRINT"  FREESPACE WAVELENGTH =";LO;" (MM)"
4650 PRINT
4660 PRINT"  EEFF ="EF
4670 PRINT
4680 PRINT"  EFF WAVELENGTH =";LEFF;" (MM)"
4690 PRINT
4700 PRINT"  EFF HALFWAVE =";LH;" (MM)"
4710 PRINT
4720 PRINT"  EFF QUARTERWAVE =";LQ;" (MM)"
4730 PRINT""'''''""
4740 PRINT"      DO YOU WISH TO RUN THE"
4750 PRINT
4760 INPUT"      PROGRAM AGAIN  (Y/N)",A$
4770 IF A$="Y" GOTO 40
4780 GOTO 4790
4790 CLS
4800 MODE 128
4810 VDU19,1,0,0,0,0
4820 VDU19,2,2,0,0,0
4830 PRINT""'''''''''''''''""
4840 PRINT "        END OF PROGRAM"
4850 PRINT "        ***************"
4860 GOTO  5980
4870 CLS
4880 MODE 128
4890 VDU19,1,0,0,0,0
4900 VDU19,2,2,0,0,0
4910 REM"AMPLIFIER DESIGN"
4920 PRINT
4930 INPUT"     INPUT the LENGTH (L2)",L12
4940 L2S=(L12*1280/225.5)
4950 PRINT
4960 INPUT"     INPUT the WIDTH (W)",W
4970 WVS=(W*1024/163.5)
4980 WHS=(W*1280/225.5)
4990 PRINT
5000 INPUT"     INPUT the LENGTH (L3)",L13
5010 L3S=(L13*1280/225.5)
5020 PRINT
5030 INPUT"     INPUT the LENGTH (L1)",L11
5040 L1S=(L11*1024/163.5)
5050 PRINT
5060 INPUT"     INPUT the LENGTH (L4)",L14
5070 L4S=(L14*1024/163.5)
5080 PRINT
5090 INPUT"     INPUT the BIAS LINE LENGTH (g)",g1
5100 gS=(g1*1200/202)
5110 PRINT
5120 INPUT"     INPUT the DISTANCE BETWEEN STUBS & CAP'S (B)",Z1
5130 ZS=(Z1*1280/225.5)
5140 PRINT
5150
5160 INPUT"     INPUT the DISTANCE BETWEEN THE CAP. &SOURCE/LOAD (A)",A1
5170 AS=(A1*1280/225.5)
5180 PRINT
5190 INPUT"     INPUT the TRANSISTOR GAP (h)",H1
5200 HS=(H1*1280/225.5)
5210 PRINT
5220 INPUT"     INPUT the CAPACITOR GAP (c)",C1
5230 CS=(C1*1280/225.5)
5240 C2S=(C1*1024/163.5)
5250 PRINT
5260 INPUT"     INPUT BIAS PADS DIMENSION (e,f)",M1,N1
5270 MS=(M1*1280/225.5)
5280 NS=(N1*1024/163.5)
5290 GOSUB 5820
5300 CLS
5310 VDU19,1,3,0,0,0
5320 MODE 128
5330 MOVE 633,512
5340 DRAW (633-L2-(WH/2)-Z),512
5350 DRAW (633-L2-(WH/2)-Z),(512-WV)
5360 DRAW (633-L2-(WH/2)),(512-WV)
5370 MOVE 633,(512-WV)
5380 DRAW (633-L2+(WH/2)),(512-WV)
5390 DRAW (633-L2+(WH/2)),((512-WV)-L1+(WV/2))
5400 DRAW (633-L2-(WH/2)),((512-WV)-L1+(WV/2))
5410 DRAW (633-L2-(WH/2)),(512-WV)
```

```
 5420 MOVE 633,512:DRAW 633,(512-WV)
 5430 O=(633+H)
 5440 MOVE O,512
 5450 DRAW(O+L3+Z+(WH/2)),512
 5460 DRAW(O+L3+Z+(WH/2)),(512-WV)
 5470 DRAW(O+L3+(WH/2)),(512-WV)
 5480 DRAW(O+L3+(WH/2)),((512-WV)-L4+(WV/2))
 5490 DRAW(O+L3-(WH/2)),((512-WV)-L4+(WV/2))
 5500 DRAW(O+L3-(WH/2)),(512-WV)
 5510 DRAW O,(512-WV)
 5520 DRAW O,512
 5530 MOVE (C+O+L3+Z+(WH/2)),512
 5540 DRAW (C+O+L3+Z+(WH/2)+A),512
 5550 DRAW (C+O+L3+Z+(WH/2)+A),(512-WV)
 5560 DRAW (C+O+L3+Z+(WH/2)),(512-WV)
 5570 DRAW (C+O+L3+Z+(WH/2)),512
 5580 MOVE (633-C-Z-L2-(WH/2)),512
 5590 DRAW (633-C-Z-L2-(WH/2)-A),512
 5600 DRAW (633-C-Z-L2-(WH/2)-A),(512-WV)
 5610 DRAW (633-C-Z-L2-(WH/2)),(512-WV)
 5620 DRAW (633-C-Z-L2-(WH/2)),512
 5630 VDU 29,633;512;
 5640 MOVE 0,0
 5650 X1=g*COS(RAD(135)):Y1=g*SIN(RAD(135))
 5660 DRAW X1,Y1
 5670 VDU 29,0;512;
 5680 MOVE 0,0
 5690 X2=g*COS(RAD(45)):Y2=g*SIN(RAD(45))
 5700 DRAW X2,Y2
 5710 OX=(g*COS(RAD(135))):OY=(g*SIN(RAD(135)))
 5720 VDU 29 ,(OX+633);(OY+512);
 5730 MOVE 0,0:DRAW M/2,0:DRAW M/2,N:DRAW -M/2,N:DRAW -M/2,0:DRAW 0,0
 5740 MOVE M/2,(N+C2):DRAW M/2,((2*N)+C2):DRAW -M/2,((2*N)+C2):DRAW -M/2,(N+C2):
DRAW M/2,(N+C2)
 5750 OX1=(g*COS(RAD(45))):OY1=(g*SIN(RAD(45)))
 5760 VDU 29 ,(OX1+O);(OY1+512);
 5770 MOVE 0,0:DRAW M/2,0:DRAW M/2,N:DRAW -M/2,N:DRAW -M/2,0:DRAW 0,0
 5780 MOVE M/2,(N+C2):DRAW M/2,((2*N)+C2):DRAW -M/2,((2*N)+C2):DRAW -M/2,(N+C2):
DRAW M/2,(N+C2)
 5790 INPUT TAB(0,29),"WOULD YOU LIKE TO TRY AOTHER SCALE(Y/N)",B$
 5800 IF B$="Y" GOTO 5290
 5810 GOTO 5980
 5820 INPUT "ENTER SCALE FACTOR",SF
 5830 L2=L2S*SF
 5840 WV=WVS*SF
 5850 WH=WHS*SF
 5860 L3=L3S*SF
 5870 L1=L1S*SF
 5880 L4=L4S*SF
 5890 g=gS*SF
 5900 Z=ZS*SF
 5910 A=AS*SF
 5920H=HS*SF
 5930C=CS*SF
 5940 C2=C2S*SF
 5950 M=MS*SF
 5960 N=NS*SF
 5970 RETURN
 5980 END
```

REFERENCES

[1] *S*-parameters . . . circuit analysis and design. HP application note 95, Sept. 1968.

[2] Transistor parameter measurements. HP application note 77–1.

[3] *S*-parameter design. HP application note 154, April 1972.

[4] Stripline component measurement with the 8410A network analyser. HP application note 117-2, Jan. 1971.

[5] NEC, Application of microwave GaAs FETs, general technical literature.

[6] Gentili, C., Microwave amplifiers and oscillators, North Oxford Academic, 1986.

[7] Tri, T. Ha, Solid state microwave amplifier design, J. Wiley and Sons, 1981.

6

Microwave filter design

INTRODUCTION

A filter is a frequency selective device with low levels of attenuation or insertion loss in its passband and specified high levels of attenuation in its stopband. An example specification for a microwave low pass filter is shown in Fig. 6.1. In this example the

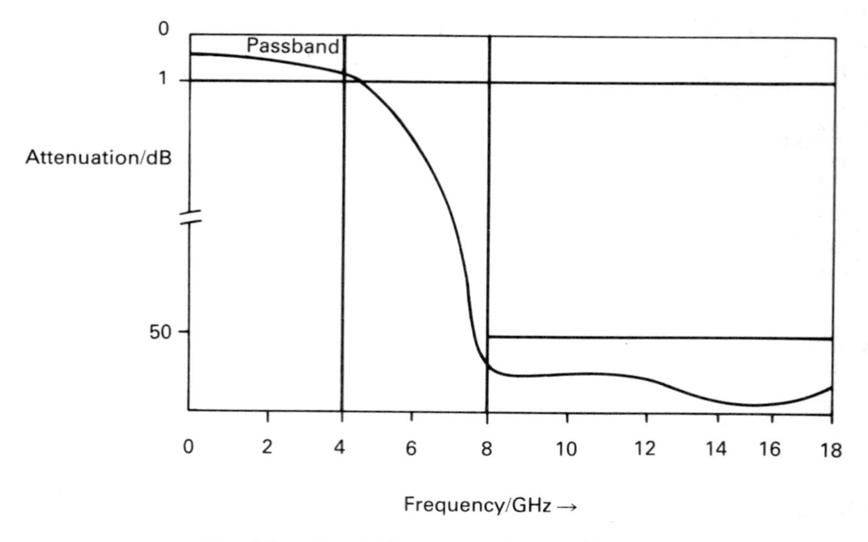

Fig. 6.1 — A typical microwave lowpass filter specification.

passband insertion loss must be less than 1 dB from d.c. to 4 GHz. The stopband attenuation must be greater than 50 dB from 8 GHz to 18 GHz. In addition there is a specification on the input return loss of greater than 20 dB in the passband. This means that any signal incident in the filter in the frequency range of its passband must be 99% transmitted or absorbed; only 1% of incident power can be reflected. A high level of return loss, typically 20 dB or greater, implies a flat low ripple insertion loss characteristic, which is very desirable from the point of view of signal distortion.

How do we design a device to produce this lowpass response, or any other filter response? There is a vast amount of published literature on the design of microwave filters, some of it highly mathematical. Fortunately, there are some relatively straight forward procedures which enable us to design certain useful classes of microwave filters. These procedures follow some basic steps.

First we decide upon the specification. It is usually presented graphically as in Fig. 6.1.

Second, we design a suitable lowpass prototype network. This is a lumped element lowpass filter operating in a 1 Ω impedance system with a cutoff frequency of 1 rad/sec. This prototype filter meets the specification in a normalized sense.

Third, we transform the lowpass prototype into a microwave network. Normally, the microwave network is a distributed circuit composed of various interconnections of transmission lines and coupled transmission lines.

Finally we convert the distributed circuit into a physical structure, using printed circuit, coaxial or waveguide technology.

In this chapter these basic steps have been applied to the design of two simple, but useful, types of microwave filter. These are the stepped impedance lowpass filter and the parallel coupled bandpass filter. An example of each type of filter has been developed from basic theory through to construction. Experimental results are presented for both filters.

LOWPASS PROTOTYPE FILTERS

The maximally flat filter

To understand lowpass prototype filters we shall start by analysing the frequency response of a simple three element LC filter (Fig. 6.2). This type of circuit is known as a ladder network.

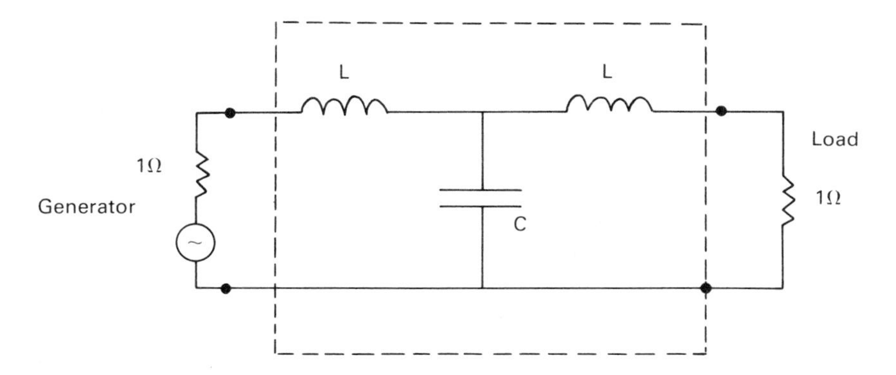

Fig. 6.2 — A three-element lowpasss filter.

The transfer matrix [T] of the filter is

$$[\mathbf{T}] = \begin{bmatrix} 1 & j\omega L \\ 0 & 1 \end{bmatrix} \times \begin{bmatrix} 1 & 0 \\ j\omega C & 1 \end{bmatrix} \times \begin{bmatrix} 1 & j\omega L \\ 0 & 1 \end{bmatrix} \tag{6.1}$$

$$= \begin{bmatrix} 1-\omega^2 LC & j(2\omega L - \omega^3 L^2 C) \\ j\omega C & 1-\omega^2 LC \end{bmatrix} \tag{6.2}$$

The power attentuation between generator and load, or insertion loss, for a lossless network with transfer [**T**] is given by

$$IL = 10 \log\left[1 + \frac{1}{4}\Big((A - D^2) + (B - C)^2\Big)\right] \text{ dB} \tag{6.3}$$

where

$$[\mathbf{T}] = \begin{bmatrix} A & jB \\ jC & D \end{bmatrix} \tag{6.4}$$

Thus for our lowpass filter

$$IL = 10 \log\left[1 + \frac{1}{4}\left(2\omega L - \omega C - \omega^3 L^2 C\right)^2\right] \text{ dB} \tag{6.5}$$

which is a sixth degree function in ω. In general an Nth degree network produces a $2N$th degree insertion loss function.

To simplify this expression we let $L = 1$ and $C = 2$, then

$$IL = 10 \log(1 + \omega^6) \text{ dB} \tag{6.6}$$

This is the insertion loss of a third degree maximally flat filter. For $\omega < 1$ then the insertion loss is small, the insertion loss is equal to 3 dB at $\omega = 1$, and for $\omega > 1$ the insertion loss rapidly increases. In general the Nth degree maximally flat filter has an insertion loss given by

$$IL = 10 \log(1 + \omega^{2N}) \text{ dB} \tag{6.7}$$

The passband of the maximally flat lowpass filter extends to $\omega = 1$; above $\omega = 1$ we rapidly extend into regions of high attenuation, i.e. the stopband.

The name 'Maximally flat' is used because of the behaviour of the derivatives of the function

$$F(\omega) = 1 + \omega^{2N} \tag{6.8}$$

differentiating with respect to ω

$$F^1(\omega) = 2N\omega^{2N-1} = 0 \text{ at } \omega = 0 \tag{6.9}$$

differentiating $2N-1$ times we obtain

$$F^{2N-1}(\omega)=2N!\omega=0 \text{ at } \omega=0 \tag{6.10}$$

The first $2N-1$ derivatives of $F(\omega)$ are zero at $\omega=0$. This obviously implies a flat response around $\omega=0$, hence the name.

In filter design, we need to know N, the degree of the insertion loss function which is required to meet a particular specification. Figure 6.3 shows a specification

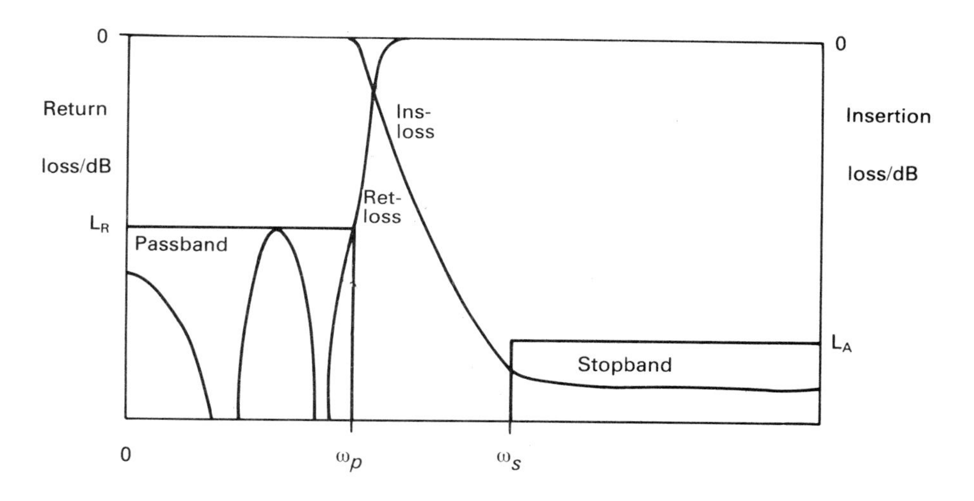

Fig. 6.3 — Specification for a lowpass filter.

for a lowpass filter. L_A is the minimum insertion loss required at frequency ω_s. L_R is the minimum passband return loss required up to a frequency ω_p. Defining S as

$$S=\omega_s/\omega_p \tag{6.11}$$

Then it can be shown that [6.2]

$$N \geqslant \frac{L_A+L_R}{20\log(S)} \tag{6.12}$$

Using the example of Fig. 6.1 where $S=2$, then N must be greater than 11.7. Since N represents the number of elements in the lowpass prototype, it must be an integer, so we round up to $N=12$.

The Chebyshev filter

The maximally flat filter response rolls off to 3 dB at $\omega = 1$ before attentuation starts increasing rapidly. This is not optimum in terms of approximating an ideal 'brickwall' type filter. A better approximation is one where the insertion loss ripples between two values in the passband prior to increasing rapidly in the stopband [6.3]. This type of 'Equiripple' behaviour is more optimum in the sense of minimizing the number of elements to meet a particular specification than the maximally flat response. An example of an equiripple or 'Chebyshev' response is shown in Fig. 6.4 for $N = 5$ and $N = 6$.

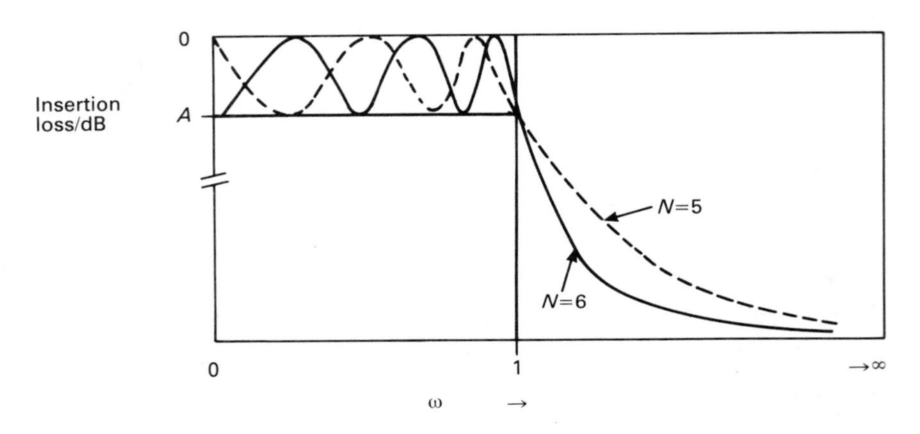

Fig. 6.4 — Equiripple lowpass filter response.

The insertion loss ripple level A is normally expressed as

$$A = 10 \log(1 + \varepsilon^2) \tag{6.13}$$

The passband ripple is thus controlled by ε.

If we now let

$$IL = 10 \log[1 + \varepsilon^2 T_N^2(\omega)] \tag{6.14}$$

then $T_N(\omega)$ is a function which must ripple between ± 1 at the maximum number of points in the region $|\omega| < 1$ (Fig. 6.5).

From Fig. 6.5 we see that all points lies in the region $|\omega| < 1$ where $|T_N(\omega)| = 1$ must be turning points except when $\omega = \pm 1$. Thus

$$\frac{dT_N(\omega)}{d\omega} = 0 \text{ when } |T_N(\omega)| = 1 \tag{6.15}$$

(except at $|\omega| = 1$)

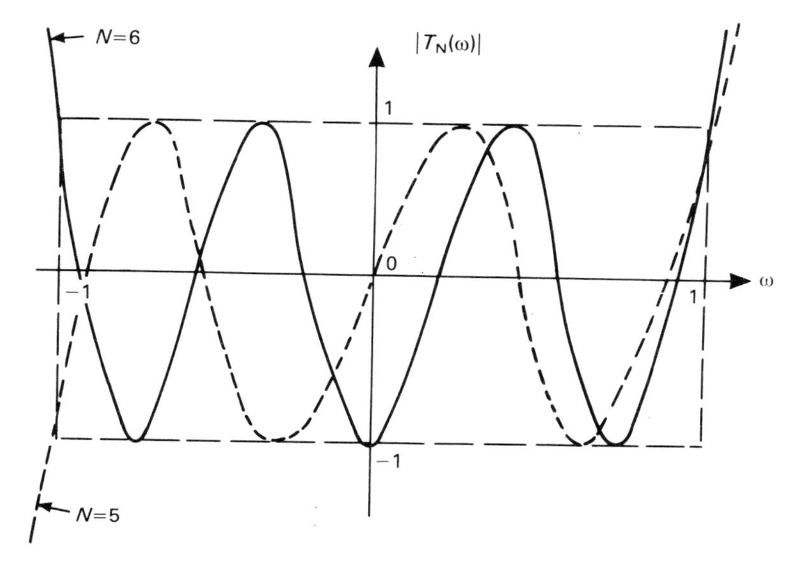

Fig. 6.5 — Plot of $|T_N(\omega)|$.

Or

$$\frac{dT_N(\omega)}{d\omega} = \frac{C_N\sqrt{1 - T_N^2(\omega)}}{\sqrt{1-\omega^2}} \tag{6.16}$$

Separating variables and integrating, we obtain

$$\cos^{-1}(T_N(\omega)) = C_N \cos^{-1}(\omega) \tag{6.17}$$

It can be shown that for $T_N(\omega)$ to possess N zeros then $C_N = N$ [6.4]. Thus

$$T(\omega) = \cos\,(N \cos^{-1}(\omega)) \tag{6.18}$$

$T(\omega)$ is the Chebyshev polynomial given by the recurrence formula [6.5]

$$T_{N+1}(\omega) = 2\omega\; T(\omega) - T_{N-1}(\omega) \tag{6.19}$$

with

$$T_0(\omega) = 1 \tag{6.20}$$

$$T_1(\omega) = \omega \tag{6.21}$$

As an example let us work out $T_3(\omega)$. From eqn. (6.19)

$$T_2(\omega) = 2\omega T_1(\omega) - T_0(\omega) = 2\omega^2 - 1 \tag{6.22}$$

Thus

$$T_3(\omega) = 2\omega T_2(\omega) - T_1(\omega) \tag{6.23}$$
$$T_3(\omega) = 4\omega^3 - 3\omega \tag{6.24}$$

Substituting $T_3(\omega)$ into the general expression (6.14) for insertion loss, we obtain

$$IL = 10 \log[1 + \varepsilon^2(4\omega^3 - 3\omega)^2] \tag{6.25}$$

It is left to the reader to choose a value of ε such that the passband ripple is small and plot eqn. (6.25) as a function of frequency to show that it is indeed an equiripple function.

It is worth noting at this point that the third degree Chebyshev function given by eqn. (6.25) could be obtained from the circuit shown in Fig. 6.2, simply by choosing the appropriate element values. Thus a Chebyshev response can be obtained from the same type of circuit used to obtain a maximally flat response.

The formula to calculate N to meet a particular specification when using a Chebyshev filter is [6.6]

$$N \geqslant \frac{L_A + L_R + 6}{20 \log(S + \sqrt{S^2 - 1})} \tag{6.26}$$

where L_A, L_R, S have the same significance as before.

If we use the same example as for the maximally flat filter then we see that $N = 7$. This compares very favourably with the value of $N = 12$ required for the maximally flat filter. For this reason the design procedures outlined in this chapter will use a Chebyshev lowpass prototype.

It is, however, worth noting that there are other types of filter functions. These include generalized Chebyshev responses with ripples in the passband and stopband [6.7], and also filter functions with specified phase response [6.8]. These are beyond the scope of this book.

Formulae for Chebyshev lowpass prototype filters

The Chebyshev lowpass prototype filter can be realized using either of the following networks. The first, shown in Fig. 6.6, is an LC ladder network. The formulae for the element values are presented below [6.9].

First calculate N from eqn. (6.26) then compute

$$\varepsilon = (10^{L_R/10} - 1)^{-1/2} \tag{6.27}$$

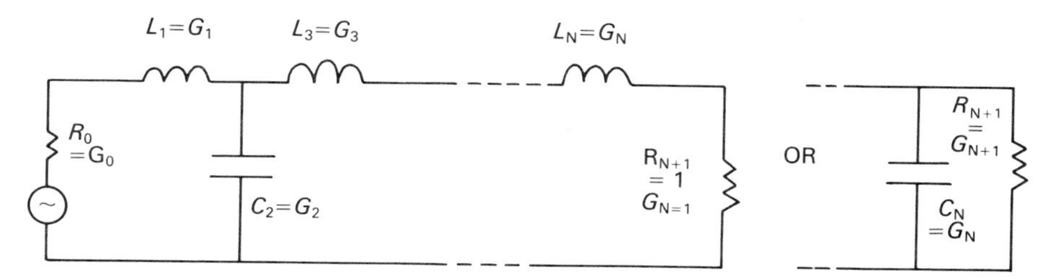

Fig. 6.6 — Ladder network lowpass prototype filter.

$$y = \sinh\left[\frac{1}{N}\sinh^{-1}(1/\varepsilon)\right] \tag{6.28}$$

$$x = \tfrac{1}{2}\sinh^{-1}(1/\varepsilon) \tag{6.29}$$

Then

$$A_R = \sin\left[\frac{(2R-1)\pi}{2N}\right] \quad (6.30$$

$$B_R = y^2 + \sin^2\left(\frac{R\pi}{N}\right) \tag{6.31}$$

Then the element values G_R are given by

$$G_1 = \frac{2A_1}{y} \tag{6.32}$$

$$G_R = \frac{4A_{R-1}\cdot A_R}{B_{R-1}\cdot G_{R-1}} \tag{6.33}$$

$(R = 2,3,\ldots,N)$

$$G_{N+1} = 1 \quad (N \text{ odd}) \tag{6.34}$$

$$G_{N+1} = \coth^2(x) \ (N \text{ even}) \tag{6.35}$$

The LC ladder network prototype filter is suitable for the design of microwave lowpass filters. If a narrowband bandpass filter is required then an alternative equivalent network is more suitable. This network uses a circuit element known as an impedance inverter (Fig. 6.7).

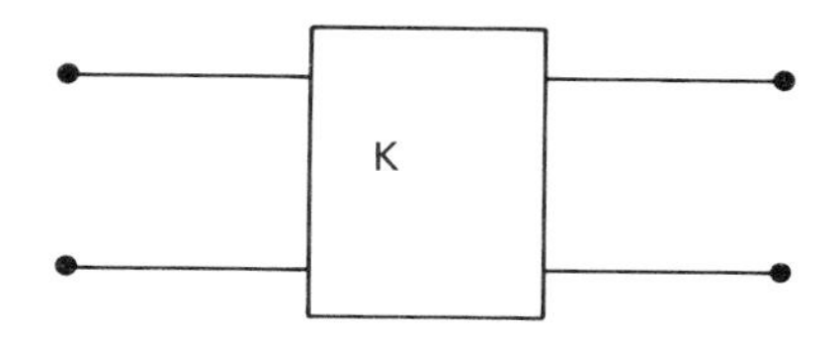

Fig. 6.7 — An impedance inverter.

The impedance inverter is defined by its transfer matrix [**T**] where

$$[\mathbf{T}] = \begin{bmatrix} 0 & jK \\ j/K & 0 \end{bmatrix} \tag{6.36}$$

where K is the characteristic impedance of the inverter. The function of the inverter can be seen by terminating the inverter in an inductor of impedance L_p. Then the input impedance of the inductively terminated inverter is given by

$$Z_{\text{in}}\,(p) = \frac{K^2}{Lp} \tag{6.37}$$

This is the impedance of a capacitor of value L/K^2. Thus a network consisting of series inductors separated by impedance inverters is equivalent to an LC ladder network. This alternative prototype network is shown in Fig. 6.8.

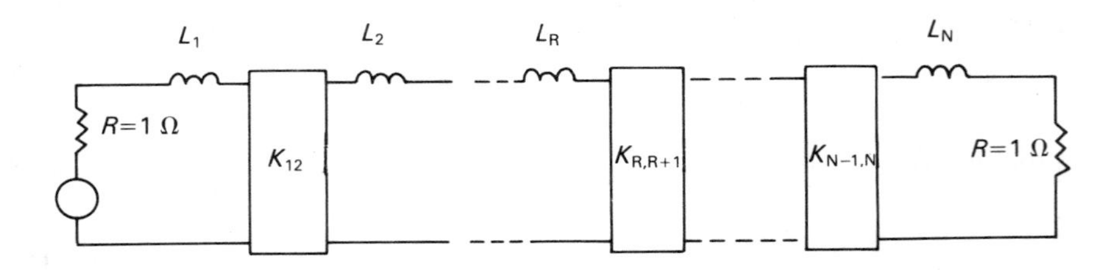

Fig. 6.8 — An impedance inverter coupled lowpass prototype filter.

The element values of the inverter coupled prototype are given by [6.10]

$$L_{\mathrm{R}} = \frac{2}{y} \sin \left[\frac{(2_{\mathrm{R}-1})\pi}{2N} \right] \tag{6.38}$$

and

$$K_{\mathrm{R.R+1}} = \frac{\sqrt{y^2 + \sin^2(R\pi/N)}}{y} \tag{6.39}$$

$$(R = 1,2,3,\ldots,N-1)$$

y is defined as before.

DESIGN OF MICROWAVE FILTERS

We can now specify a filter response and calculate the required degree and element values of the Chebyshev lowpass prototype filter. It now remains to convert the prototype into a microwave structure. Two types of filter will be discussed.

The first type, the stepped impedance lowpass filter, is designed from the *LC* ladder network prototype filter. We shall see that a short section of high impedance transmission line can approximate a series inductor. Similarly, a short section of low impedance transmission line can approximate a shunt capacitor. Thus the *LC* ladder network prototype can be transformed into a microwave structure composed entirely of alternating low and high impedance transmission lines. This type of structure is easily manufactured by printing the lines on a microstrip circuit.

The second type of filter is a narrowband bandpass filter. In this case the inverter coupled prototype is used. A microwave structure known as the parallel coupled bandpass filter will be used. This structure consists on an array of open circuited coupled lines. The lines are equivalent to series resonant circuits and the coupling gaps between them correspond to the impedance inverters in the low pass prototype.

The stepped impedance lowpass filter

Consider a section of transmission line of characteristic impedance Z_o and length l. This line has a transfer matrix [**T**] where:

$$[T] = \begin{bmatrix} \cos(\theta) & jZ_o\sin(\theta) \\ j\dfrac{\sin(\theta)}{Z_o} & \cos(\theta) \end{bmatrix} \tag{6.40}$$

where

$$\theta = \frac{\omega l}{\upsilon} \tag{6.41}$$

and v is the velocity of signal propagation along the line.

We now find the π section equivalent circuit of the line shown in Fig. 6.9. The transfer matrix of this equivalent is

$$[T] = \begin{bmatrix} 1 & 0 \\ jB & 1 \end{bmatrix} \times \begin{bmatrix} 1 & jX \\ 0 & 1 \end{bmatrix} \times \begin{bmatrix} 1 & 0 \\ jB & 1 \end{bmatrix} \tag{6.42}$$

$$= \begin{bmatrix} 1-XB & jX \\ jB(2-XB) & 1-XB \end{bmatrix} \tag{6.43}$$

Thus

$$\cos(\boldsymbol{\theta}) = 1 - XB \tag{6.44}$$

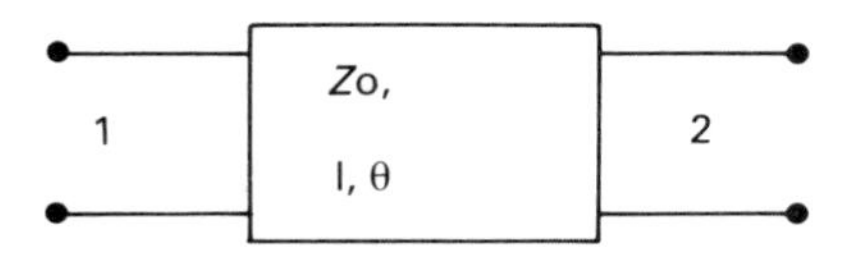

=

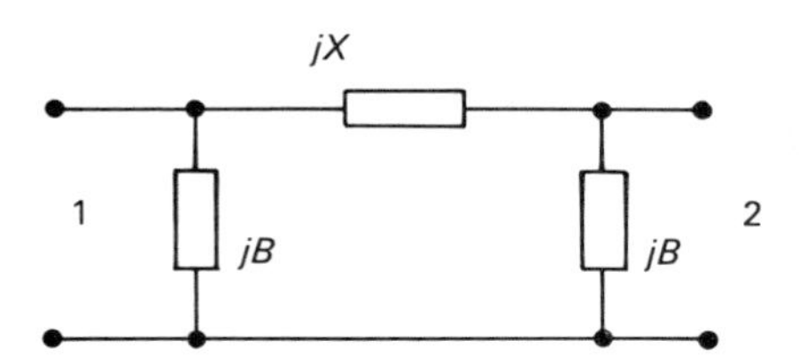

Fig. 6.9 — π equivalent of a transmission line.

and

$$X = Z_o \sin(\theta) \tag{6.45}$$

therefore

$$B = \frac{(1 - \cos(\theta))}{Z_o \sin(\theta)} \tag{6.46}$$

$$= Y_o \tan(\theta/2) \tag{6.47}$$

where

$$Y_o = 1/Z_o \tag{6.48}$$

Now assume that the line length is significantly less than one quarter wavelength at the operating frequency and that the characteristic impedance of the line is greater than unity, i.e.

$$\theta < \pi/2 \tag{6.49}$$

and

$$Z_o > 1 \tag{6.50}$$

Under these conditions, X is positive and B is small. Thus the short section of high impedance line approximates a series inductive impedance.

Let the impedance of the inductive section of the line be Z_L and the length of the Rth section by l_R. Then we can apply the transformation

$$\omega L_R \rightarrow Z_L \sin\left(\frac{\omega l_R}{v}\right) \tag{6..51}$$

where L_R is the Rth series inductor in the ladder network prototype filter.

We must map the band edge at $\omega = 1$ in the lowpass prototype to the band edge in the microwave filter at ω_c. Thus

$$L_R = Z_L \sin\left(\frac{\omega_c l_R}{v}\right) \tag{6.52}$$

or

$$l_R = \frac{v}{\omega_c} \sin^{-1}\left(\frac{L_R}{Z_L}\right) \tag{6.53}$$

Next we wish to obtain the transmission line equivalent of the shunt capacitors in the lowpass prototype filter. In this case we take the T section equivalent circuit of a transmission line (Fig. 6.10). The transfer matrix of this T section is

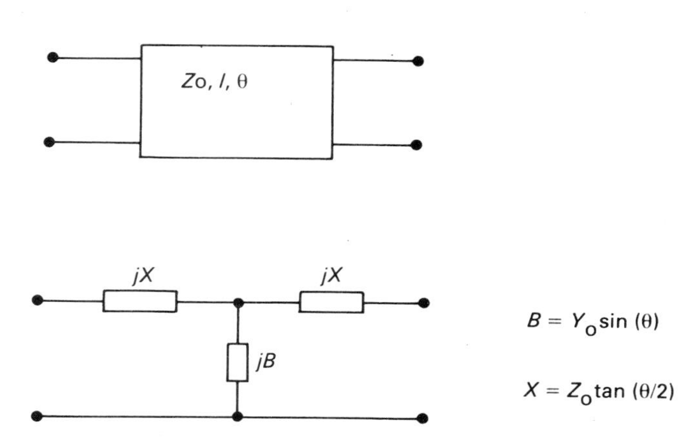

Fig. 6.10 — **T equivalent of a transmission line.**

$$[\mathbf{T}] = \begin{bmatrix} 1 & jX \\ 0 & 1 \end{bmatrix} \times \begin{bmatrix} 1 & 0 \\ jB & 1 \end{bmatrix} \times \begin{bmatrix} 1 & jX \\ 0 & 1 \end{bmatrix} \tag{6.54}$$

Equating with eqn. (6.40) we obtain

$$B = Y_o \sin(\theta) \tag{6.55}$$

and

$$X = Z_o \tan(\theta/2) \tag{6.56}$$

Now let

$$\theta < \pi;/2 \text{ and } Z_o < 1 \tag{6.57}$$

Then X becomes small and the line is predominantly a shunt capacitive impedance. Thus a short low impedance section of line approximates a shunt capacitor. Let the admittance of this line by Y_c with length l_R. Thus we transform the shunt capacitors in the lowpass prototype into lines by the transformation

$$\omega C_R \rightarrow Y_c \sin\left(\frac{\omega l_R}{v}\right) \tag{6.58}$$

Mapping $\omega = 1$ in the prototype to ω_c

$$C_R = Y_c \sin\left(\frac{\omega_c l_R}{v}\right) \tag{6.59}$$

or

$$l_R = \frac{\text{v}}{\omega_c} \sin^{-1}(C_R/Y_c) \tag{6.60}$$

Next we must account for the small residual impedances and admittances in the π and T sections. If we cascade a series of alternating high and low impedance lines then we obtain the equivalent circuit shown in Fig. 6.11.

It is reasonable to absorb the series impedances X from the capacitive T sections into the series inductive sections. The series impedance of the Rth inductive section is then

$$Z_{ser} = Z_L \sin\left(\frac{\omega l_R}{v}\right) + Z_c \left[\tan\left(\frac{\omega l_{R-1}}{2v}\right) + \tan\left(\frac{\omega l_{R+1}}{2v}\right)\right] \tag{6.61}$$

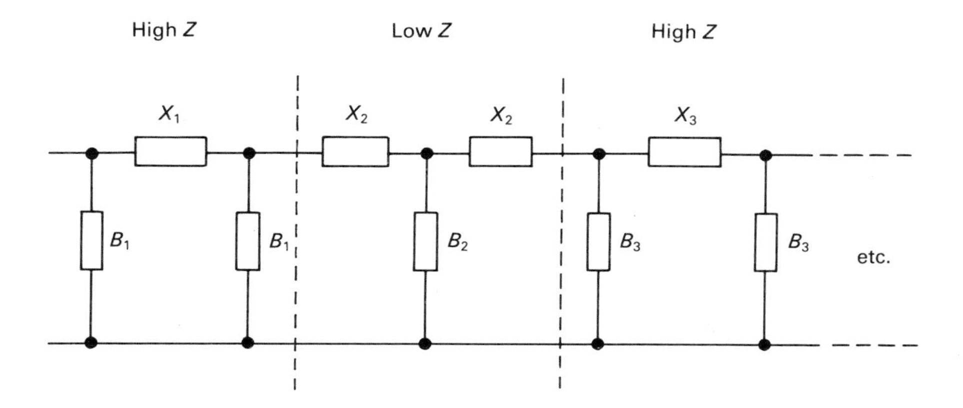

Fig. 6.11 — Equivalent circuit of a cascade of high and low impedance transmission lines.

For short line lengths we have the approximation

$$\sin(\theta) \simeq \tan(\theta) \simeq \theta \tag{6.62}$$

Thus

$$Z_{\text{ser}} = \frac{Z_L \omega l_R}{v} + Z_c \left(\frac{\omega l_{R-1}}{2v} + \frac{\omega l_{R-1}}{2v} \right) \tag{6.63}$$

The Rth series inductive impedance is increased by loading from its adjacent capacitive sections. If we modify the length of the Rth section to compensate for this effect we obtain

$$\frac{Z_L \omega l_R}{v} = \frac{Z_L \omega l'_R}{v} + Z_c \left[\frac{\omega l_{R-1}}{2v} + \frac{\omega l_{R+1}}{2v} \right] \tag{6.64}$$

or

$$l'_R = l_R - \frac{Z_c}{Z_L} \left(\frac{l_{R-1}}{2} + \frac{l_{R+1}}{2} \right) \tag{6.65}$$

Thus the new length l'_R has been shortened to compensate for the loading.

In a similar way we can absorb the shunt admittances from the inductive sections into the admittance of the capacitive sections. The loaded admittance of the Rth capacitive section Y_{shR} is then given by

$$Y_{\text{shR}} = Y_c \sin\left(\frac{\omega l_R}{v}\right) + Y_L\left[\tan\left(\frac{\omega l_{R-1}}{2v}\right) + \tan\left(\frac{\omega l_{R+1}}{2v}\right)\right] \tag{6.66}$$

or for short line lengths

$$Y_{\text{shR}} = \frac{Y_c \omega l_R}{v} + Y_L\left(\frac{\omega l_{R-1}}{2v} + \frac{\omega l_{R+1}}{2v}\right) \tag{6.67}$$

Modifying l_R to l'_R to compensate for the loading then

$$Y_c l_R = Y_c l'_R + Y_L\left(\frac{l_{R-1}}{2} + \frac{l_{R+1}}{2}\right) \tag{6.68}$$

or

$$l'_R = l_R - \frac{Y_L}{Y_c}\left(\frac{l_{R-1}}{2} + \frac{l_{R+1}}{2}\right) \tag{6.69}$$

We have now formulated all the design equations for the lowpass filter. To summarize, first we obtain the lowpass prototype filter. Let us consider the odd degree case starting and finishing with an inductor. Then the microwave filter will start and finish with a high impedance section. We choose the impedances Z_L and Z_c of the high and low sections for our convenience. Typically let $Z_L = 2$ and $Z_c = 1/2$. The lengths of the sections are then calculated as follows:

$$l_R = \frac{v}{\omega_c}\sin^{-1}\left(\frac{L_R}{Z_L}\right) \tag{6.70}$$

$(R = 1,3,5,\ldots,N)$

$$l_R = \frac{v}{\omega_c}\sin\left(\frac{C_R}{Y_c}\right) \tag{6.71}$$

$(R = 2,4,6,\ldots,N-1)$

Then calculate the modified values from

$$l'_R = l_R - \frac{Z_c}{Z_L}\left(\frac{l_{R-1}}{2} + \frac{l_{R+1}}{2}\right) \tag{6.72}$$

$(R = 1,2,3,\ldots,N)$

A BASIC computer program has been written to design stepped impedance lowpass filters. This program starts with an arbitrary specification of band edge frequency, passband return loss and stopped insertion loss. It then calculates the element values of the lowpass prototype and then determine the lengths of the high and low impedance lines. Microstrip equations are then used to determine the dimensions of the metallization pattern on any microstrip substrate. Finally, the program performs an analysis of the filter by multiplying the transfer matrices of the individual line sections together and working out the insertion loss and return loss at various frequencies.

A design of a stepped impedance lowpass filter has been performed using the computer program.

The specification of the filter was:

Band edge frequency = 4 GHz
Passband return loss = 20 dB
Stopband insertion loss ≡ ≥40 dB at 5 GHz
System impedance level = 50 Ω

From eqn. (6.26) this specification implies $N = 11$.

The line impedances Z_L and Z_c were chosen to be

$$Z_L = 100\ \Omega$$
$$Z_c = 25\ \Omega$$

Thus in a 1 Ω system for which the design equations are valid we must divide Z_L and Z_c by 50. Thus

$$Z_L = 2$$
$$Z_c = 0.5$$

The computer program works out the dimensions of the lines on a microstrip substrate. In our case the substrate was Rogers 5880 material with a thickness of 0.75 mm and a dielectric constant of 2.33. The dimensions of the metallization pattern on the microstrip substrate are shown in Fig. 6.12.

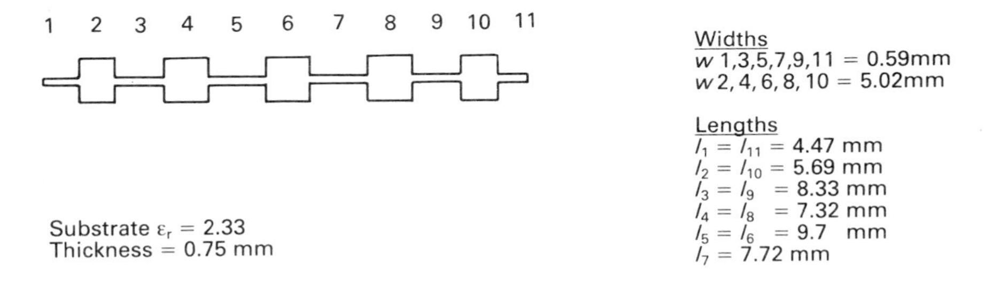

Fig. 6.12 — Metallization pattern for lowpass filter.

The filter was constructed in a shielded enclosure to avoid spurious couplings and radiation losses. The substrate material was soldered directly to the base of a brass housing with interfaces via SMA flange mounted connectors.

The measured performance of the filter is shown in Fig. 6.13.

The parallel coupled bandpass filter

This type of filter is composed of a cascade of pairs of coupled parallel open circuited lines which all one quarter wavelength long at the centre frequency of the filter (Fig. 6.14).

Consider a pair of coupled lines with even and odd mode impedances Z_{oe} and Z_{oo} respectively. The equivalent circuit of this structure [eqn. (6.11)] is composed of a pair of series open circuited lines separated by a section of 'unit element' of transmission line (Fig. 6.15). The transfer matrix of the unit element is

$$[\mathbf{T}] = \begin{bmatrix} \cos(\theta) & \dfrac{j(Z_{oe} - Z_{oo}]}{2} \sin(\theta) \\ \dfrac{j2\sin(\theta)}{[Z_{oe} - Z_{oo})} & \cos(\theta) \end{bmatrix} \tag{6.73}$$

which can be decomposed into

$$[\mathbf{T}] = \begin{bmatrix} 1 & \dfrac{-j(Z_{oe} - Z_{oo}}{2\tan(\theta)} \\ 0 & 1 \end{bmatrix} \times \begin{bmatrix} 0 & \dfrac{j[Z_{oe} - Z_{oo}]}{2\sin(\theta)} \\ \dfrac{j2\sin(\theta)}{(Z_{oe} - Z_{oo})} & 0 \end{bmatrix} \times \begin{bmatrix} 1 & \dfrac{-j(Z_{oe} - Z_{oo})}{2\tan(\theta)} \\ 0 & 1 \end{bmatrix} \tag{6.74}$$

which is the matrix of a pair of open circuited stubs separated by an impedance inverter (Fig. 6.16)of characteristic impedance

$$K = \frac{Z_{oe} - Z_{oo}}{2\sin(\theta)} \tag{6.75}$$

Since we are interested in narrow band filters the variation in K with θ can be ignored, for example in an octave band filter θ would vary from 60° to 120° over the entire passband. This produces a variation in $\theta \sin(\theta)$ of 0.866:1 which will cause a relatively small degradation in passband return loss. Thus the equivalent circuit of the pair of coupled lines over a narrowband of frequencies around $\theta = 90°$ can be represented in Fig. 6.17. This equivalent circuit consists of two resonant open circuited stubs separated by an impedance inverter.

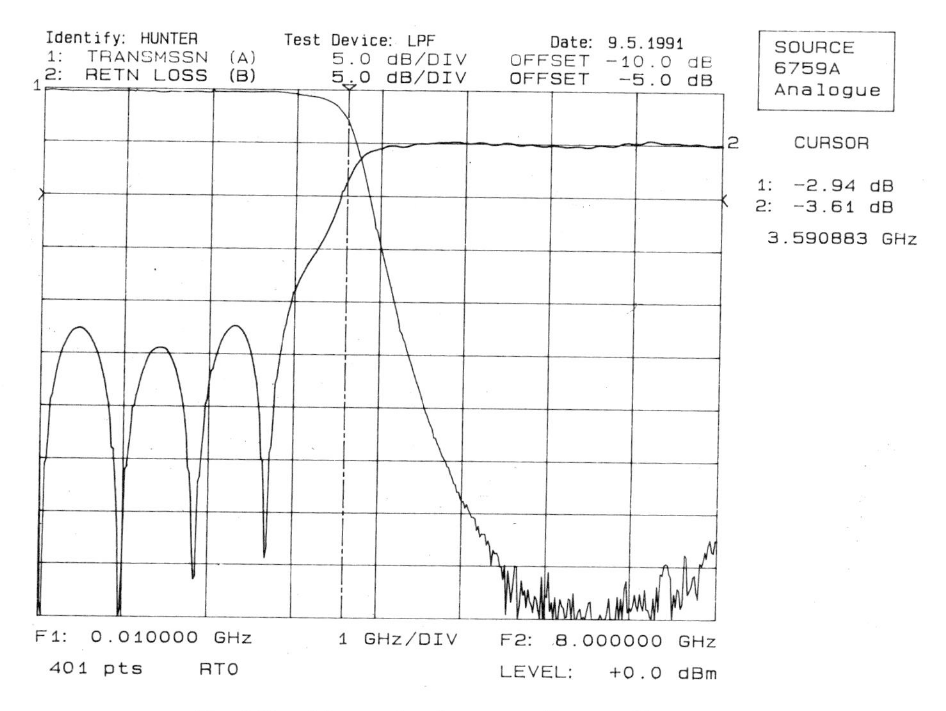

Fig. 6.13 — Measured performance of lowpass filter.

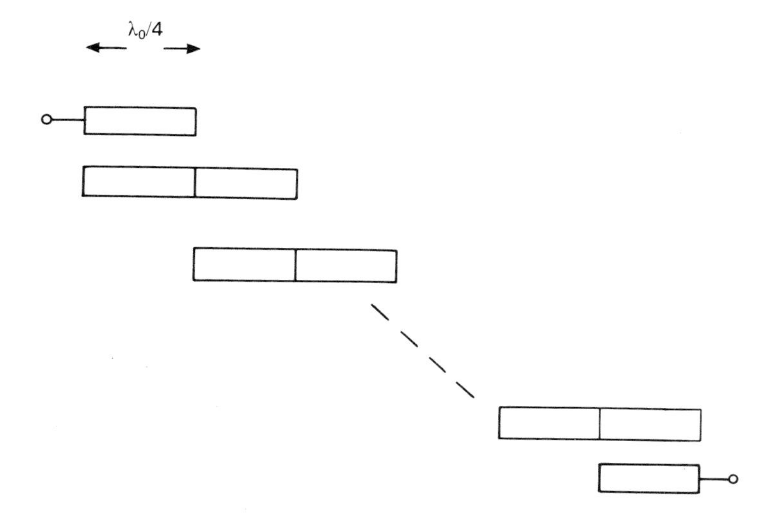

Fig. 6.14 — Parallel coupled bandpass filter.

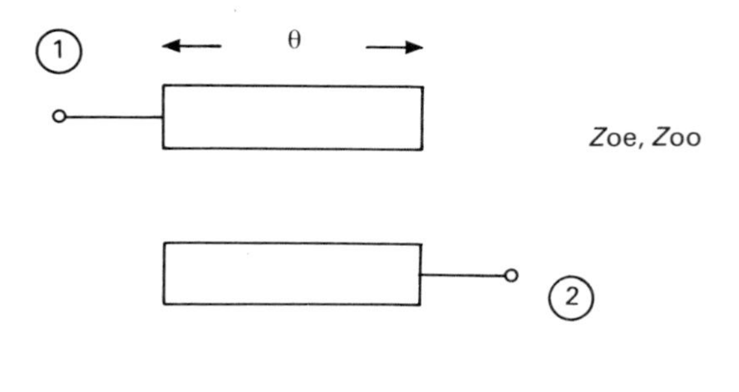

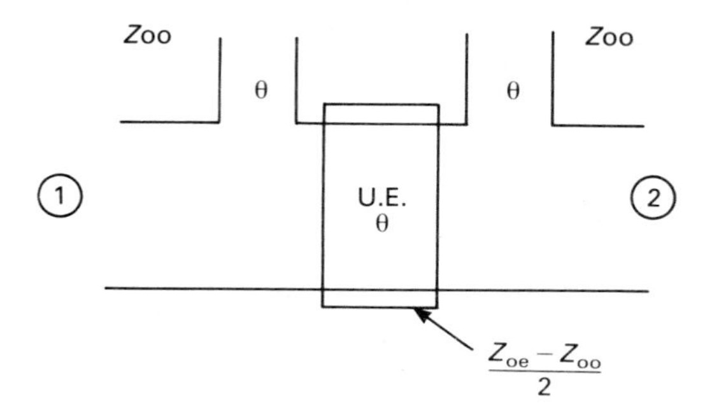

Fig. 6.15 — Equivalent circuit of a pair of coupled lines.

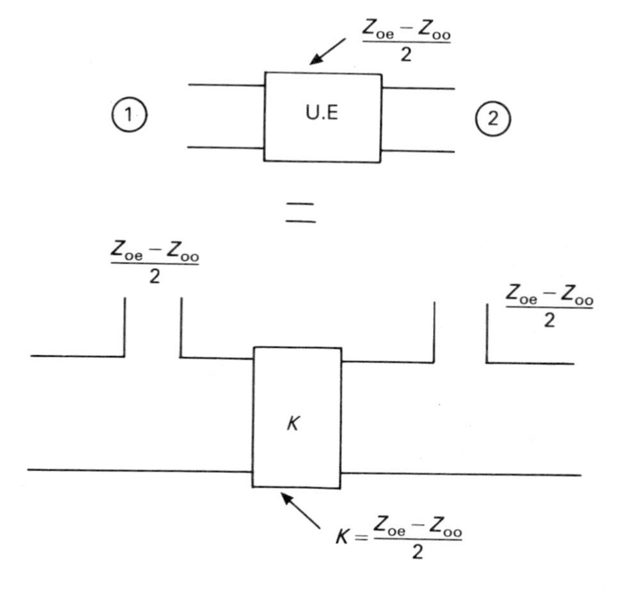

Fig. 6.16 — Equivalent circuit of a transmission line.

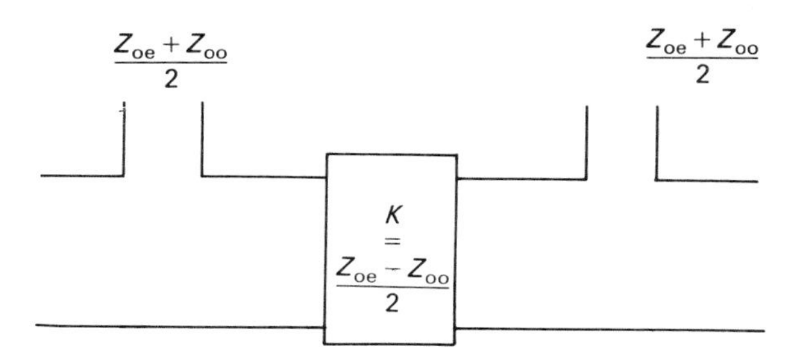

Fig. 6.17 — Final equivalent circuit of the pair of coupled lines.

If we now cascade a series of coupled line pairs together we obtain the equivalent circuit shown in Fig. 6.18. Here the even and odd mode impedances of the Rth coupled line pair are Z_{oeR} and Z_{ooR} respectively.

The impedance of the Rth series element is

$$Z_R = \frac{-j}{2\tan(\theta)}(Z_{oeR-1} + Z_{ooR-1} + Z_{oeR} + Z_{ooR}) \tag{6.76}$$

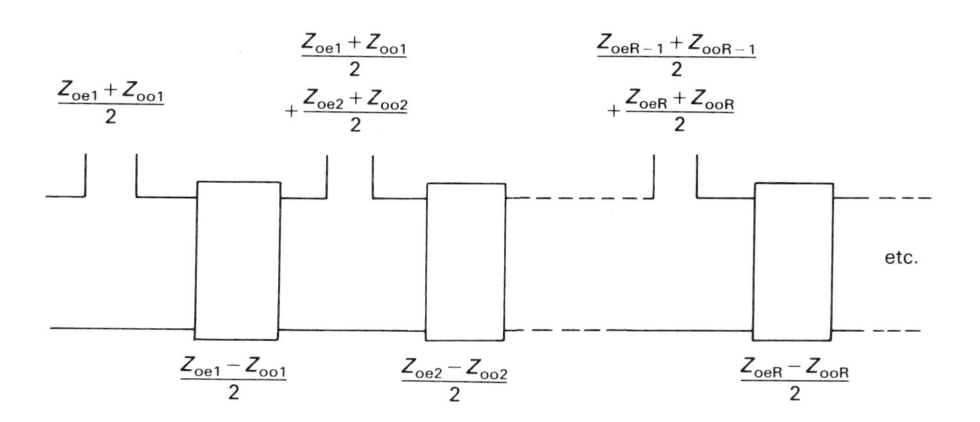

Fig. 6.18 — Equivalent circuit of a cascade of coupled lines.

For as narrow band filter response around $\theta = 90°$ Z_R must change rapidly with θ. From eqn. (6.78) we see that this would require very large even and odd mode line impedances which would be physically difficult to realize. To avoid this situation we scale the impedance of all the series elements; the Rth impedance is scaled by a factor n_R^2, resulting in transformers either side of Z_R of $1{:}n_R$ respectively (Fig. 6.19).

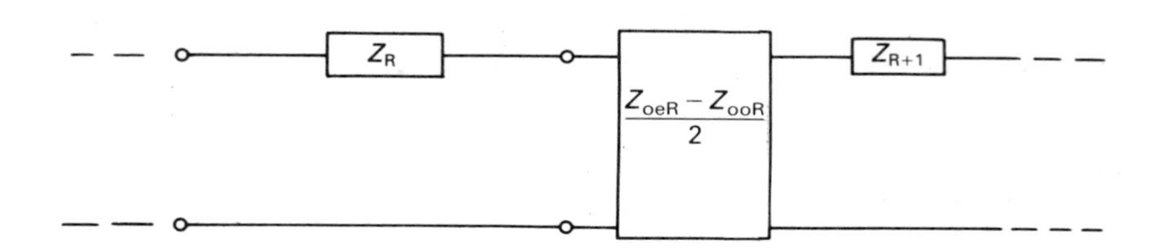

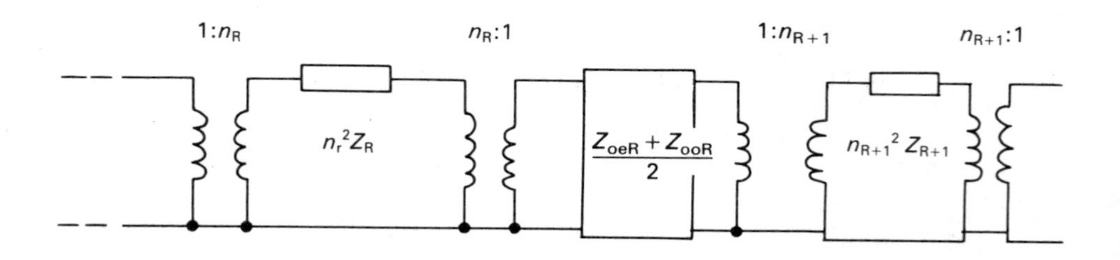

Fig. 6.19 — Scaling the impedance of the resonators.

These ideal transformers can be absorbed into the inverters. The transfer matrix of the R to $R+1$th inverter then becomes

$$[\mathbf{T}] = \begin{bmatrix} n_R & 0 \\ 0 & 1/n_R \end{bmatrix} \times \begin{bmatrix} 0 & \dfrac{j(Z_{oeR} - Z_{ooR})}{2} \\ \dfrac{j2}{(Z_{oeR} - Z_{ooR})} & 0 \end{bmatrix} \times \begin{bmatrix} 1/n_{R+1} & 0 \\ 0 & n_{R+1} \end{bmatrix} \tag{6.77}$$

$$[\mathbf{T}] = \begin{bmatrix} 0 & \dfrac{j n_R \cdot n_{R+1}}{2} (Z_{oeR} - Z_{ooR}) \\ \dfrac{j2}{n_R \cdot n_{R+1} (Z_{oeR} - Z_{ooR})} & 0 \end{bmatrix} \tag{6.78}$$

The matrix must equate to the matrix of the R to $R+1$th inverter in the lowpass prototype filter. Thus

$$K_{R.R+1} = \frac{n_R . n_{R+1}}{2} (Z_{oeR} - Z_{ooR}) \tag{6.79}$$

and the impedance of the Rth series element in the bandpass filter is

$$Z_R = -jn_R^2 \frac{(Z_{oeR-1} + Z_{ooR-1} + Z_{oeR} + Z_{ooR})}{2 \tan(\theta)} \tag{6.80}$$

Now let the sum of the even and odd mode impedances of every pair of coupled lines by equal to a constant.

$$\frac{Z_{oeR} + Z_{ooR}}{2} = X \tag{6.81}$$

then

$$Z_R = \frac{-jn_R^2 2X}{\tan(\theta)} \tag{6.82}$$

Z_R represents the impedance of the Rth series resonant circuit in the band pass filter. Thus Z_R must have the same impedance as the series inductors in the lowpass prototype filter at the band edge frequencies.

Thus we use the lowpass to bandpass transformation

$$\omega L_R \rightarrow \frac{-n_R^2 2X}{\tan(\theta)} \tag{6.83}$$

or

$$\omega \rightarrow -\alpha/\tan(\theta) \tag{6.84}$$

where

$$\alpha = \frac{2n_R^2 X}{L_R} \tag{6.85}$$

We equate the band edges of the lowpass prototype filter at $\omega = \pm 1$ to the band edges of the band pass filter at ω_1 and ω_2. This is a mapping from the lowpass prototype to the bandpass filter (Fig. 6.20).

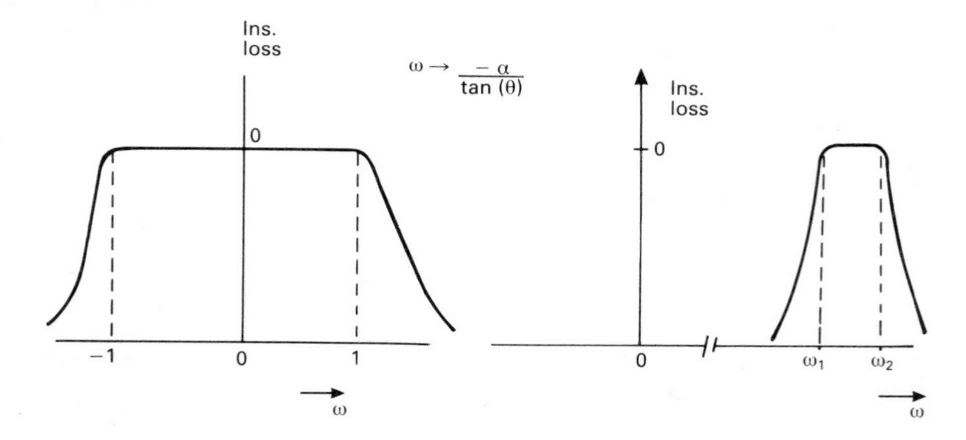

Fig. 6.20 — Lowpass to bandpass transformation.

Let the centre frequency of the bandpass filter be ω_0 and the passband bandwidth be $\Delta\omega$. Then

$$\omega_1 = \omega_0 - \frac{\Delta\omega}{2} \tag{6.86}$$

and at ω_1

$$\theta_1 = \frac{90^\circ \times \omega_1}{\omega_0} \tag{6.87}$$

Then from eqn. (6.84)

$$-1 = -\alpha/\tan(\theta_1) \tag{6.88}$$

or

$$\alpha = \tan(\theta_1) \tag{6.89}$$

hence from eqn. (6.85)

$$n_R = \sqrt{\frac{\tan(\theta_1)L_R}{2X}} \tag{6.90}$$

Now let

$$\frac{Z_{oeR} - Z_{ooR}}{2} = Y_R \tag{6.91}$$

Then from eqn. (6.79)

$$Y_R = \frac{K_{R.R+1}}{n_R . n_{R+1}} \tag{6.92}$$

and from eqns (6.81) and (6.91)

$$Z_{oeR} = X + Y_R \tag{6.93}$$

$$Z_{ooR} = X - Y_R \tag{6.94}$$

Equations (6.93) and (6.94) give us the impedances of all the coupled line pairs. However, we have not quite finished. The process of scaling all the resonators resulted in the introduction of ideal transformers into the network. These were all absorbed into the inverters apart from the $1:N1$ and $N1:1$ transformers at each end of the filter.

These are removed by introducing an extra coupled line section with even and odd mode impedances Z_{oe0} and Z_{oo0} at the input and output of the filter. If the coupled line section is represented by its equivalent circuit the input to the filter is as shown in Fig. 6.21. By moving the ideal transformer to the left of this network it can be absorbed into the inverter.

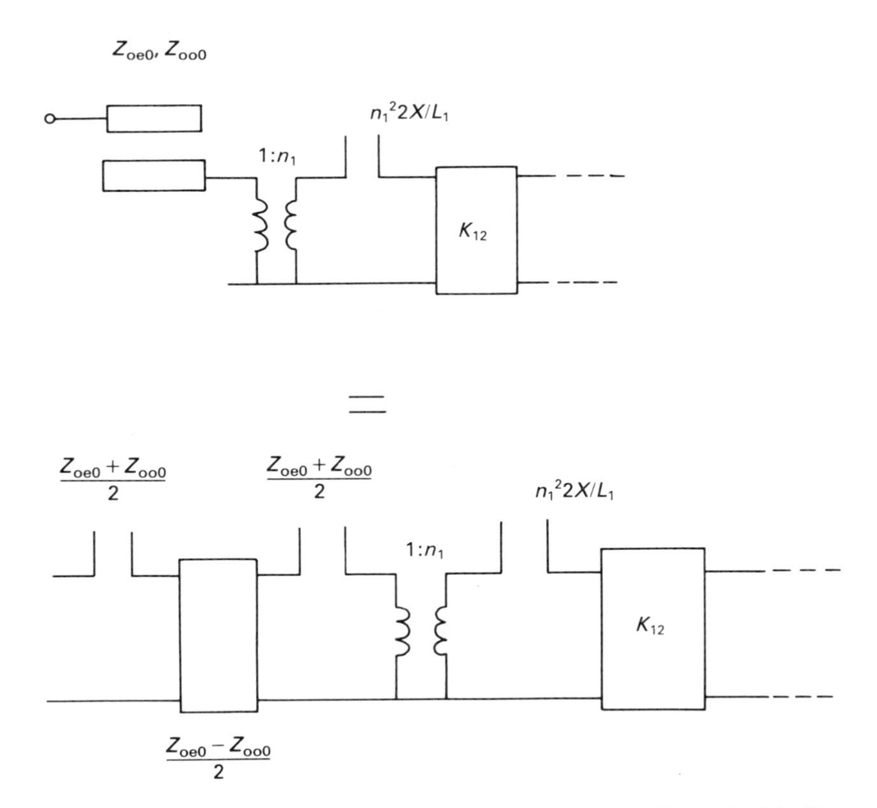

Fig. 6.21 — Introduction of an extra coupled line section at the input (and output) of the filter.

Now let

$$\frac{Z_{oe0} + Z_{oo0}}{2} = 1 \tag{6.95}$$

and

$$\frac{n_1(Z_{oe0} - Z_{oo0})}{2} = 1 \tag{6.96}$$

Remember that the equivalent circuit for a unit element is a pair of open circuit stubs separated by an inverter. Then the equivalent circuit of the filter can be represented as Fig. 6.22. Here the transformer has been cancelled and we are left with a unity impedance unit element. The impedance level of the network is unity thus the unit element will not affect the amplitude response of the filter.

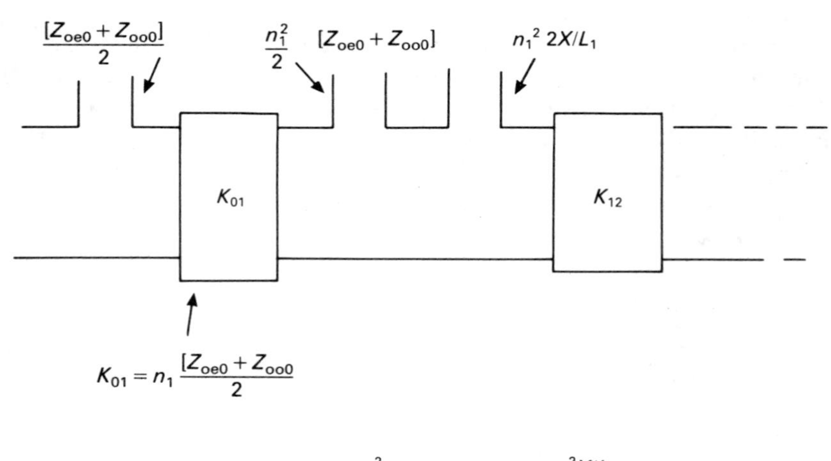

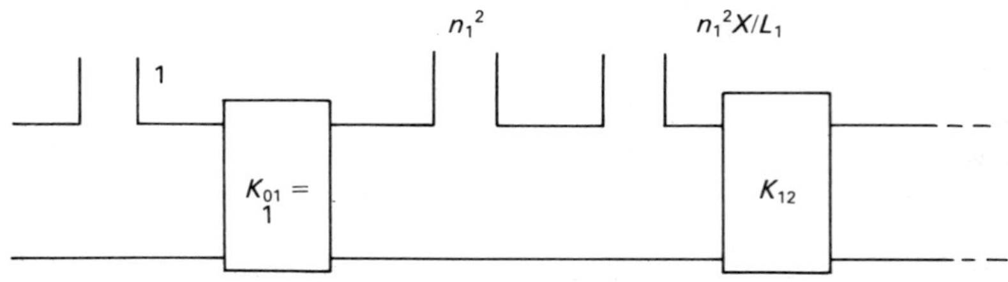

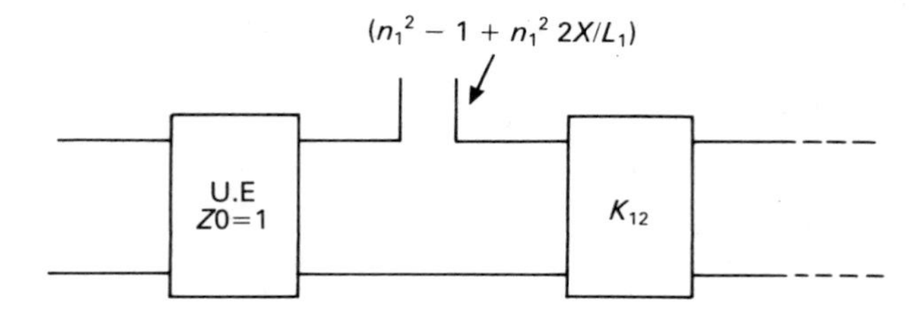

Fig. 6.22 — Final equivalent circuit of the filter.

The impedance of the first resonator is now

$$Z_1 = -j\,\frac{(n_1^2 + n_1^2 X - 1)}{\tan(\theta)} \tag{6.97}$$

and since

$$\omega \rightarrow -\alpha/\tan(\theta) \tag{6.98}$$

Then for the first resonator

$$\alpha = \frac{n_1^2\,(1 + X)}{L_1} - 1 \tag{6.99}$$

or

$$n_1 = \sqrt{\frac{\alpha L_1 + 1}{1 + X}} \tag{6.100}$$

This finishes the development of the filter design procedure which can now be summarized.

First specify the filter in terms of its passband edges ω_1 and ω_2, and the frequencies ω_3 and ω_4, where the stopband attenuation of L_A dB must be achieved. The passband return loss is L_R dB (Fig. 6.23).

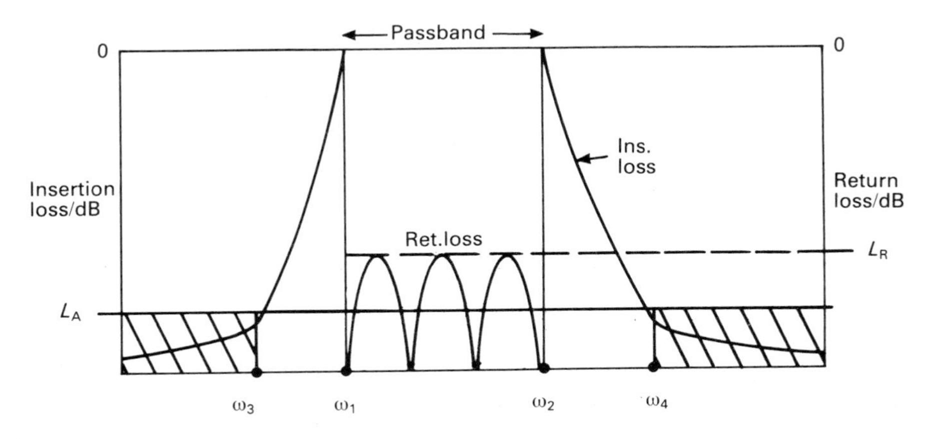

Fig. 6.23 — Specification of a bandpass filter.

Now assuming that the attenuation characteristic of the filter is symmetrical around its centre frequency, calculate S, the selectivity from

$$S = \frac{\omega_4 - \omega_3}{\omega_2 - \omega_1} \tag{6.101}$$

We can then calculate N, the degree of the filter from eqn. (6.20) and the values of the series inductors and inverters in the prototype from eqns (6.38) and (6.39).

Now we calculate ω_0, the centre frequency from

$$\omega_0 = \frac{\omega_1 + \omega_2}{2} \tag{6.102}$$

and the passband bandwidth $\Delta\omega$ from

$$\Delta\omega = \omega_2 - \omega_1 \tag{6.103}$$

Then choose a suitable value of X, typically around 1 in a 1 Ω system.

Next calculate α from

$$\alpha = \tan(\theta_1) \tag{6.104}$$

where

$$\theta_1 = \frac{90° \times \omega_1}{\omega_0} \tag{6.105}$$

Then calculate the scaling factors n_R from

$$n_R = \sqrt{\frac{\alpha\ L_R}{2X}} \tag{6.106}$$

$(R = 2,3,\ldots,N-1)$

and

$$n_R = \sqrt{\frac{\alpha\ L_1 + 1}{1 + X}} \tag{6.107}$$

$(R = 1 \text{ and } R = N)$

Then find Y_R from

$$Y_R = \frac{K_{R.R+1}}{n_R . n_{R+1}} \quad (R = 1 \text{ to } N - 1) \tag{6.108}$$

Finally, we can find the even and odd mode coupled line section impedances from

$$Z_{oeR} = X + Y_R \tag{6.109}$$
$$Z_{ooR} = X - Y_R \tag{6.110}$$
$$(R = 1 \text{ to } N - 1)$$

(Remember that there are N resonant circuits produced by $N - 1$ coupled line sections.)

Then calculate the even and odd mode impedances of the first (and last) coupled line sections from

$$Z_{oeo} = 1 + 1/n_1 \tag{6.111}$$
$$Z_{ooo} = 1 - 1/n_1 \tag{6.112}$$

Next multiply all the coupled line impedances by the impedance level of the system in which the filter is to operate. Typically this is 50 Ω.

Finally, work out the lengths of the coupled line sections; they are one quarter wavelength long at the centre frequency ω_0.

A computer program was written to design and analyse parallel coupled line filters. The program starts with the filter specifications and generates the dimensions of the coupled line sections for a microstrip realization.

A design example has been done to the following specification.

Centre frequency	8 GHz
Passband bandwidth	200 MHz
Passband return loss	20 dB
Stopband attentuation at centre frequency ±	40 dB
System impedance	50 Ω

The dimensions of the filter are shown in Fig. 6.24. These are for a substrate thickness of 0.75 mm and dielectric constant of 2.33.

The filter was constructed in a similar manner to the lowpass filter, i.e. the substrate was soldered directly to the base of a brass housing. Input and output interfaces were SMA connectors.

The measured performance of the filter is shown in Fig. 6.25.

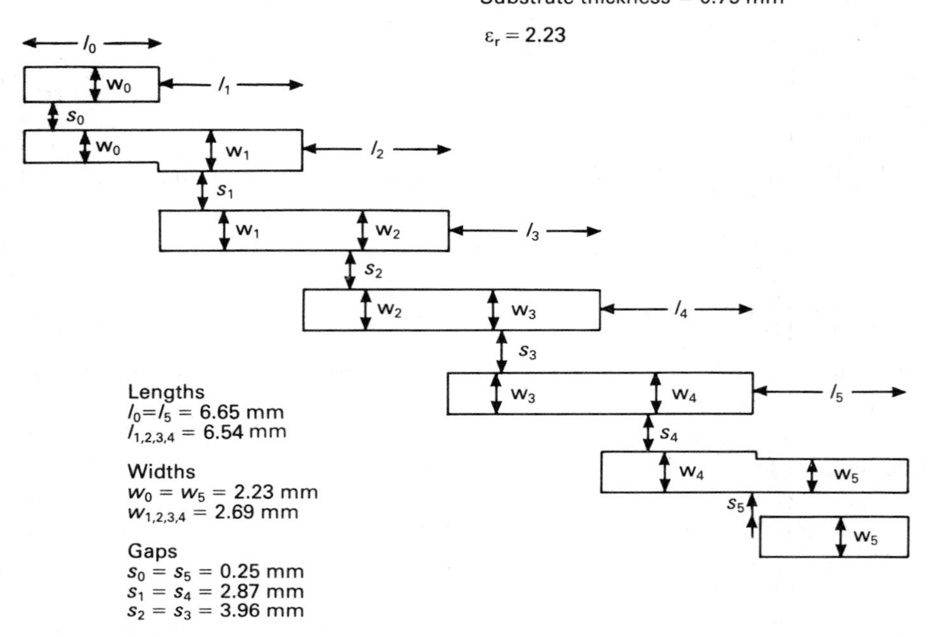

Fig. 6.24 — Metallization pattern for bandpass filter.

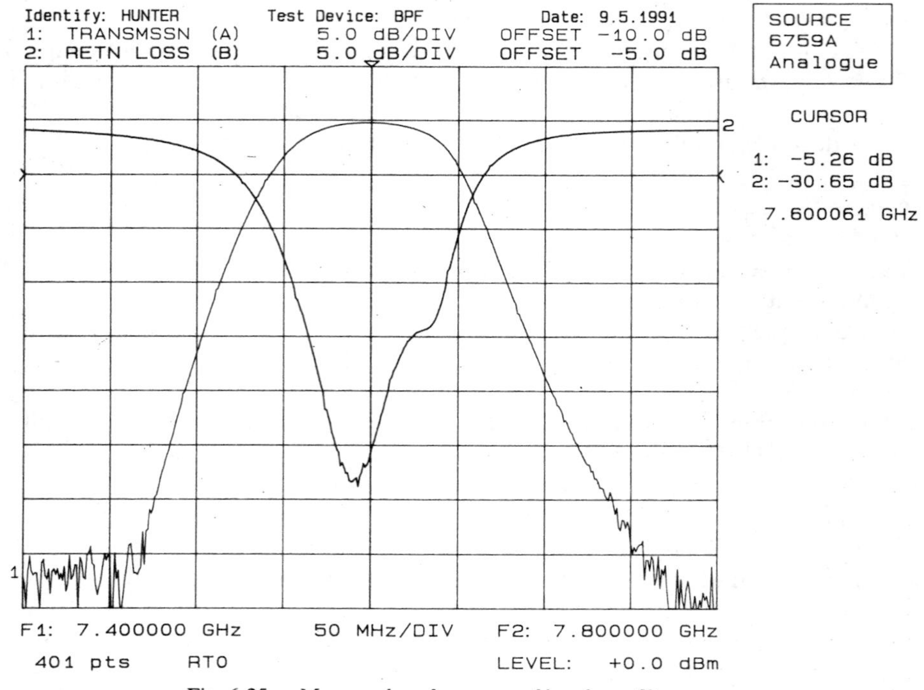

Fig. 6.25 — Measured performance of bandpass filter.

```
STEPPED IMPEDANCE LOWPASS FILTER
DIM  IL(500),RL(500),S(20),L(20),C(20),A(20),B(20),G(20),Z(20),L1(20),GD(500),
      PHASE(500),DELP(500) , LN(20),ER(20)
DIM LX(20)
SCREEN 2:CLS
WINDOW (0,-30)-(100,100)
DEF FNA(X) = ATN(X/SQR(1-X2))
INPUT ‘ RL,IL,S’;RL,IL,S
CLS
N = (RL + IL + 6)/8.686/LOG(S + SQR(S2-1))
FG = INT(N)
IF N < FG THEN N = FG
IF N > FG THEN N = FG + 1
PRINT ‘DEGREE = ’;N
PRINT
PI = 4*ATN(1)
LA = 4.343*LOG(1/(1-10(-RL/10)))
X = LA/17.37
BETA =  LOG((EXP(X) + EXP(-X))/(EXP(X)-EXP(-X)))
ETA = .5*(EXP(BETA/2/N)-EXP(-BETA/2/N))
FOR R =  1 TO N
A(R) = SIN((2*R-1)*PI/2/N)
B(R) = ETA2 + (SIN(R*PI/N))2
IF R = 1 THEN 10
G(R) = 4*A(R)*A(R-1)/B(R-1)/G(R-1)
10 G(1) = 2*A(1)/ETA
NEXT R
FOR R =  1 TO N STEP 2
L(R) = G(R)
NEXT R
FOR R =  2 TO N-1 STEP 2
C(R) = G(R)
NEXT R
INPUT ‘BANDEDGE OF FILTER IN GHZ’;FC
PRINT
WC = 2*PI*1E9*FC
REM SERIES INDUCTIVE SECTIONS
INPUT ‘INDUCTIVE LINE IMPEDANCE’;ZL
PRINT
ZL = ZL/50
FOR R =  1 TO N STEP 2
S(R) = 3E8*FNA(L(R)/ZL)/WC
NEXT R
REM SERIES CAPACITIVE SECTIONS
INPUT ‘CAPACITIVE LINE IMPEDANCE’;ZC
PRINT
ZC = ZC/50
FOR R = 2 TO N-1 STEP 2
S(R) = 3E8/WC*FNA(ZC*C(R))
NEXT R
```

```
FOR R = 3 TO N-2 STEP 2
L1(R) = S(R)-ZC/ZL*(S(R-1) + S(R + 1))/2
NEXT R
L1(1) = S(1)-ZC/ZL/2*(S(R)*.15 + S(R + 1))
L1(N) = L1(1)
FOR R =  2 TO N-1 STEP 2
L1(R) = S(R)-ZC/ZL/2*(S(R-1) + S(R + 1))
NEXT R
FOR R = 1 TO N
S(R) = L1(R)
NEXT R
FOR R = 1 TO N STEP 2
Z(R) = ZL
NEXT R
FOR R =  2 TO N-1 STEP 2
Z(R) = ZC
NEXT R
REM MICROSTRIP DIMENSION CALCULATION
INPUT 'BOARD THICKNESS ,ER';H,ER
PRINT
H = H/1000/39.3*100
REM INDUCTIVE SECTIONS
ZL = ZL*50
HL = ZL*SQR(2*(ER + 1))/119.9 + .5*((ER-1)/(ER + 1))*(LOG(PI/2) +
1/ER*LOG(4/PI))
WL = H*(EXP(HL)/8-1/4/EXP(HL))(-1)
EL = (ER + 1)/2*((1-1/2/HL*(ER-1)/(ER + 1)*(LOG(PI/2) + 1/ER*LOG(4/PI)))(-2))
REM CAPACITIVE SECTIONS
ZC = 50*ZC
DC = 59.95*PI2/ZC/SQR(ER)
WC = H*2/PI*((DC-1)-LOG(2*DC-1))  + H*(ER-1)/PI/ER*(LOG(DC-1) +
  .293-.517/ER)
EC = ER/(.96 + ER*(.109-.004*ER)*(.434*LOG(10 + ZC)-1))
PRINT 'THIN W = ';WL/2.54*1000,'EL = ';EL
PRINT 'THICK W = ';WC/2.54*1000,'EC = ';EC
PRINT
REM DIELECTRIC LENGTH CONTRACTION
FOR R =  1 TO N STEP 2
ER(R) = EL
LN(R) = S(R)/SQR(EL)
NEXT R
FOR R  =  2 TO N-1 STEP 2
ER(R) = EC
LN(R) = S(R)/SQR(EC)
NEXT R
REM DISCONTINUITY CALCULATION
LEO = .412*(EC + .3)/(EC-.258)*(WC/H + .262)/(WC/H + .813)*H
LES = 2*LEO*(1-WL/WC)
PRINT ' LES =  ';LES
FOR R =  2 TO N-1 STEP 2
```

```
LX(R) = LN(R)-LES/100
NEXT R
FOR R = 1 TO N STEP 2
LX(R) = LN(R)
NEXT R
PRINT
PRINT ' LINE LENGTHS IN DIEL IN THOU ARE'
PRINT
FOR R = 1 TO N
PRINT R,LX(R)*100000/2.54
NEXT R
REM ANALYSIS SECTION
PRINT
INPUT 'ANALYSIS FREQS AND STEP WIDTH ';F1,F2,S
CLS
PRINT 'FREQ (GHZ) INS LOSS (DB) RET LOSS (DB) GROUP DELAY (NS)
PRINT
FOR F = F1 TO F2 STEP S
FOR IND = 0 TO 1 STEP 1
W = 2*PI*F
W = W + .0001*IND
A1 = 1
B1 = 0
C1 = 0
D1 = 1
FOR R = 1 TO N
A2 = COS(W*LN(R)/.3*SQR(ER(R)))
B2 = Z(R)*SIN(W*LN(R)/.3*SQR(ER(R)))
C2 = 1/Z(R)*SIN(W*LN(R)/.3*SQR(ER(R)))
D2 = A2
A3 = A1*A2-B1*C2
B3 = A1*B2 + B1*D2
C3 = C1*A2 + D1*C2
D3 = D1*D2-C1*B2
A1 = A3
B1 = B3
C1 = C3
D1 = D3
NEXT R
IL = 4.343*LOG(1 + .25*((A1-D1)2 + (B1-C1)2))
RL = 4.343*LOG(1 + 4/(B1-C1)2)
REA = A1 + D1
IMA = B1 + C1
MAG = SQR(REA2 + IMA2)
IF IND = 1 THEN 2000
PHASE = -2*ATN(IMA/(REA + MAG))
2000 NEXT IND
DELP = -2*ATN(IMA/(REA + MAG))-PHASE
KK = KK + 1
IL(KK) = IL:RL(KK) = RL
```

```
GD = DELP/.0001
GD(KK) = -GD
PRINT F,IL,RL,-GD
NEXT F
REM GRAPHICS SECTION
N = KK
CLS
FOR I =  0 TO 10
LINE (0,I*10)-(100,I*10)
LINE (I*10,0)-(I*10,100)
NEXT I
IF RL(1) > 25 THEN RL(1) = 25
LINE (0,100)-(0,100-2*IL(1))
FOR I =  2 TO N
XP = I*100/N
IF IL(I) > 50 THEN IL(I) = 50
YP = 100-2*IL(I)
LINE -(XP,YP)
NEXT I
LINE (0,100)-(0,100-4*RL(1))
FOR I =  2 TO N
XP = I*100/N
IF RL(I) > 25 THEN RL(I) = 25
YP = 100-4*RL(I)
LINE -(XP,YP)
NEXT I
IF GD(1) > 2 THEN GD(1) = 2
LINE (0,50*GD(1))-(0,50*GD(1))
FOR I = 2 TO N
XP = I*100/N
IF GD(I)  >  2 THEN GD(I) = 2
YP = 50*GD(I)
LINE-(XP,YP)
NEXT I
VIEW PRINT 22 TO 24
PRINT ' INSERTION LOSS AND RETURN LOSS OF LOWPASS FILTER '
```

```
IP BANDPASS FILTER DESIGN PROGRAM
PI = 4*ATN(1)
V = 3E8
E = 2.7182818
E0 = 1E-9/36/PI
DIM IL(500),RL(500),GD(500),PHASE(500),DELP(500)
SCREEN 2 : CLS
WINDOW (0,-30)-(100,100)
11 REM LOWPASS PROTOTYPE SYNTHESIS
DIM C(20),L(20),N(20),ZO(20)
INPUT 'RL,STOPBAND REJ SELECTIVITY ';LR,LA,S
PRINT
PRINT
N = (LA + LR + 6)/8.686/LOG(S + SQR(S2-1))
NB = INT(N)
IF N-NB > 0 THEN N = NB + 1
IF N-NB < 0 THEN N = NB
PRINT 'DEGREE = ';N
PRINT
EP = (10(LR/10)-1)(-.5)
DEF FNG(X) = LOG(X + SQR(X2 + 1))
U = FNG(1/EP)/N
U = (EXP(U)-EXP(-U))/2
FOR R = 1 TO N
O = (2*R-1)*PI/2/N
L(R) = 2*SIN(O)/U
K(R) = SQR(U2 + (SIN(R*PI/N))2)/U
NEXT R
REM IMPEDANCE MATRIX CALCULATION
PRINT
INPUT 'F0,DEL(GHZ)';F0,DEL
PRINT
PRINT
W0 = 2*PI*1E9*F0
DEL = 2*PI*1E9*DEL
W1 = W0-DEL/2
O1 = W1*PI/2/W0
AL = TAN(O1)
X = .9
FOR R = 1 TO (N + 1)/2
IF R = 1 THEN 600
N(R) = SQR(AL*L(R)/2/X)
600 N(1) = SQR((AL*L(1) + 1)/(X + 1))
NEXT R
FOR R = 2 TO (N + 1)/2
Y(R-1) = K(R-1)/N(R-1)/N(R)
NEXT R
FOR R = 1 TO (N-1)/2
ZE(R) = X + Y(R)
ZO(R) = X-Y(R)
```

```
NEXT R
FOR R = (N + 1)/2 TO N-1
ZE(R) = ZE(N-R)
ZO(R) = ZO(N-R)
NEXT R
ZE(0) = 1 + 1/N(1)
ZO(0) = 1-1/N(1)
ZE(N) = ZE(0)
ZO(N) = ZO(0)
PRINT ' R ZOE ZOO '
PRINT
FOR R = 0 TO N
ZOE(R) = ZE(R)
ZOO(R) = ZO(R)
PRINT R,ZOE(R)*50,ZOO(R)*50
NEXT R
PRINT
PRINT
REM MICROSTRIP COUPLED LINE DIMENSION ITERATIVE SYNTHESIS
INPUT ' SUBSTRATE THICKNESS AND DIELECTRIC CONSTANT ',H,ER
CLS
JJ = ER
FOR Z = 0 TO (N-1)/2
ER = JJ
EP = ER
ZOE = ZOE(Z)*50
ZOO = ZOO(Z)*50
FOR T = 1 TO 2
Z(Z,1) = ZOE/2
Z(Z,2) = ZOO/2
NEXT T
FOR T = 1 TO 2
IF Z(Z,T) < (44-2*ER) THEN 2
H11 = Z(Z,T)*SQR(2*(ER + 1))/119.9 + .5*(ER-1)/(ER + 1)*(LOG(PI/2) +
1/ER*LOG(4/PI))
SH(T) = (EXP(H11)/8-1/4/EXP(H11))-1
GOTO 3
2 DE = 59.95*PI2/Z(Z,T)/SQR(ER)
SH(T) = 2/PI*((DE-1)-LOG(2*DE-1)) + (ER-1)/PI/ER*(LOG(DE-1) + .293-.517/ER)
3 NEXT T
WHE = SH(1)
WHO = SH(2)
DEF FNF(X) = (EXP(X) + EXP(-X))/2
DEF FNE(X) = LOG(X + SQR(X2-1))
J9 = FNF(PI/2*WHE)
J8 = FNF(PI/2*WHO)
J = (J9 + J8-2)/(J8-J9)
S = H*4/PI*FNE(J)
G = FNF(PI*S/2/H)
DD = .5*((G + 1)*FNF(PI/2*WHE)-1 + G)
```

```
W = H/PI*(FNE(DD)-PI*S/2/H)
WX = W : S0 = S
REM ANALYSIS OF COUPLED LINE STRUCTURE
FOR Q = 1 TO 6
FOR IND = 0 TO 2
ER = EP
IF IND = 2 THEN 7
W = (1 + .001*IND)*WX
GOTO 8
7 W = WX
S = (1 + .001*IND)*S0
8 FOR R = 1 TO 2
REM SINGLE LINE CALCULATION
IF W/H > 3.3 THEN 10
SCD = LOG(4*H/W + SQR(16*(H/W)2 + 2))-.5*(ER-1)/(ER + 1)*(LOG(PI/2) +
1/ER*LOG(4/PI))
Z0 = 119.9/SQR(2*(ER + 1))*SCD
GOTO 20
10 CCD = W/2/H + LOG(4)/PI + LOG(E*PI2/16)/2/PI*((ER-1)/ER2)
CCD = CCD + (ER + 1)/2/PI/ER*(LOG(PI*E/2) + LOG(W/2/H + .94))
Z0 = 119.9*PI/2/SQR(ER)/CCD
20 IF W/H < 1.3 THEN 30
EF = (ER + 1)/2 + (ER-1)/2*(1 + 10*H/W)-.555
GOTO 40
30 H1 = LOG(4*H/W + SQR(16*(H/W)2 + 2))
XD = 1-.5/H1*(ER-1)/(ER + 1)*(LOG(PI/2) + 1/ER*LOG(4/PI))
EF = (ER + 1)/2/XD2
40 REM COUPLED LINE ANALYSIS
CP = ER*E0*W/H
CF = .5*(SQR(EF)/V/Z0-CP)
DEF FNA(X) = (EXP(X)-EXP(-X))/(EXP(X) + EXP(-X))
DEF FNB(X) = EXP(-.1*EXP(2.33-2.53*X))
IF S > H THEN 999
GOTO 1000
999 CF1 = CF/(1 + H/S*FNB(W/H))
GOTO 1010
1000 CF1 = CF/(1 + H/S*FNB(W/H)*FNA(8*S/H))
1010 K = S/H/(S/H + 2*W/H)
K1 = SQR(1-K2)
DEF FNC(X) = PI/LOG(2*(1 + SQR(X))/(1-SQR(X)))
DEF FND(X) = 1/PI*LOG(2*(1 + SQR(X))/(1-SQR(X)))
IF K2 > .5 THEN 100
CGA = E0*FND(K1)
GOTO 200
100 CGA = E0*FNC(K)
200 CGD = E0*ER/PI*LOG(1/FNA(PI*S/4/H)) + .65*CF*(.02/S*H*SQR(ER) +
1-1/ER/ER)
CE(R) = CP + CF + CF1
CO(R) = CP + CF + CGA + CGD
ER = 1
```

```
NEXT R
ZE = 1/V/SQR(CE(1)*CE(2))
ZO = 1/V/SQR(CO(1)*CO(2))
EE = CE(1)/CE(2)
EO = CO(1)/CO(2)
ZE(IND) = ZE
ZO(IND) = ZO
NEXT IND
DEW = (ZE(0)-ZE(1))/.001/W
DES = (ZE(0)-ZE(2))/.002/S
DOW = (ZO(0)-ZO(1))/.001/W
DOS = (ZO(0)-ZO(2))/.002/S
IF ABS(DES) > ABS(DOS) THEN S = S-(ZOE-ZE(0))/DES
IF ABS(DOS) > ABS(DES) THEN S = S-(ZOO-ZO(0))/DOS
IF ABS(DEW) > ABS(DOW) THEN W = W-(ZOE-ZE(0))/DEW
IF ABS(DOW) > ABS(DEW) THEN W = W-(ZOO-ZO(0))/DOW
WX = W : S0 = S
NEXT Q
W(Z) = W
S(Z) = S
ZOE(Z) = ZE:ZOO(Z) = ZO:EE(Z) = EE:EO(Z) = EO
PRINT 'LOOP = ';Z
NEXT Z
FOR Z = (N + 1)/2 TO N
ZOE(Z) = ZOE(N-Z): ZOO(Z) = ZOO(N-Z)
EE(Z) = EE(N-Z): EO(Z) = EO(N-Z)
W(Z) = W(N-Z):S(Z) = S(N-Z)
NEXT Z
CLS
DEM = 1000
PRINT:PRINT:PRINT
PRINT ' ZE ZO EE EO'
PRINT
FOR R = 0 TO N
PRINT USING '###.## '; ZOE(R), ZOO(R), EE(R), EO(R)
NEXT R
FOR AA = 1 TO 10000
NEXT AA
CLS
PRINT:PRINT:PRINT
PRINT ' WIDTH SPACING LINE LENGTH'
PRINT
FOR R = 0 TO N
REM LENGTH CALCULATION
LENTH = 30/F0/2.54/4
EA = SQR(EE(R)*EO(R))
LL(R) = LENTH/SQR(EA)
REM LM(R) = PHYSICAL LENGTH OF QUARTER WAVE SECTION IN METRES
LM(R) = LL(R)/39.3
PRINT USING ' ####.## ' ; W(R), S(R), LL(R)*DEM
```

```
NEXT R
PRINT
PRINT
REM CALCULATION OF Q
LG = 30/F0/SQR(EE(2))/2.54
Q = 237*W(2)/1000*ZOE(2)/SQR(F0)/LG
PRINT 'Q = ';Q
V = Q*2
PRINT
PRINT 'DIEL THICKNESS =  ';H,' ER = ';JJ
REM ANALYSIS SECTION IL RL TD WITH INHOMOGENEOUS SECTIONS AND
DISSIPATION LOSS
DEF FNH(X) = (EXP(X)-EXP(-X))/2
PRINT
PRINT
INPUT 'F1,F2,DEL';F1,F2,S
CLS
PRINT ' FREQ INS.L RET.L GP.DEL
PRINT
FOR F =  F1 TO F2 STEP S
FOR INC = 0 TO 1 STEP 1
W = 2*PI*F  + .0001*INC
Q = V*SQR(F/F0)
A1 = 1: A2 = 0: B1 = 0: B2 = 0 : C1 = 0: C2 = 0: D1 = 1: D2 = 0
FOR R =  0 TO (N-1)/2
ZOE = ZOE(R)/50
ZOO = ZOO(R)/50
EE = EE(R)
EO = EO(R)
AE = W*LM(R)*SQR(EE)/.3
BE = -W*LM(R)*SQR(EE)/.3/Q
AO = AE*SQR(EO)/SQR(EE)
BO = BE*SQR(EO)/SQR(EE)
CTRE = TAN(AE)*(1-FNA(BE)2)/(TAN(AE)2 + FNA(BE)2)
CTRO = TAN(AO)*(1-FNA(BO)2)/(TAN(AO)2 + FNA(BO)2)
CTIE = -(FNA(BE) + TAN(AE)2*FNA(BE))/(TAN(AE)2 + FNA(BE)2)
CTIO = -(FNA(BO) + TAN(AO)2*FNA(BO))/(TAN(AO)2 + FNA(BO)2)
NLLE = SIN(AE)2*FNF(BE)2 + COS(AE)2*FNH(BE)2
NLLO = SIN(AO)2*FNF(BO)2 + COS(AO)2*FNH(BO)2
CSRE = SIN(AE)*FNF(BE)/NLLE
CSRO = SIN(AO)*FNF(BO)/NLLO
CSIE = -COS(AE)*FNH(BE)/NLLE
CSIO = -COS(AO)*FNH(BO)/NLLO
DENR = ZOE*CSRE-ZOO*CSRO
DENI = ZOE*CSIE-ZOO*CSIO
NUMR = DENR/(DENR2 + DENI2)
NUMI = -DENI/DENR*NUMR
A3 = NUMR*(ZOE*CTRE + ZOO*CTRO) -NUMI*(ZOE*CTIE + ZOO*CTIO)
A4 = NUMI*(ZOE*CTRE + ZOO*CTRO)  + NUMR*(ZOE*CTIE + ZOO*CTIO)
GAM = CSRE*CSRO-CSIE*CSIO + CTRE*CTRO-CTIE*CTIO
```

```
GIM = CSRE*CSIO + CSIE*CSRO + CTRE*CTIO + CTIE*CTRO
B3 = -NUMI/2*(ZOE2 + ZOO2-2*ZOE*ZOO*GAM) + NUMR*ZOE*ZOO*GIM
B4 = NUMI*ZOE*ZOO*GIM + NUMR/2*(ZOE2 + ZOO2-2*ZOE*ZOO*GAM)
C3 = -2*NUMI
C4 =  2*NUMR
D3 = A3
D4 = A4
A5 = A1*A3-A2*A4 + B1*C3-B2*C4
A6 = A1*A4 + A2*A3 + B1*C4 + B2*C3
B5 = A1*B3-A2*B4 + B1*D3-B2*D4
B6 = A1*B4 + A2*B3 + B1*D4 + B2*D3
C5 = C1*A3-C2*A4 + D1*C3-D2*C4
C6 = C1*A4 + C2*A3 + D1*C4 + D2*C3
D5 = C1*B3-C2*B4 + D1*D3-D2*D4
D6 = C1*B4 + C2*B3 + D1*D4 + D2*D3
A1 = A5: A2 = A6: B1 = B5: B2 = B6: C1 = C5: C2 = C6: D1 = D5: D2 = D6
NEXT R
A7 = A1*D1-A2*D2 + B1*C1-B2*C2
A8 = A1*D2 + A2*D1 + B1*C2 + B2*C1
B7 = A1*B1-A2*B2 + B1*A1-B2*A2
B8 = A1*B2 + A2*B1 + B1*A2 + B2*A1
C7 = C1*D1-C2*D2 + D1*C1-D2*C2
C8 = C1*D2 + C2*D1 + D1*C2 + D2*C1
D7 = C1*B1-C2*B2 + D1*A1-D2*A2
D8 = C1*B2 + C2*B1 + D1*A2 + D2*A1
REA = A7 + B7 + C7 + D7
IMA = A8 + B8 + C8 + D8
IL = 4.343*LOG(.25*(REA2 + IMA2))
RL = IL-4.343*LOG(.25)-4.343*LOG((A7 + B7-C7-D7)2 + (D8 + B8-C8-A8)2)
MAG = SQR(REA2 + IMA2)
IF INC = 1 THEN 20000
PHASE = -2*ATN(IMA/(REA + MAG))
20000 NEXT INC
DELP = -2*ATN(IMA/(REA + MAG))-PHASE
KK = KK + 1
IL(KK) = IL
RL(KK) = RL
GD = DELP/.0001
GD(KK) = -GD
PRINT USING '###.### '; F,IL,RL,-GD
NEXT F
CLS
REM GRAPHICS
N = KK
FOR I =  0 TO 10
LINE (0,I*10)-(100,I*10)
LINE (I*10,0)-(I*10,100)
NEXT I
IF IL(1) > 50 THEN IL(1) = 50
LINE (0,0)-(0,100-2*IL(1))
```

```
FOR I =  1 TO N
XP = I*100/N
IF IL(I) > 50 THEN IL(I) = 50
YP = 100-2*IL(I)
LINE -(XP,YP)
NEXT I
RL(1) = 0
LINE (0,0)-(0,100-2*RL(1))
FOR I = 2 TO N
XP = I*100/N
IF RL(I) > 50 THEN RL(I) = 50
YP = 100-2*RL(I)
LINE -(XP,YP)
NEXT I
LINE (0,2*GD(1))-(0,2*GD(1))
FOR I =  2 TO N
XP = I*100/N
IF GD(I) > 50 THEN GD(I) = 50
YP = 2*GD(I)
LINE-(XP,YP)
NEXT I
VIEW PRINT 22 TO 24
PRINT ' INS LOSS RET LOSS AND GROUP DELAY OF BANDPASS FILTER '
```

7

Dielectric resonators

INTRODUCTION

The term 'dielectric resonator' (DR) was introduced by Richtmyer in 1939 when he demonstrated that a suitably shaped dielectric object could be used to behave as an electro-magnetic resonator [1]. In 1960 Okaya [2] 'rediscovered' the DR by observing that a piece of high dielectric material (rutile) used in the investigation of paramagnetic resonance acted as a resonator. About two years later he and Barash furnished the first analysis of modes for a DR [3]. Subsequently other researchers analysed and experimented with DR's, amongst them Cohn [4] who showed that rutile ceramic (high-purity $Ti\ O_2$) had a high dielectric constant in the order of 100 with an unloaded Q-factor around 10 000. Despite all this the DR was not widely used owing to poor temperature stability.

In early 1970 Raytheon [5] and then the Bell laboratories [6] reported low-loss, high dielectric constant and temperature stable barium tetratitanates of the type $BaTi_4O_9$ and $Ba_2Ti_9O_{20}$. Although these materials were not commercially available, DRs were implemented in some microwave designs.A major breakthrough was initiated by Murata Manufacturing Co., who produced $(Zr,Sn)TiO_4$ ceramics and made it commercially available [7].

The following sections describe some of the features, uses and measurements of DRs as well as their analysis.

Dielectric resonator

A dielectric resonator is a piece of high dielectric constant ceramic material which is comparable in characteristics , but much smaller in size, to the metallic resonant cavity. The high dielectric constant, typically around 40, causes most of the electromagnetic energy to be confined within the DR. Resonance in the DR is produced by reflection of the energy at the boundary or interface between the dielectric of the DR and its surrounding air or metallic wall. DRs can resonate in various modes. The resonance frequency is governed by the dimensions of the DR and its immediate suroundings as well as the dielectric constant of material.

A major problem with previously available high Q materials was the poor temperature stability of the dielectric constant and the resulting instability of the resonance frequency of the dielectric resonators. To overcome this, however, a number of temperature stable dielectric materials have been developed during the last decade. These new materials have dielectric constants between 30 and 40 and have minimal resonance frequency variation with temperature. Basic properties of some high quality ceramics developed for DRs are presented in Table 7.1.

Table 7.1 — Some DR compositions [8]

Material composition	Manufacturer	ε_r	Loss tan. at 4 GHz	Temperature coefficient (ppm/°C)
$BaTi_4O_9$	Raytheon, Transtech	38	0.0001	+4
$Ba_2Ti_9O_{20}$	Bell Labs. Murata	40	0.001	+2
$(Zr,Sn)TiO_4$	Thomson-CSF Siemens Transtech NTK	38	0.0001	−4 to +10 adj.
$Ba(Zn_{1/3}Nb_{2/3})O_2$ $Ba(Zn_{1/3}Ta_{2/3})O_2$	Panasonic Murata	~30	0.00004	0 to +10 adj.

For a fundamental mode response, the dimensions of a DR are in the order of one guided wavelength λ_g given by $\lambda_o/\sqrt{\varepsilon_r}$. Since the dielectric constant is high, most of the electromagnetic energy is confined to the DR and therefore radiation losses are very small. The unloaded quality factor Q_u is limited mainly by losses within the DR. Hence, using a low-loss, high dielectric material results in a small resonator of high Q.

At about 1 GHz the size of the DR becomes large, so it is rarely used below this frequency, whilst the maximum frequency limit is imposed by the lowest Q_u, at about 100 GHz [9]. The present optimum frequency application range of the DR is 4 to 30 GHz.

DRs have many advantages over the traditional cavity resonators. They are small in size, lightweight, high Q, temperature stable, low-cost and easily integrated with MICs. These properties make its application very popular in both active and passive microwave circuits such as MIC oscillators [10] and MIC filters [11].

FEATURES OF DR CIRCUITS

There are basically two types of microwave circuits in which DRs are used: oscillators and filters. Although known for a long time, it was the progress in semiconductor technology and DR material science that propelled the circuits into the present high ranking position. Some of the advantageous features of DR circuits and in particular DROs are as follows:

— small physical size;
— rugged;
— low-cost;
— tunable;
— oscillate over large frequency range;
— low d.c. supply power;
— operates over wide temperature range;
— low phase–noise characteristics;
— good efficiency.

In Table 7.2 a general comparison is made between the DRO and two other types of microwave oscillators.

Table 7.2 — Comparison of some microwave oscillators

Feature	DRO	*X*-tal stabilized mutliplier oscillator	Cavity stabilized Gunn oscillator
Size	Small	Large	Small
f-stability	High	Very high	Good
Efficiency	High	Medium	Low
Tunable	Yes	No	Yes
Production cost	Very low	Medium	Low

APPLICATIONS

Because of some of the features outlined in the previous section DR filters and oscillators are widely used. Amongst others they find application in

— terrestrial communication systems;
— space communication systems;
— ECM receivers;
— radars;
— missile transponders.

PHYSICAL ASPECTS OF DRs

DRs are available in different configurations and some examples are given in Fig. 7.1. Special shapes can be manufactured to cater for specific needs. Table 7.3 provides a typical set of DR material characteristics [12].

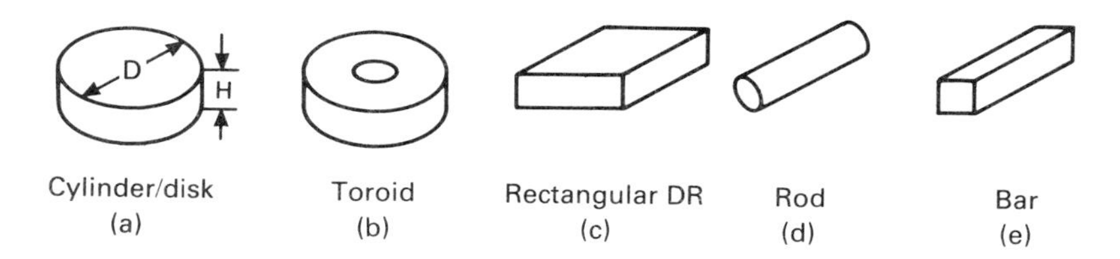

Fig. 7.1 — Some DR shapes.

Table 7.3 — Typical DR material characteristics

Material designation	U(R-O4C) material	S-series material
Dielectric constant ε_r	38	29
Temperature coefficient (p.p.m./°C)	− 4–10	0–6
$Q(1/\tan\delta)$ min.	6000 (at 7 GHz)	13 000 (at 7 GHz)
Insulation resistance (Ω cm)	10^{14}	$>10^{14}$
Thermal coefficient of expansion (p.p.m./°C)	6.5	10.2
Thermal conductivity (cal/cm s °C)	0.0046	0.0063
Specific heat (cal/g°C)	0.15	0.07
Density (g/cm^3)	5.0	7.7
Water absorption (%)	<0.01	<0.01
Vicker's hardness no. (kg/mm)	900	700
Flexural strength (kg/cm)	1000	800
Operating frequency range (GHz)	1.5–10	8–50

The disk or puck shape shown in Fig. 7.1 is most frequently used in microwave circuits since it can be more easily manufactured than other shapes. Furthermore, the disks are mostly operated in the $TE_{O1\delta}$ mode which represents the lowest possible resonance frequency. This resonance frequency can be calculated to an accuracy of about 1%. The DR is usually specified by its physical dimensions, i.e. its height or length H and diameter D. The DR aspect ratio is thus D/H.

A cylindrical DR (disk) supports resonances which may be classified in terms of the following transverse and hybrid modes: $TE_{np\upsilon}$, $TM_{np\upsilon}$, $HE_{np\upsilon}$ and $EH_{np\upsilon}$.

The suffixes n, p and υ describe the standing wave pattern in the azimuthal (n), radial (p) and axial (υ) direction of the cylinder. For the fundamental transverse electric mode TE the suffix υ is usually replaced by δ, namely $TE_{O1\delta}$, where δ lies in the range 0.5 to unity. Some simple modes of a DR are contrasted in Fig. 7.2 and other modes are illustrated in references [8] and [13].

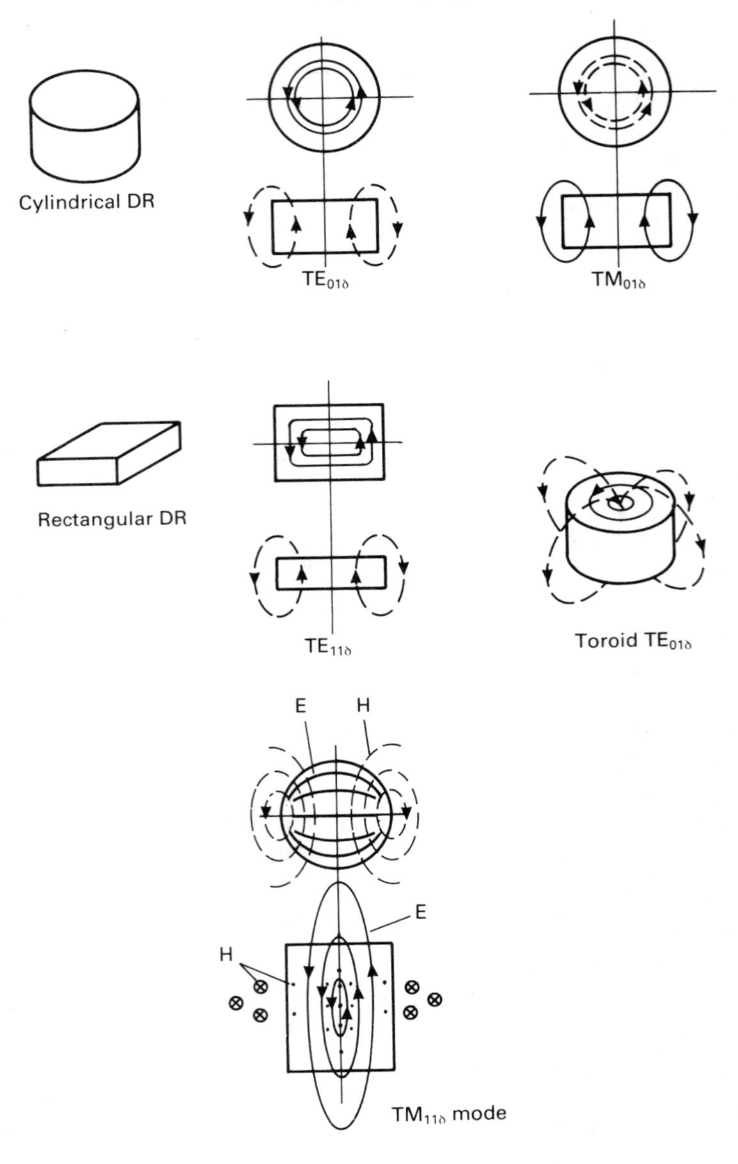

Fig. 7.2 — Examples of DR modes. —— *E*-field; --- *H*-field.

If a DR is placed with its circular sides between two metallic walls, with one or two of these sides having an airgap as shown in Fig. 7.3, then it can be made to resonate in its fundamental mode $TE_{01\delta}$ or $HE_{11\delta}$. Most applications use one or the

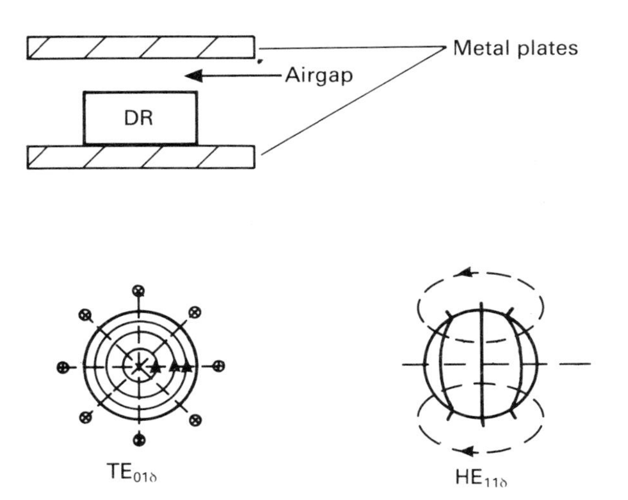

Fig. 7.3 — DR-structure and $TE_{01\delta}$ and $HE_{11\delta}$ mode.

other of those two fundamental modes. Which of those modes predominates depends on the totality of the DR system, e.g. the aspect ratio of the DR, the presence of any tuning elements,the type of substrate used and the enclosure/ shielding employed. The $TE_{01\delta}$ mode usually prevails for aspect ratios which are larger than about 1.5.

A more detailed drawing of the $TE_{01\delta}$ mode is given in Fig. 7.4. The electric field lines E are perpendicular to the *z*-axis. If a DR is cut in half as shown in the top part of Fig. 7.4, then the E-lines enter the right half of the DR, go around the *z*-axis and then emanate on the left half. Seen from the top of the DR, the E-lines thus form concentric circles. Naturally, again referring to the same figure, the magnetic field lines H are at right angles to the E-lines, and are emanating from the top of the DR at the high dielectric/air boundary, go through the air and then enter the bottom of the DR. The H-lines are thus also forming closed loops. Of course, as the DR has a high dielectric constant, the intensity of the field is highest within the DR and with the field diminishing rapidly on the outside. But there is still a small percentage of lines both, electric and magnetic, outside the DR. It is because of this that coupling from the DR to outside microwave circuitry is possible as will be discussed later.

EXCITATION AND TUNING OF A DR

Once a DR has been placed onto a suitable substrate and into a housing or similar enclosure, it needs to be excited.This may be done in different ways, three of which

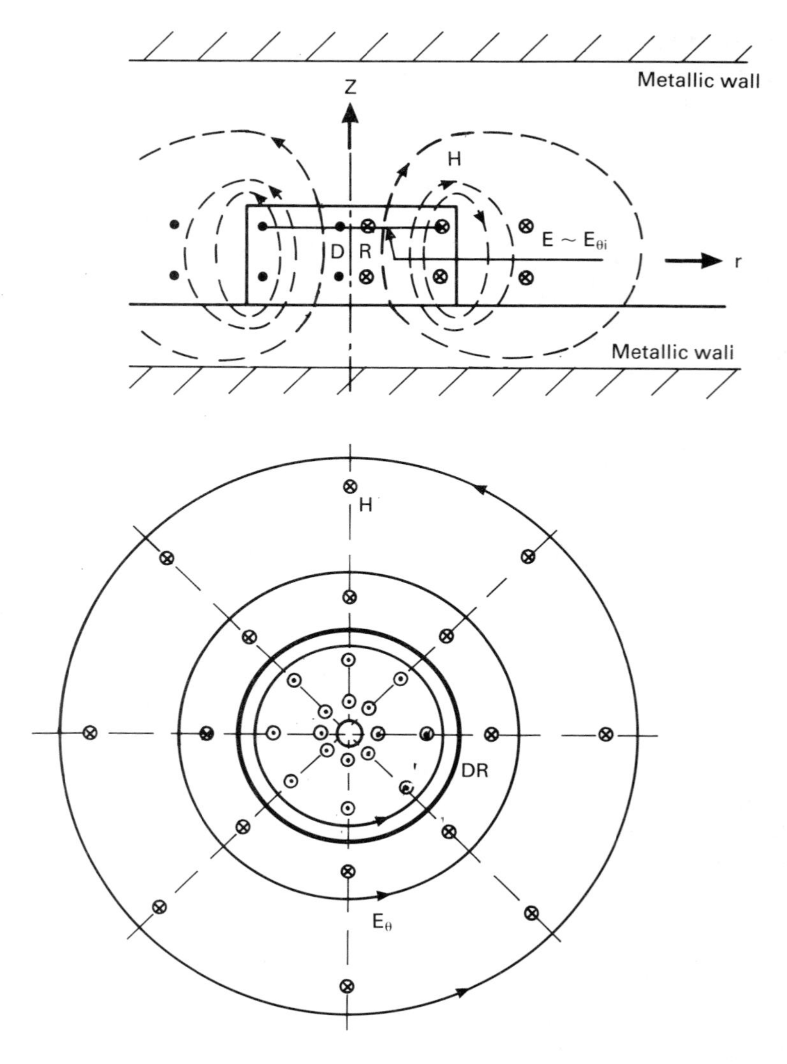

Fig. 7.4 — A more detailed drawing of the $TE_{01\delta}$ mode.

are illustrated here. The technique or way in which one couples energy into or out of a DR depends on factors such as:

— amount of coupling required;
— mode of excitation;
— type of transmission medium: microstrip, coaxial line, waveguide.

The three techniques (a,b,c) are shown in Fig. 7.5 which is self-explanatory. One or the other technique will be discussed in more detail later on.

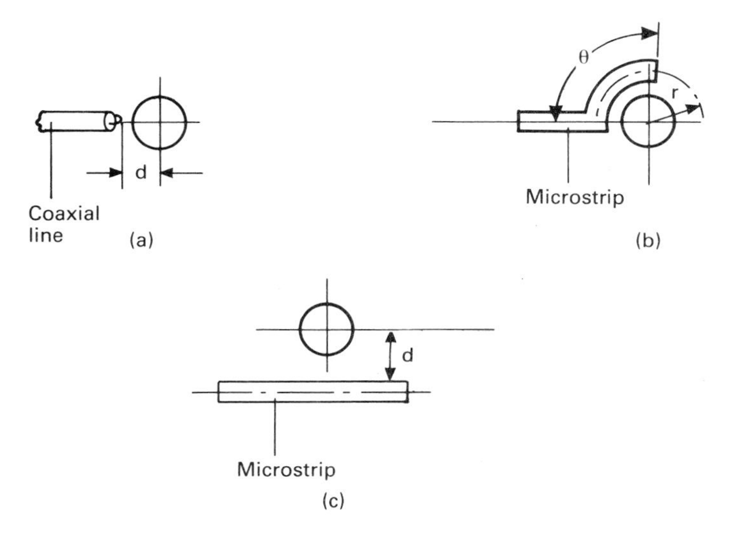

Fig. 7.5 — Structure for coupling to a DR.

DRs can be tuned or made to meet a desired resonance frequency by changing its diameter, length or both. A DR structure can also be tuned by perturbing the fringing (external) fields of the DR either on its circular side or at some point of its circumference. The first case is shown in Fig. 7.6(a) . If a metallic disk or plunger is brought close to one circular side of the DR, then its fringe field will be disturbed, resulting in a frequency change. As the metal plunger comes closer to the DR (airgap reducing), the frequency goes up and vice versa. If the metal plunger is replaced by an arrangement, usually a screw, made from the same material as the DR, then the frequency of the DR structure will go down as the airgap reduces (Fig. 7.6(b)). Using a metal side screw as shown in Fig. 7.6(c) will also decrease the oscillating frequency. Plunger and side screw tuning have different effects on different DR oscillating modes. Please note that the direction of frequency change does not only depend on the tuning arrangement, but also on the type of material employed for the tuning element, i.e. the plunger or screw.

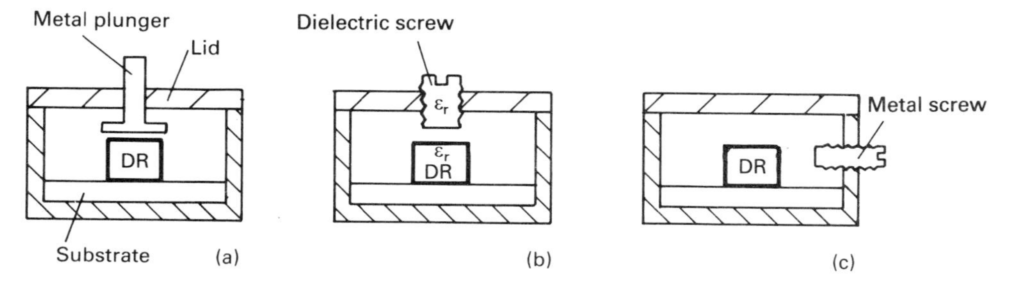

Fig. 7.6 — Examples of DR tuning structures.

DRs IN AN ENCLOSURE

Any microwave circuit employing a DR will eventually have to be put into a metal enclosure (box/housing) to facilitate circuit handling, prevent both, the radiation and degradation of the resonator Q and the interference of electromagnetic waves from outside. Aspects relating to the effect of box size on microwave circuit operation are discussed in Chapter 9. The calculations given in that chapter may also be applied to DR enclosures. Here it is sufficient to say that a DR in a housing may be represented as in Fig. 7.7. Apart from the DR there will be some kind of coupling structure associated with the DR, as discussed in the previous section. Most frequently this is a microstrip type configuration.

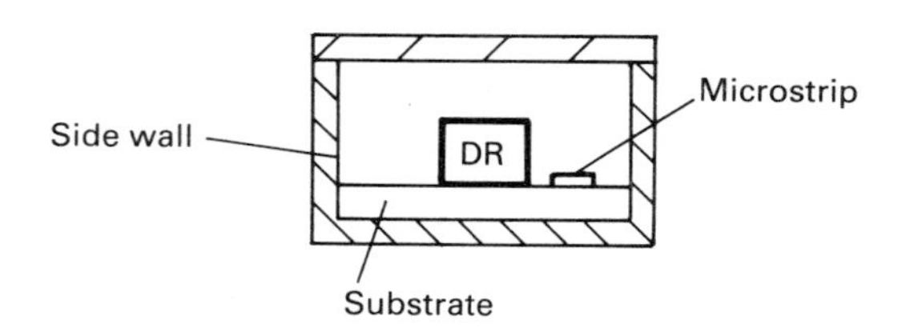

Fig. 7.7 — Enclosed DR structure.

In practice the coupling structure is small and only negligibly disturbs the fields external to the DR. This is convenient when analysing a DR structure as that shown in Fig. 7.7. Furthermore, let us assume that the DR is placed sufficiently far away from the walls of the box, so as to eliminate major field distortions.

QUALITY FACTOR

An important parameter of any resonant system or component is the quality factor, Q. The quality factor of a component is a measure of its ability to behave as a pure reactance. In the context of filters and oscillators employing a DR (or cavity) one has to consider three different factors, namely the unloaded quality factor Q_u, the loaded quality factor Q_L and the external quality factor Q_{ext} also often written as Q_e for brevity.

The loaded quality factor of a resonance system is defined as:

$$Q_L = \frac{\omega_r W}{P} = 2\pi \frac{W}{PT_r} \tag{7.1}$$

where

ω_r $= 2\pi f_r$ is the resonance frequency in radians,

W = total stored energy of the electromagnetic field in the resonator system,
P = total power loss in the resonator and load,
T_r = $1/f_r$ is the period of oscillation,
$P \times T_r$ = total energy loss during each period of oscillation.

The total power loss can be mathematically expressed as the loss in the DR system (P_r) and the load (P_L) namely:

$$P = P_r + P_L \tag{7.2}$$

The losses associated with the DR system can be expressed as:

$$P_r = P_R + P_D + P_C \tag{7.3}$$

where

P_R = radiation loss,
P_D = dielectric losses,
P_C = conduction loss, mainly in the 'skin' of the metal wall.

Hence

$$\frac{1}{Q_L} = \frac{PT_r}{2\pi W} = \frac{(P_r + P_L)T_r}{2\pi W} = \frac{P_r T_r}{2\pi W} + \frac{P_L T_r}{2\pi W}$$

$$= \frac{1}{Q_u} + \frac{1}{Q_e} \tag{7.4}$$

$$Q_u = 2\pi \frac{W}{P_r T_r} \tag{7.5}$$

$$Q_e = 2\pi \frac{W}{P_L T_r} \tag{7.6}$$

where Q_u is the unloaded quality factor and Q_e the external quality factor. In order to obtain a high Q_u, it is essential to make P_R, P_D and P_C as low as possible.

The energy stored in this DR system can be described by means of the electromagnetic field theory as

$$W = \int_{v} \frac{\mu |H|^2}{2} dV \tag{7.7}$$

or

$$W = \int_{v} \frac{\varepsilon |E|^2}{2} \mathrm{d}V \tag{7.8}$$

where $\int_{v}$ denotes an integral over a volume. From lumped circuit theory we obtain for the electromagnetic energy stored in an inductor or capacitor

$$W = 0.5 L_r I_r^2 \tag{7.9}$$

and

$$W = 0.5 C_r V_r^2 \tag{7.10}$$

with the unloaded quality factor Q_u defined as

$$Q_u = \frac{\omega_r L_r}{R_r} \tag{7.11}$$

where R_r is the series resistance representing the resistive loss and L_r the inductance of the circuit. The DR coupled to a microstrip and its equivalent circuit are shown in Fig. 7.8. Sometimes the quality factor is expressed in terms of the loss tangent tan δ, where $Q = 1/\tan \delta$. Reference [14] gives a detailed theoretical analysis. A computer

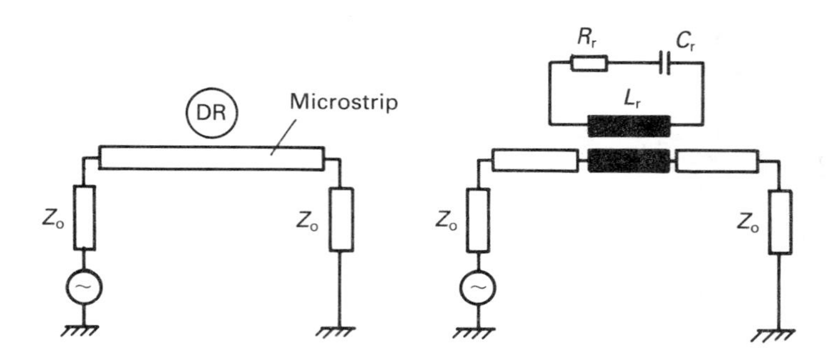

Fig. 7.8 — DR coupled to a microstrip and its equivalent circuit.

program is also given in this book. Some lines of the program have been re-written as DR10 for calculating the resonance frequency and unloaded quality factor (only taking into account the conduction loss, P_c, of the DR-substrate-metal shielding system).

RESONANCE FREQUENCY OF A SHIELDED TE MODE SINGLE PIECE DR

Previous sections have given a general introduction to dielectric resonators. We shall now deal with the theoretical analysis of a DR. The magnetic wall model can be used to analyse the field problem of a DR. In the magnetic wall model the tangential magnetic field component and the normal component of the electric field vanish at the DR/air boundary. Some of the field leaks out of the DR and if not taken into account in an analysis results in discrepancies with measured results.

In order to improve analysis the two lateral magnetic walls of the DR were removed and replaced by a magnetic wall waveguide model working below the cut-off frequency [4,26]. Subsequently the circular dielectric waveguide model was developed by removing the circular magnetic wall of the DR [27,28]. This improved the theoretical accuracy to about 2% .The best accuracy can be obtained by using the variational method [29]. The frequency calculation accuracy for the $TE_{01\delta}$ mode of DR operation is about 1%.

The dielectric waveguide model is now being used for analysis. For a good resonator the practical conduction and dielectric losses are small and are thus neglected in this analysis. The TE mode ,which is called transversal electric mode ($E_z = 0$) with no variation in the azimuthal direction, is only discussed here. Based on Helmholz's equation

$$\nabla^2 H_z + k^2 H_z = 0 \qquad (7.12)$$

and the appropriate boundary conditions an approximation for the field distribution inside and outside the DR is introduced.The composite structure to be analysed here is shown in Fig. 7.9. The structure is sub-divided into six regions. Since we are

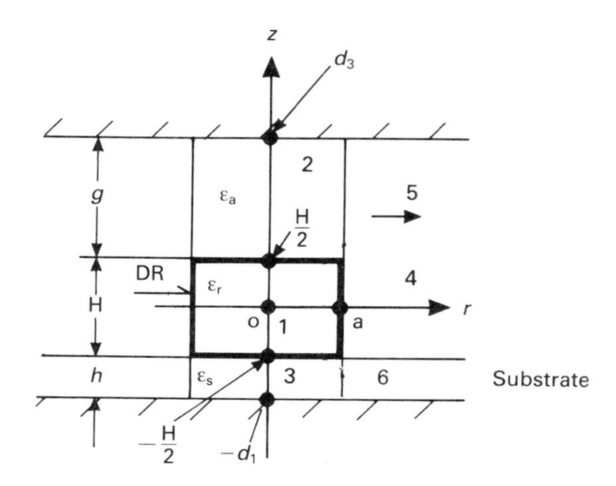

Fig. 7.9 — Regions associated with an enclosed DR system. O = reference point.

only interested in the resonance frequency, regions 1 to 4 need only to be analysed. The expressions for regions 5 and 6 are omitted. The longitudinal magnetic components in regions 1 to 4 are as follows:

$$
\begin{aligned}
H_{z1} &= H_o J_0(k_r r)\cos(\beta z + \varphi) \\
H_{z2} &= H_o \frac{\cos(\beta H/2 + \varphi)}{\sinh \alpha_a g} J_0(k_r r)\sinh \alpha_a (d_3 - z) \\
H_{z3} &= H_o \frac{\cos(\beta H/2 - \varphi)}{\sinh \alpha_s h} J_0(k_r r)\sinh \alpha_s (z + d_1) \\
H_{z4} &= H_o \frac{J_0(k_r a)}{K_0(k_a a)} K_0(k_a r)\cos(\beta z + \varphi)
\end{aligned}
\tag{7.13}
$$

and

$$
\begin{aligned}
\beta^2 &= \varepsilon_r(\omega_r/c)^2 - k_r^2 = \varepsilon_a(\omega_r/c)^2 + k_a^2 \\
\alpha_a^2 &= k_r^2 - \varepsilon_a(\omega_r/c)^2 \\
\alpha_s^2 &= k_r^2 - \varepsilon_s(\omega_r/c)^2
\end{aligned}
\tag{7.14}
$$

After solving Maxwell's equations:

$$
\begin{aligned}
E_r &= \frac{-j\omega_r\mu_o}{k^2 - k_z^2} \times \frac{\partial H_z}{r\partial\theta} \\
H_r &= \frac{1}{k^2 - k_z^2} \times \frac{\partial^2 H_z}{\partial r \partial z} \\
E_\theta &= \frac{j\omega_r\mu_o}{k^2 - k_z^2} \times \frac{\partial H_z}{\partial r} \\
H_\theta &= \frac{1}{k^2 - k_z^2} \times \frac{\partial^2 H_z}{r\partial\theta\partial z}
\end{aligned}
\tag{7.15}
$$

The other field components can be obtained in a similar manner if necessary. The symbols used are as follows:

H_θ = amplitude of the field component
J_0 = Bessel function of the first kind
K_0 = modified Bessel function of the second kind

k = wave number of each region
$k_z \ldots k_z = \beta$ in regions 1 and 4
$k_z = j\alpha_a$ in regions 2 and 5
$k_z = j\alpha_s$ in regions 3 and 6
k_r, k_a, β, α_a, α_s = propagation or distribution constants of each region
φ = phase
r = radial coordinate
z = axial coordinate
θ = azimuthal coordinate
a = radius of the DR
d_1 = coordinate point, thickness of the substrate is $h = d_1 - H/2$
d_3 = coordinate point, distance between the top of the DR and upper metal shield is $g = d_3 - H/2$
H = height of the DR
ω_r = $2\pi f_r$
c = 3×10^{10} [cm/s]
μ_0 = permeability of free space 1.257×10^{-6} [H/m]
ε_a = relative permittivity of regions 2, 4, 5 (air)
ε_s = relative permittivity of regions 3, 6 (substrate)
ε_r = relative permittivity of region 1 (DR)

The azimuthal components of the electric field are deduced as follows:

$$E_{\theta 1} = \frac{-j\omega_r\mu_o}{k_r} H_o J_1(k_r r)\cos(\beta z + \varphi)$$

$$E_{\theta 2} = \frac{-j\omega_r\mu_0}{k_r} H_o \frac{\cos[(\beta H/2) + \varphi)]}{\sinh\alpha_a(d_3 - H/2)} J_1(k_r r)\sinh\alpha_a(d_3 - z)$$

$$E_{\theta 3} = \frac{-j\omega_r\mu_0}{k_r} H_o \frac{\cos[(\beta H/2) - \varphi]}{\sinh\alpha_s h} J_1(k_r r)\sinh\alpha_s(z + d_1) \qquad (7.16)$$

$$E_{\theta 4} = \frac{j\omega_r\mu_0}{k_a} H_o \frac{J_0(k_r a)}{K_0(k_a a)} K_1(k_a r)\cos(\beta z + \varphi)$$

The radial components of magnetic field are given below:

$$H_{r1} = \frac{\beta H_o}{k_r} J_1(k_r r)\sin(\beta z + \varphi)$$

$$H_{r2} = \frac{\alpha_a H_o \cos[(\beta H/2) + \varphi]}{k_r \sinh\alpha_a g} J_1(k_r r)\cosh\alpha_a(d_3 - z) \qquad (7.17)$$

$$H_{r3} = -\frac{\alpha_s H_o}{k_r}\frac{\cos[(\beta H/2) - \varphi]}{\sinh \alpha_s h} J_1(k_r r) \cosh \alpha_s (z + d_1)$$

$$H_{r4} = -\frac{\beta H_o}{k_a}[J_0(k_r a)/K_0(k_a a)] K_1(k_a r) \sin(\beta z + \varphi)$$

where

J_1 = Bessel function of the first kind,
K_1 = modified Bessel function of the second kind.

At the circular wall of a dielectric resonator, $r = a$, the E_θ is continuous over the interface between regions 1 and 4, that is $E_{\theta 1} = E_{\theta 4}$; thus one obtains the transcendental equation

$$\frac{J_1(k_r a)}{(k_r a) J_0(k_r a)} = -\frac{K_1(k_a a)}{(k_a a) K_0(k_a a)} \tag{7.18}$$

Solving the above equation a series of separate roots, $k_{r1}, k_{r2} \ldots k_{rn}$, are obtained, where k_{r1} is the lowest root which is associated with the $TE_{01\delta}$ mode.

With the upper boundary of the dielectric resonator at $z = H/2$, the magnetic field component H_r is continuous over the interface between regions 1 and 2, that is $H_{r1} = H_{r2}$. Thus the following equation can be derived:

$$\beta \tan[(\beta H/2) + \varphi] = \alpha_a \coth(\alpha_a g) \tag{7.19}$$

With the lower boundary of the dielectric resonator at $z = -H/2$, where $H_{r1} = H_{r3}$, one obtains eqn. (7.20)

$$\beta \tan[(\beta H/2) - \varphi] = \alpha_s \coth(\alpha_s h) \tag{7.20}$$

From eqns (7.19) and (7.20) the following equations can be derived

$$\beta H = q\pi + \tan^{-1}[(\alpha_a/\beta)\coth(\alpha_a g)] + \tan^{-1}[(\alpha_s/\beta)\coth(\alpha_s h)] \tag{7.21}$$

where $q = 0, 1, 2, 3 \ldots$.

$$\varphi = 0.5\{\tan^{-1}[(\alpha_a/\beta)\coth(\alpha_a g)] - \tan^{-1}[(\alpha_s/\beta)\coth(\alpha_s h)]\} \tag{7.22}$$

Equations (7.14), (7.18) and (7.21) are the Eigen equations for calculating the resonance frequencies of an enclosed DR system operating in the $TE_{op(q+\delta)}$, mode in which the lowest one, $p = 1, q = 0$, is the $TE_{O1\delta}$ mode. A computer program with the filename DR10C was written with the help of the above equations, in order to calculate the resonance frequency.

The resonance frequency can be calculated using programme DR10 and this will be shown in a later section.The same results can also be obtained with program DR10A.

RESONANCE FREQUENCY OF A SHIELDED TE MODE STACKED DR

The stacked-type resonator, composed of two cylinders of the same diameter but different material compositions was proposed in references [15,16]. The temperature coefficients of the two materials can be chosen in such a manner that they compensate each other over a large temperature range . With this kind of structure a very high temperature stability of ± 0.06 ppm/K from -50°C to $+100$°C at 11 GHz was reported [16]. The same structural arrangement can be used for optical tuning of a DRO [17,18]. The composite structure is shown in Fig. 7.10, which is sub-divided into eight regions. Regions 1 (ε_r) and 7 (ε_{top}) represent the DR itself.

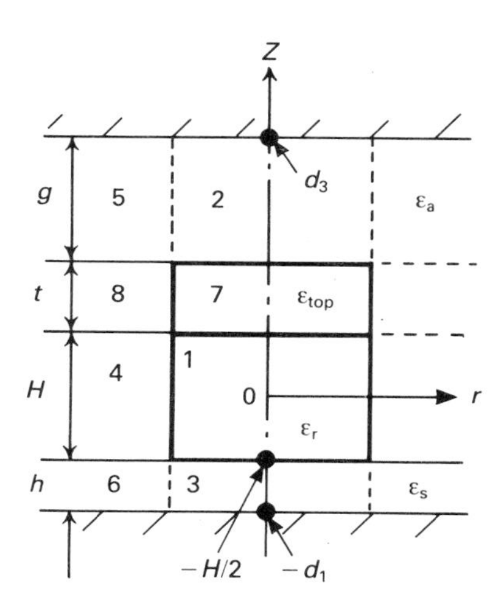

Fig. 7.10 — Regions associated with a stacked DR system.

For the stacked configuration as employed in an optically controlled DR system, with $\varepsilon_{top} < (\sim 0.5\varepsilon_r)$, the longitudinal magnetic components in regions 1 to 8 can be expressed mathematically as:

$$H_{z1} = H_o J_0(k_r r)\{\cos(\beta z) + A_{21}\sin(\beta z)\}$$

$$H_{z2} = 2H_o J_0(k_r r) A_{31} e^{-\alpha_a d_3} \sinh[\alpha_a(d_3 - z)]$$

$$H_{z3} = 2H_o J_0(k_r r) A_{41} e^{-\alpha_s d_1} \sinh[\alpha_s(d_1 + z)]$$

$$H_{z4} = H_o A_{51} K_0(k_a r)\{\cos(\beta z) + A_{65}\sin(\beta z)\}$$

$$H_{z5} = 2H_o A_{81} K_0(k_a r) e^{-\alpha_a d_3} \sinh[\alpha_a(d_3 - z)] \tag{7.23}$$

$$H_{z6} = 2H_o A_{71} K_0(k_r r) e^{-\alpha_s d_1} \sinh[\alpha_s(d_1 + z)]$$

$$H_{z7} = 2H_o A_{91} J_0(k_r r)\{\cosh[\alpha_\xi(z - H/2)] + A_{92}\sinh[\alpha_\xi(z - H/2)]\}$$

$$H_{z8} = 2H_o A_{101} K_0(k_a r)\{\cosh[\alpha_\xi(z - H/2)] + A_{102}\sinh[\alpha_\xi(z - H/2)]\}$$

The equations for k_r, k_a, β, α_a and α_s are the same as those given in eqn. (7.14), where

$$\alpha_\xi^2 = k_r^2 - \varepsilon_{top}(\omega_r/c)^2 \tag{7.24}$$

By matching the fields at each boundary one obtains the following coefficients for A[19]:

$$A_{21} = A_{65} = \frac{\alpha_s - \beta\tan(\beta H/2)\tanh(\alpha_s h)}{\alpha_s\tan(\beta H/2) + \beta\tanh(\alpha_s h)}$$

$$A_{41} = \frac{\cos(\beta H/2) - A_{21}\sin(\beta H/2)}{2e^{-\alpha_s d_1}\sinh(\alpha_s h)}$$

$$A_{51} = \frac{J_0(k_r a)}{K_0(k_r a)}$$

$$A_{91} = \frac{\cos(\beta H/2) + A_{21}\sin(\beta H/2)}{2} \tag{7.25}$$

$$A_{92} = A_{102} = \frac{\beta}{\alpha_\xi} \times \frac{A_{21} - \tan(\beta H/2)}{1 + A_{21}\tan(\beta H/2)} = -\frac{\alpha_a + a_\xi\tanh(\alpha_\xi t)\tanh(\alpha_a g)}{\alpha_\xi\tanh(\alpha_a g) + \alpha_a\tanh(\alpha_\xi t)}$$

$$A_{31} = \frac{A_{91}\{\cosh(\alpha_\xi t) + A_{92}\sinh(\alpha_\xi t)}{e^{-\alpha_a d_3}\sinh(\alpha_a g)}$$

$$A_{71} = A_{41}A_{51}$$

$$A_{81} = A_{31}A_{51}$$

$$A_{101} = A_{91}A_{51}$$

and obtains

$$\frac{J_1(k_r a)}{(k_r a)J_0(k_r a)} = -\frac{K_1(k_a a)}{(k_a a)K_0(k_a a)} \tag{7.26}$$

This is the same as eqn. (7.18). In a similar way to eqn. (7.21) we obtain the following equation for the $TE_{O1(q+\delta)}$ mode:

$$\beta H = q\pi + \tan^{-1}\left[\frac{\beta[\alpha_s - \alpha_\xi A_{92}\tanh(\alpha_s h)]}{\beta^2\tanh(\alpha_s h) + \alpha_s\alpha_\xi A_{92}}\right] \tag{7.27}$$

where $q = 0, 1, 2, \ldots$

For the $TE_{01\delta}$ mode DR we have $0.5\pi < \beta H < \pi$, $\tan(\beta H) < 0$.

For the stacked configuration employed in highly frequency stable DROs, $\varepsilon_{top} > (\sim 0.5\,\varepsilon_r)$. The formulas for H_{z1} to H_{z6} are the same as those in eqn. (7.23), whilst the longitudinal magnetic components in regions 7 and 8 can be expressed as:

$$H_{z7} = 2H_o A_{91} J_0(k_r r)\{\cos[\beta_t(z - H/2)] + A_{92}\sin[\beta_t(z - H/2)]\} \tag{7.28}$$

$$H_{z8} = 2H_o A_{101} K_0(k_a r)\{\cos[\beta_t(z - H/2)] + A_{102}\sin[\beta_t(z - H/2)]\}$$

where

$$\beta_t^2 = \varepsilon_{top}(\omega_r/c)^2 - k_r^2 \tag{7.29}$$

The coefficients A_{92}, A_{102} and A_{31} become:

$$A_{92} = A_{102} = \frac{\beta}{\beta_t}\times\frac{A_{21} - \tan(\beta H/2)}{1 + A_{21}\tan(\beta H/2)} = -\frac{\alpha_a - \beta_t\tan(\beta_t t)\tanh(\alpha_a g)}{\beta_t\tanh(\alpha_a g) + \alpha_a\tan(\beta_t t)}$$

$$A_{31} = \frac{A_{91}\{\cos(\beta_t t) + A_{92}\sin(\beta_t t)\}}{e^{-\alpha_a d_3}\sinh(\alpha_a g)} \tag{7.30}$$

The other coefficients, namely A_{21}, A_{65}, A_{41}, A_{51}, A_{91}, A_{71}, A_{81} and A_{101} remain the same and do not change. Thus we obtain the following Eigen equation:

$$\tan(\beta H)\{\beta^2\beta_t\tanh(\alpha_a g)\tanh(\alpha_s h) - \alpha_a\alpha_s\beta_t\}$$

$$+ \tan(\beta H)\tan(\beta_t t)\{\alpha_a\beta^2\tanh(\alpha_s h) + \alpha_s\beta_t^2 + \tanh(\alpha_a g)\}$$

$$+ \tan(\beta_t t)\{\beta\beta_t^2\tanh(\alpha_a g)\tanh(\alpha_s h) - \alpha_a\alpha_s\beta\}$$

$$= \beta\beta_t\{\alpha_a\tanh(\alpha_s h) + \alpha_s\tanh(\alpha_a g)\} \tag{7.31}$$

Alternatively this may be expressed in concise form as

$$\beta H + \beta_t t = q\pi + f(\varepsilon_r, \varepsilon_s, \varepsilon_a, \varepsilon_{top}, a, H, g, t, h) \tag{7.32}$$

where $q = 0, 1, 2, \ldots$

For the $TE_{01\delta}$ mode the following condition must be satisfied:

$$0.5\pi < (\beta H + \beta_t t) < \pi$$

Based on the equations in this section computer program DR10B was written to allow for a speedy calculation of the resonance frequency of a stacked DR. Examples of calculated curves using this program are given in the following.

1. DR COUPLING TO MICROSTRIP

Field analysis

In order to be of practical use, a DR must be coupled by some means to the microwave circuit of which it forms a part.One way of coupling is via a microstrip. Mathematical expressions of the coupling between a dielectric resonator and a microstrip have been derived in [20] and an improved approach has been given in [21]. In [21], rather than employing coupling in one plane, the concept of using an effective coupling length, which is a function of an angle, has been introduced in this analysis.

The following assumptions were made in the analysis of the coupling between a cylindrical DR and microstrip:

(a) The main oscillation mode in the cylindrical DR is $TE_{01\delta}$ and the main mode of propagation in the microstrip is quasi-TEM.
(b) The field disturbances owing to the presence of the microstrip and radiation losses including that owing to field mismatch can be neglegted.
(c) The electromagnetic waves cannot penetrate the microstrip and only the radial magnetic field components of the DR and microstrip can match. The coupling between the DR ($TE_{01\delta}$ mode) and the microstrip (quasi-TEM) is mainly due to the radial magnetic component H_r and is not due to the component H_z. The field distribution of the DR-microstrip system is illustrated in Fig. 7.11.
(d) The losses in the conductor and dielectric material are all considered negligible.

As was shown earlier on, a resonator composite structure can be subdivided into six regions. This is again shown in Fig. 7.12, for the purpose of a mathematical analysis. The azimuthal electric field components E_θ in each region of the structure for the $TE_{01\delta}$ mode are given in eqn. (7.33)

$$E_{\theta 1} = \frac{-j\omega_0\mu_0}{k_r} H_o J_1(k_r r)\left[\cos\beta z + A_{21}\sin\beta z\right]$$

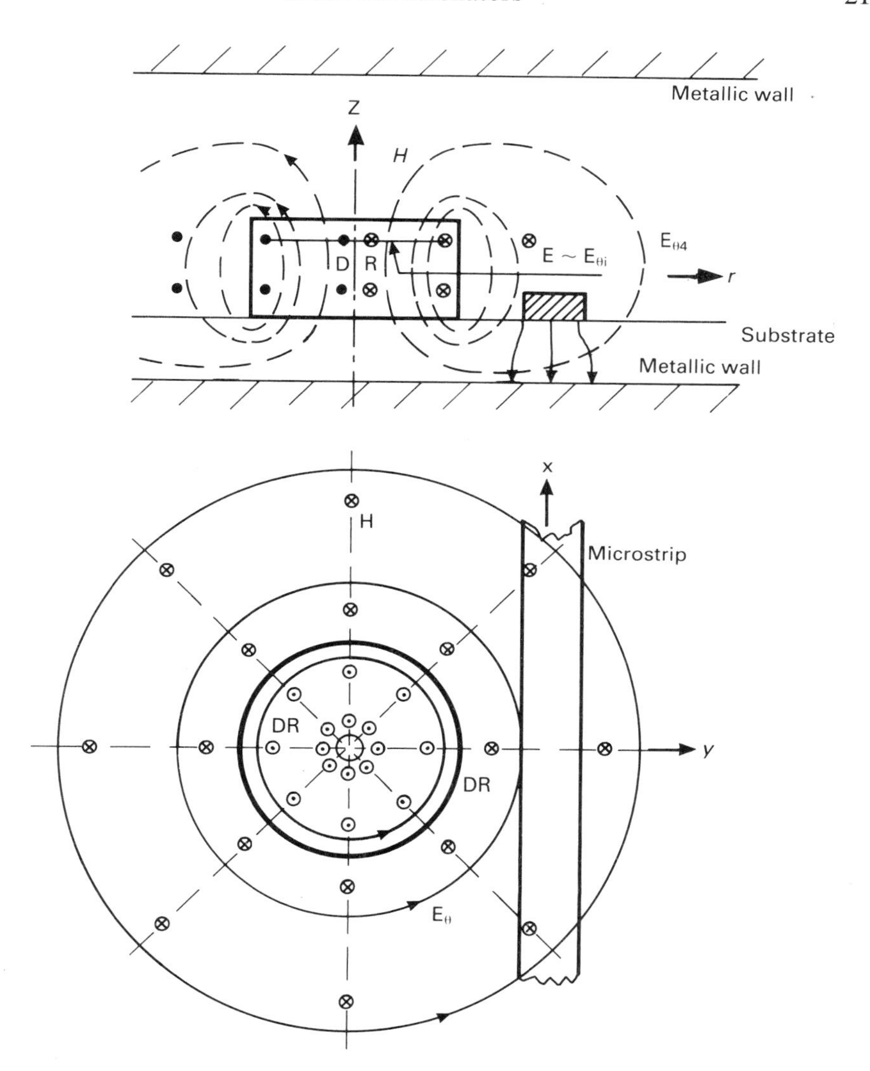

Fig. 7.11 — Field distribution of a DR-microstrip system.

$$E_{\theta 2} = \frac{-j\omega_r\mu_0}{k_r} 2H_o A_{31} e^{-\alpha_a d_3} J_1(k_r r) \sinh[\alpha_a(d_3 - z)]$$

$$E_{\theta 3} = \frac{-j\omega_r\mu_0}{k_r} 2H_o A_{41} e^{-\alpha_s d_1} J_1(k_r r) \sinh[\alpha_s(d_1 + z)]$$

$$E_{\theta 4} = \frac{j\omega_r\mu_0}{k_a} H_o A_{51} K_1(k_a r) [\cos\beta z + A_{65} \sin\beta z]$$

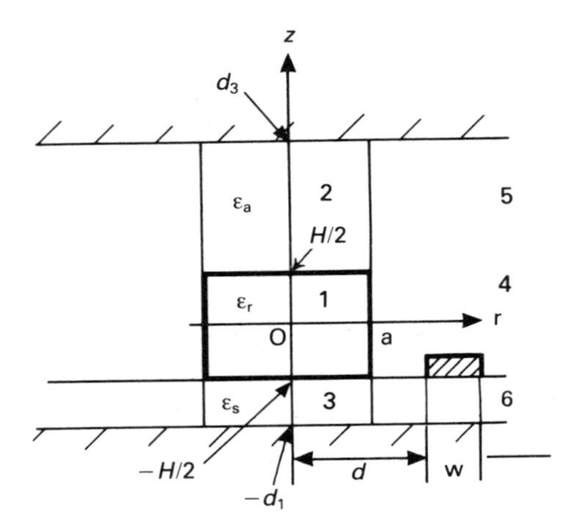

Fig. 7.12 — Regions of a DR-microstrip structure.

$$E_{\theta 5} = \frac{j\omega_r\mu_0}{k_a} 2H_oA_{81}e^{-\alpha_a d_3}K_1(k_a\mathrm{r})\sinh[\alpha_a(d_3 - z)]$$

$$E_{\theta 6} = \frac{j\omega_r\mu_0}{k_a} 2H_oA_{71}e^{-\alpha_s d_1}K_1(k_a r)\sinh[\alpha_s(d_1 + z)] \tag{7.33}$$

where

$$\frac{J_1(k_r a)}{(k_r a)J_0(k_r a)} = \frac{-K_1(k_a a)}{(k_a a)K_0(k_a a)} \tag{7.34}$$

And the equations which are satisfied by propagation constants of each region are given in eqn. (7.35) (Note. They are the same as that given in eqn. (7.14))

$$\begin{aligned} k_r^2 &= \varepsilon_r(\omega_r/c)^2 - \beta^2 \\ \alpha_a^2 &= k_r^2 - \varepsilon_a(\omega_r/c)^2 \\ \alpha_s^2 &= k_r^2 - \varepsilon_s(\omega_r/c)^2 \\ k_a^2 &= \beta^2 - \varepsilon_a(\omega_r/c)^2 \end{aligned} \tag{7.35}$$

The values of the constant A_{ij} are given below:

$$A_{21} = A_{65} = \frac{\alpha_s - \beta\tanh\alpha_s[d_1 - (H/2)]\tan(\beta H/2)}{\alpha_s\tan(\beta H/2) + \beta\tanh\alpha_s(d_1 - H/2)}$$

$$A_{31} = \frac{\cos(\beta H/2) + A_{21}\sin(\beta H/2)}{2e^{-\alpha_a d_3}\sinh\alpha_a(d_3 - H/2)}$$

$$A_{41} = \frac{\cos(\beta H/2) - A_{21}\sin(\beta H/2)}{2e^{-\alpha_s d_1}\sinh\alpha_s(d_1 - H/2)} \tag{7.36}$$

$$A_{51} = \frac{J_0(k_r a)}{K_0(k_a a)}$$

$$A_{71} = A_{41}A_{51}$$

$$A_{81} = A_{31}A_{51}$$

The symbols used in the analysis were defined in the previous section.

The electromagnetic energy stored in a DR structure can be expressed in terms of the electric field as:

$$\bar{W}_e = \frac{1}{2}\int_v \varepsilon_0\varepsilon_i|E_i|^2\,dV \tag{7.37}$$

where ε_i is the relative permittivity of each of the six regions with $i = 1, 2, \ldots 6$.

$$W_e = \frac{\omega_r^2\mu_0^2\varepsilon_o\pi H_o^2}{2}\left\{\frac{a^2}{k_r^2}[J_1^2(k_r a) - J_0(k_r a)J_2(k_r a)]\right. \tag{7.38}$$

$$\times\left\{\frac{H\varepsilon_r}{2}\left[1 + \frac{\sin\beta H}{\beta H} + A_{21}^2\left(1 - \frac{\sin\beta H}{\beta H}\right)\right]\right.$$

$$+ \frac{\varepsilon_a A_{31}^2}{\alpha_a}e^{-2\alpha_a d_3}[\sinh 2\alpha_a(d_3 - H/2) - 2\alpha_a(d_3 - H/2)]$$

$$\left. + \frac{\varepsilon_s A_{41}^2}{\alpha_s}e^{-2\alpha_s d_1}[\sinh 2\alpha_s(d_1 - H/2) - 2\alpha_s(d_1 - H/2)\right\}$$

$$+ \frac{1}{k_a^2}\left\{b^2\left[K_1^2(k_a b) - K_0^2(k_a b) - 2\frac{K_0(k_a b)K_1(k_a b)}{k_a b}\right]\right.$$

$$\left. - a^2\left[K_1^2(k_a a) - K_0^2(k_a a) - 2\frac{K_0(k_a a)K_1(k_a a)}{k_a a}\right]\right\}$$

$$\times\left\{\frac{\varepsilon_a A_{51}^2 H}{2}\left[1+\frac{\sin\beta H}{\beta H}+A_{21}^2\left(1-\frac{\sin\beta H}{\beta H}\right)\right]\right.$$

$$+\frac{\varepsilon_a A_{81}^2}{\alpha_a}e^{-2\alpha_a d_3}[\sinh(2\alpha_a)(d_3-H/2)-2\alpha_a(d_3-H/2)]$$

$$\left.+\frac{\varepsilon_s A_{71}^2}{\alpha_s}e^{-2\alpha_a d_1}[\sinh 2\alpha_s(d_1-H/2)-2a_s(\mathrm{d}_1-H/2)]\right\}\Bigg\}$$

where b is the radius of effective action of the DR. This is shown in Fig. 7.13. It is sufficient to make $b=5a$ for the purpose of calculation. Note that J_2 is the Bessel function of the first kind.

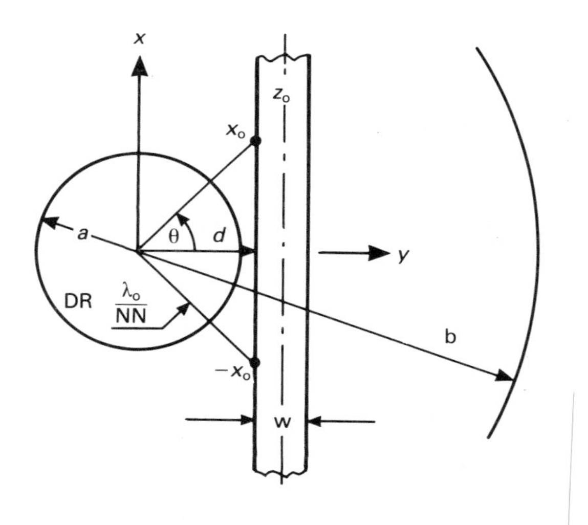

Fig. 7.13 — Effective coupling length of a DR-microstrip structure.

In order to obtain the effective coupling length $-x_0$ to x_0, a computer program was written, which forms one part of program DR 100. This shows that there is no appreciable contribution to coupling for NN<3, i.e. beyond $-x_0$ and x_0. The important radial magnetic field component H_{r6} can be derived from Maxwell's equations as

$$H_{r6}=-\frac{\alpha_s}{k_a}2H_oA_{71}e^{-\alpha_s d_1}K_1(k_a r)\cosh\alpha_s(d_1+z) \tag{7.39}$$

The external quality factor Q_e is defined as

$$Q_e = 2\pi W_h / W_s \tag{7.40}$$

where
W_h = stored energy in resonance system,
W_s = dissipated energy in load per cycle.

Figure 7.13 shows a DR coupled to a microstrip. If the microstrip to which the DR is coupled is terminated at both ends in its characteristic impedance as in Fig. 7.14, that is double loading, then one obtains for the external quality factor which is given in equation (7.41).

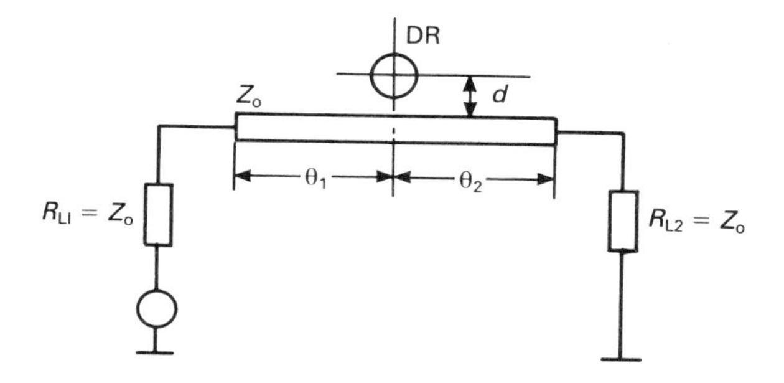

Fig. 7.14 — DR microstrip structure terminated in loads, $R_{L1} = R_{L2} = Z_0$. θ = electrical length of microstrip.

$$Q_e = \frac{4Z_o}{\omega\mu_0^2} \times \frac{\overline{W}_e}{\left\{\int\!\!\int_S \overline{H}\cdot d\overline{s}\right\}^2} \tag{7.41}$$

where

$$\int\!\!\int_S \overline{H}\cdot d\overline{s} = -\frac{4H_o A_{71} e^{-\alpha_s d_1}}{k_a}$$

$$\times (d + w/2)[\sinh \alpha_s(d_1 - H/2)] \tag{7.42}$$

$$\times \int_0^{\sqrt{[(\lambda_o/NN)^2 - d^2]}} \frac{K_1\{k_a[x^2 + (d + w/2)^2]^{0.5}\}}{[x^2 + (d + w/2)^2]^{0.5}} dx$$

and $NN = 3$, λ_0 = free space wavelength.

It is easy to calculate the integral in equation (7.42) with the computer program DR100 which is given later in this chapter.

The external quality factor Q_e, as a function of separation d, between the DR and microstrip is plotted in Fig. 7.15(a). This shows the theoretical curve using the analysis given above and compares it with that obtained in reference [20]. It was assumed that the original paper utilized a 50 Ω microstrip. Figure 7.15(b) shows the results of one of the experiments, which was given in reference [21]. As can be seen, this again shows very close agreement between theory and experiment.

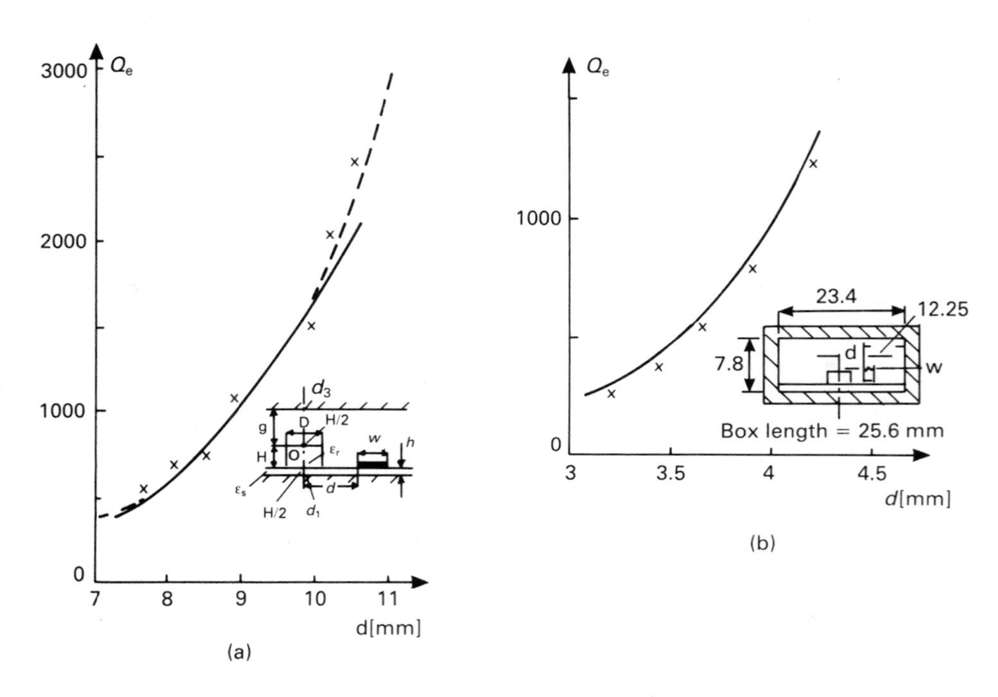

Fig. 7.15 — Variation of Q_e against coupling distance. Substrate (a): $h = 0.8$ mm, $w = 50\,\Omega$, $\varepsilon_s = 2.54$. DR: $H = 7.5$ mm, $D = 14$ mm, $\varepsilon_r = 35$, $f_r = 4$ GHz. Substrate (b): $h = 0.7874$ mm, $w = 2.25$ mm, oak laminate. DR: $H = 2.4$ mm, $D = 6.1$ mm, $\varepsilon_r = 38.9$, $f_r = 9.4$ GHz., type BT47-(IRA)TMOTA. + + + : experimental. ———: theoretical; ----: extended theory.

Note that this is the first way to discuss the coupling between a DR and microstrip. The second approach is given later.

2. THE EQUIVALENT LUMPED CIRCUIT PARAMETERS (L,C,R) OF A DR MICROSTRIP SYSTEM

The equivalent lumped circuit parameters of a DR are useful when designing microwave circuits which contain DRs coupled to microstrip. The equivalent magnetic dipole presentation is used for analysis [22]. The lumped circuit which is employed to describe the DR working in a single mode is illustrated in Fig. 7.16

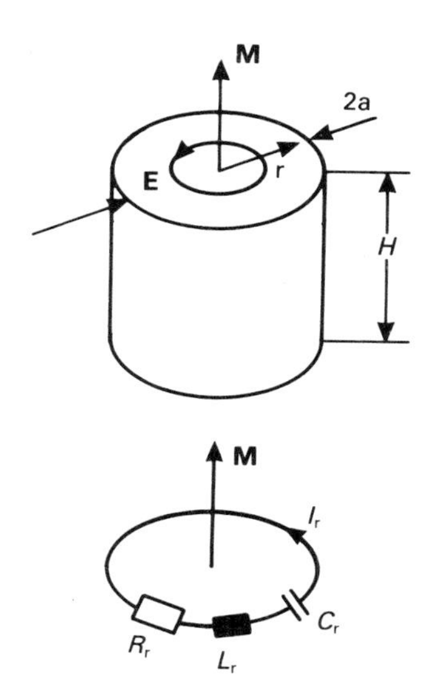

Fig. 7.16 — Equivalent lumped circuit of a DR.

where:
M = magnetic dipole,
E = electric field,
I_r = current,
I_r = equivalent inductance of DR,
C_r = equivalent capacitance of DR,
R_r = equivalent resistance of DR,
H = height of DR,
2a = diameter of DR.

The magnetic moment of a magnetic dipole is expressed by the vector product

$$\mathbf{M} = 1/2 \iiint \mathbf{r} \times \mathbf{J}\,\mathrm{d}V \tag{7.43}$$

where $\mathbf{J}$ = current density and $\mathbf{r}$ = position vector of field point.

It is assumed that there is no current flowing, i.e. there exists displacement current only in the DR. The field is assumed to have a sinusoidal time dependance e^{jwt}. The current density is then given by

$$\mathbf{J} = \varepsilon_0 \varepsilon_i \frac{\partial \mathbf{E}}{\partial t} = j\omega_r \varepsilon_o \varepsilon_i \mathbf{E} \tag{7.44}$$

Then

$$\mathbf{M} = \frac{j\omega_r \varepsilon_0}{2} \iiint \varepsilon_i \mathbf{r} \times \mathbf{E}\, \mathrm{d}V$$

$$= \mathbf{i}_z \frac{j\omega_r \varepsilon_0}{2} \iiint \varepsilon_i r \mathbf{E}_\theta\, \mathrm{d}V = \mathbf{i}_z \mathbf{M} \tag{7.45}$$

where $\mathbf{i}_z$ is the unit vector in the z direction and the permittivity of free space $\varepsilon_o = 8.854 \times 10^{-12}$ F/m.

Substituting $E_{\theta i}$, which are given in equation (7.33), for E_θ of each region in equation (7.45) and integrating then the value of **M** is obtained as:

$$\mathbf{M} = \frac{\omega_r^2 \mu_0 \varepsilon_o}{2} H_0 \left\{ \frac{J_2(k_r a)}{k_r^2} \left[\varepsilon_r \pi a^2 \frac{\sin(\beta H/2)}{\beta} \right.\right.$$

$$+ \varepsilon_a 4\pi a^2 A_{31} \frac{\cosh \alpha_a (d_3 - H/2) - 1}{\alpha_a} e^{-\alpha_a d_3}$$

$$\left. + \varepsilon_s 4\pi a^2 A_{41} \frac{\cosh \alpha_s (d_1 - H/2) - 1}{\alpha_s} e^{-\alpha_s d_1} \right]$$

$$+ \frac{(k_a b)^2 K_2(k_a b) - (k_a a)^2 K_2(k_a a)}{k_a^4} \left[\varepsilon_a 4\pi A_{51} \frac{\sin(\beta H/2)}{\beta} \right.$$

$$+ \varepsilon_a 4\pi A_{81} \frac{\cosh \alpha_a (d_3 - H/2) - 1}{\alpha_a} e^{-\alpha_a d_3}$$

$$\left.\left. + \varepsilon_s 4\pi A_{71} \frac{\cosh \alpha_s (d_1 - H/2) - 1}{\alpha_s} e^{-\alpha_s d_1} \right] \right\} \tag{7.46}$$

where K_2 is the modified Bessel function of the second kind. According to the equivalent lumped circuit theory the magnetic dipole can be expressed as

$$M = S_M I_r \tag{7.47}$$

where S_M is the equivalent area of magnetic dipole. Because the permittivity of the material of the DR is relatively high,and almost all of the energy ($\sim 98\%$ for $\varepsilon_r = 39$) of the electromagnetic field is concentrated in the DR ,the following assumption can be made using Fig. 7.16:

$$S_M \simeq \pi a^2 \tag{7.48}$$

The resonance frequency of the DR is

$$w_r = 1/\sqrt{L_r C_r} \tag{7.49}$$

The stored electromagnetic energy in the equivalent circuit of the DR is

$$\bar{W}_e = \frac{1}{2} L_r I_r^2 \tag{7.50}$$

From eqns (7.50) and (7.47) we obtain

$$L_r = \frac{2\bar{W}_e S_M^2}{M^2} \tag{7.51}$$

From eqn. (7.49) we obtain

$$C_r = \frac{1}{\omega_r^2 L_r} \tag{7.52}$$

Finally, from the definition of the unloaded quality factor Q_u one gets

$$R_r = \frac{\omega_r L_r}{Q_u} \tag{7.53}$$

The coupling between a DR and microstrip is magnetic coupling, and thus it can be represented approximately by the mutual inductance, L_m. The DR coupled to a microstrip and its equivalent circuit are illustrated in Fig. 7.17. The coupling can be expressed by the external quality factor Q_e:

$$Q_e = \frac{\omega_r L_r}{\mathrm{Re}(Z_{se})} \tag{7.54}$$

where Z_{se} is the real part of the equivalent impedance of microstrip which is coupled to a DR resonant circuit by L_m. Z_{se} is given by

$$Z_{se} = \frac{\omega_r^2 \, L_m^2}{Z_s} \tag{7.55}$$

From Fig. 7.17b one obtains

$$Z_s = Z_1 + Z_2 \tag{7.56}$$

For the input impedance Z_i we obtain (Chapter 4, eqn. 4.68):

$$Z_i = Z_o \frac{R_{Li} + j\, Z_o \tan \theta_i}{Z_o + j\, R_{Li} \tan \theta_i} \qquad \text{with } i = 1, 2 \tag{7.57}$$

Thus

$$Q_e = \frac{L_r \, \mathrm{Re}\,(Z_s)}{\omega_r \, L_m^2} \tag{7.58}$$

If $Z_s = 2Z_o$ then this is known as double loading.

$$Q_e = \frac{2\, Z_o \, L_r}{\omega_r L_m^2} \tag{7.59}$$

$$L_m = \sqrt{\frac{2Z_o \, L_r}{\omega_r \, Q_e}} \tag{7.60}$$

And if $Z_s = Z_o$, then this is single loading and

$$Q_e = \frac{Z_o L_r}{\omega_r L_m^2} \tag{7.61}$$

$$L_m = \sqrt{\frac{Z_o L_r}{\omega_r Q_e}} \tag{7.62}$$

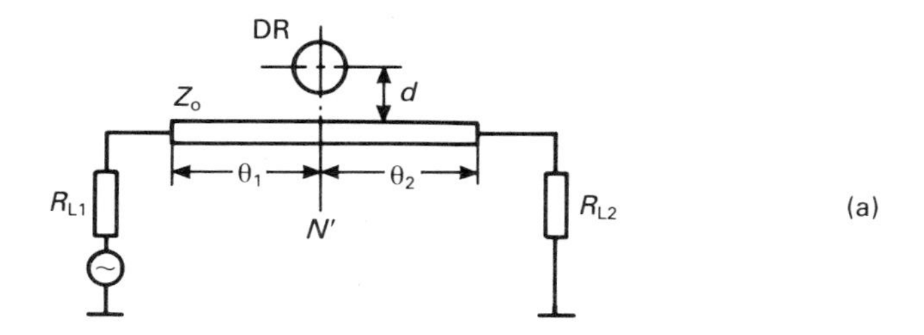

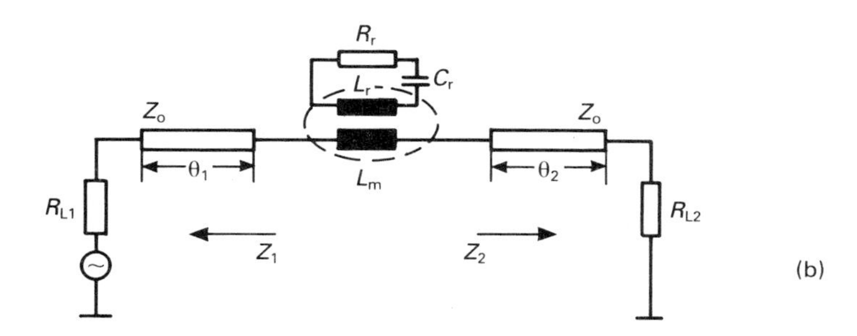

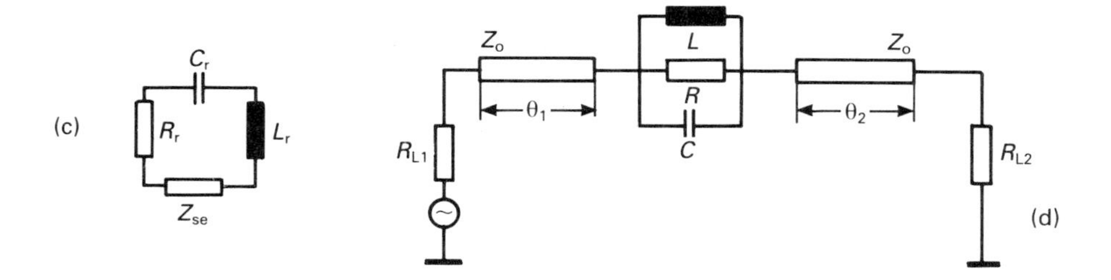

Fig. 7.17 — DR-microstrip structure and its equivalent circuit in the symmetry plane NN′. (a) DR-microstrip structure; (b) presentation of coupling between a DR and microstrip; (c) equivalent circuit in which the microstrip (Zse) is coupled 'into' the DR equivalent circuit by the mutual inductance L_m; (d) equivalent circuit in which the DR equivalent circuit is coupled 'into' the microstrip by the mutual inductance L_m.

Thus the mutual inductance L_m can be calculated by means of equations (7.60) or (7.62).

The simple equivalent circuit of a DR coupled to a microstrip is given in Fig. 7.18. In essence it is a two-port as denoted by A and B. The loss of the junction between A and B is represented by a resistance R.

The coupling factor is defined as an energy relationship, namely:

$$\beta_c' = \frac{R}{2Z_o} \tag{7.63}$$

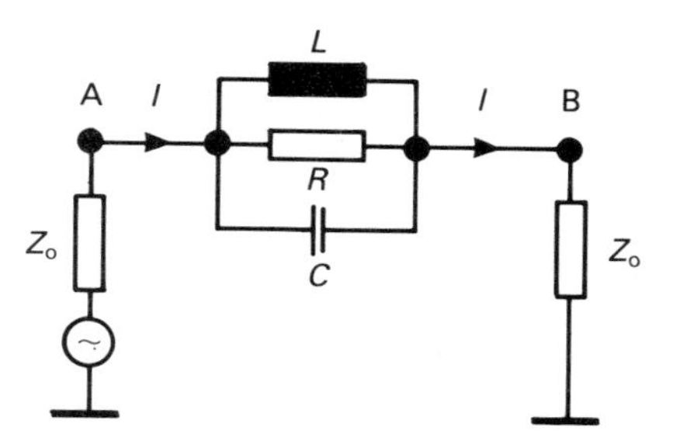

Fig. 7.18 — Equivalent circuit of a DR coupled to microstrip.

Alternatively matters can be viewed the other way around, i.e. the quality factors are employed to express the coupling factor, namely:

$$\beta_c = \frac{Q_u}{Q_e} \tag{7.64}$$

Form equation (7.59), for double loading the coupling factor may be expressed as:

$$\beta_c = \frac{\omega_r L_m^2 Q_u}{2Z_o L_r} \tag{7.65}$$

The coupling between a DR and a microstrip is reciprocal, so $\beta_c = \beta_c'$.

Thus

$$R = \frac{\omega_r L_m^2 Q_u}{L_r} \tag{7.66}$$

Through the transformation of impedance we obtain

$$C = \frac{L_r^2 C_r}{L_m^2} \tag{7.67}$$

$$L = \frac{L_m^2}{L_r} \tag{7.68}$$

It thus has been shown how by this second approach R, L and C can be obtained for a DR coupled to a microstrip.

3. THE *S*-PARAMETERS OF A DR COUPLED TO MICROSTRIP

Now, as a third approach, the scattering parameters, the basics of which were discussed in Chapter 5, are used to discuss the coupling mechanism of a DR-microstrip structure. Referring to Fig. 7.19, we have:

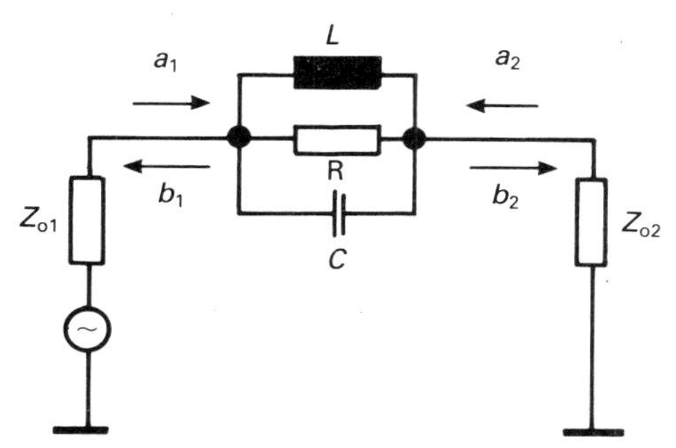

Fig. 7.19 — *S*-parameters in a DR structure.

$$b_1 = S_{11}a_1 + S_{12}a_2 \tag{7.69}$$

$$b_2 = S_{21}a_1 + S_{22}a_2 \tag{7.70}$$

where

S_{11} = reflection coefficient at the input,
S_{21} = transmission coefficient, input to output,
S_{12} = transmission coefficient, output to input,
S_{22} = reflection coefficient at the output.

If the ouput line is terminated in a matched load, Z_{o2}, that is $a_2 = 0$, then

$$b_1 = S_{11}a_1 \tag{7.71}$$

$$b_2 = S_{21}a_1 \tag{7.72}$$

A practical DR-microstrip junction is a lossy junction at resonance frequency. If there is no radiation loss, we have

$$|a_1|^2 = |b_1|^2 + |b_2|^2 + P_d \tag{7.73}$$

where

$|a_1|^2$ = incident power at the junction DR/microstrip,
$|b_1|^2$ = power reflected from the junction,

$|b_2|^2$ = power dissipated in load Z_{o2},
P_d = power dissipated at the junction.

In Fig. 7.19 if $Z_{01} = Z_{02} = Z_o$, that is the power dissipated in load Z_{02} is equal to the power dissipated in Z_{01}. Thus the coupling factor is given as:

$$\beta_c = \frac{P_d}{2|b_2|^2} \tag{7.74}$$

Using equations (7.73), (7.71) and (7.72), then the coupling factor can also be expressed in terms of the S-parameters, namely

$$\beta_c = \frac{1 - |S_{11}|^2 - |S_{21}|^2}{2|S_{21}|^2} \tag{7.75}$$

From a practical point of view it would be more convenient to express the coupling factor in terms of measured parameters. The insertion loss L_{ins} in dB and return loss L_{ret} in dB of a DR-microstrip structure can easily be measured on a network analyser. Hence, the coupling coefficient may also be expressed as

$$\beta_c = \frac{1 - 10^{-L_{ret/10}} - 10^{-L_{ins/10}}}{2 \times 10^{-L_{ins/10}}} \tag{7.76}$$

At the resonance frequency, the equivalent circuit of a DR coupled to a microstrip in a metal box is given in Fig. 7.20. The reflection coefficient is expressed as:

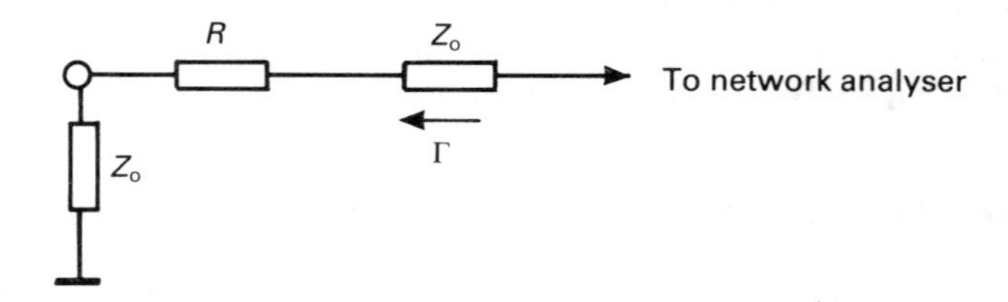

Fig. 7.20 — Equivalent circuit of a DR coupled to microstrip at resonance.

$$\Gamma_{in} = \frac{Z_L - Z_o}{Z_L + Z_o} = \frac{R}{(R + 2Z_o)} \tag{7.77}$$

or

$$|S_{11}| = \frac{R}{R + 2Z_o} \tag{7.78}$$

From eqn. (7.75) we obtain

$$|S_{21}| = \sqrt{\frac{1 - |S_{11}|^2}{1 + 2\beta_c}} \tag{7.79}$$

The experimental curves of $|S_{11}|^2$, $|S_{21}|^2$ and $|S_{11}|^2 + |S_{21}|^2$ versus distance d are shown in Fig. 7.21 and [23]. One can see that $|S_{11}|^2 + |S_{21} \neq 1$ in a practical case. The $|S_{11}|^2 + |S_{21}|^2$ curve is U-shaped and has a minimum at $|S_{11}|^2 + |S_{21}|^2 \simeq 0.5$. This value can only become approximately unity for strongest coupling.

The theoretical curves obtained with eqns (7.78) and (7.79) together with the experimental curve are shown in Fig. 7.22. As can be seen, both curves agree well, as long as the DR is in close vicinity of the microstrip. This also explains why the calculated results for R based on eqn. (7.66) are accurate within the same range. For a larger coupling distance d agreement between the two curves deteriorates. The reason is that the energy coupled from the DR to the microstrip (or vice versa) is very small compared with the losses associated with light coupling. As the coupling distance increases, the losses of the conductors and dielectric materials can no longer be neglected if one wants to compare the theoretical results with those obtained from experimentation. In the analysis of Fig. 7.22 all losses associated were neglected.

DR COMPUTER CALCULATIONS

Computer programs have been written based on the formulae derived in the previous two paragraphs.

Computer program DR10 (DR10C choose option: 1)

Figure 7.23 shows an example of a computed family of curves based on computer program DR10. These curves show the resonance frequency and unloaded quality factor versus the distance between the DR and lid of the DR enclosure with the substrate thickness h as parameter. Observing the frequency as a function of distance g, we conclude that increasing the distance beyond 3 mm does not affect the frequency of the system in an appreciable way. The dimensions and parameters of the system are inset in Fig. 7.23. When the distance g is reduced to less than about 1 mm, frequency starts to increase rapidly. At the same time as frequency increases , the unloaded quality factor is lowered. It is also seen that the Q-factor is affected apparently by the distance between the DR and lid.

The calculation of Q_u here takes only the conduction losses of the shielding into account. In practical cases both, radiation and dielectric losses are present and should thus be included for an accurate calculation of Q_u. It is often easier to measure this value.

DR theory shows that the following inequality has to be satisfied

$$1.5 < 2a/H < 2.5 \tag{7.80}$$

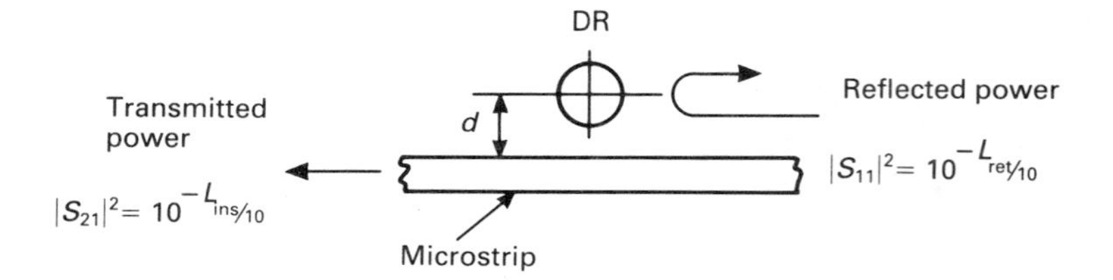

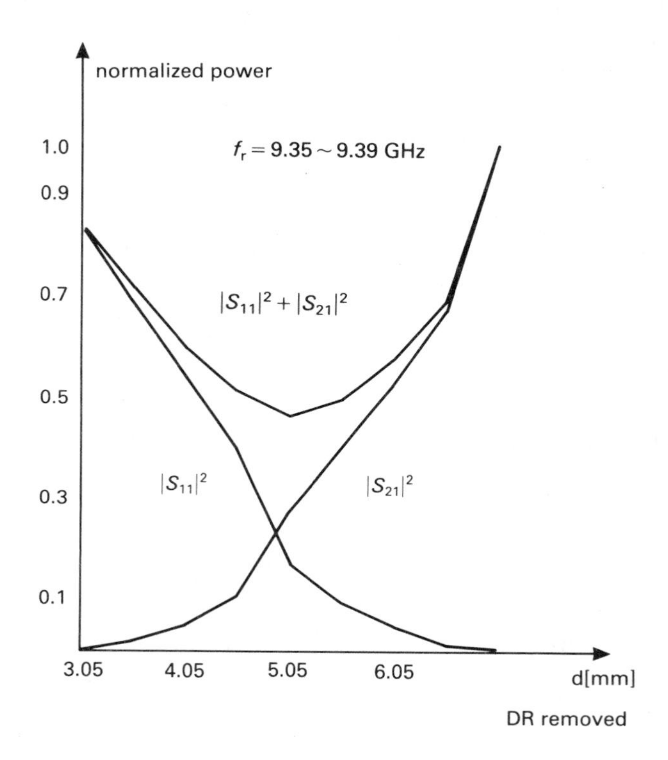

Fig. 7.21 — The experimental curves of S parameters versus coupling distance d of a DR-microstrip system. Box: length = 25.6 mm, height = 7.8 mm, width = 23.4 mm, material, aluminium. Substrate: oak laminate, $h = 0.7874$ mm, $\varepsilon_s = 2.5$. Microstrip: $Z_0 = 50\,\Omega$, $w = 2.25$ mm. DR: BTN47-IR(A)TMOTA, $H = 2.4$ mm, $a = 3.05$ mm, $\varepsilon_r = 38.9$. Air: $\varepsilon_r = 1$. Resonance frequency $f_r \sim 9.37$ GHz.

for obtaining a high unloaded quality factor, Q_u, and for preventing the development of a $TM_{11\delta}$ mode.

Computer program DR10B (DR10C choose option: 2)

The resonance frequency of a shielded stacked DR-substrate system can be calculated using program DR10B. Figure 7.24 shows the resonance frequency *fr* as a function of the dielectric constant (ε_{top}) of the top cylinder with the thickness (t) of

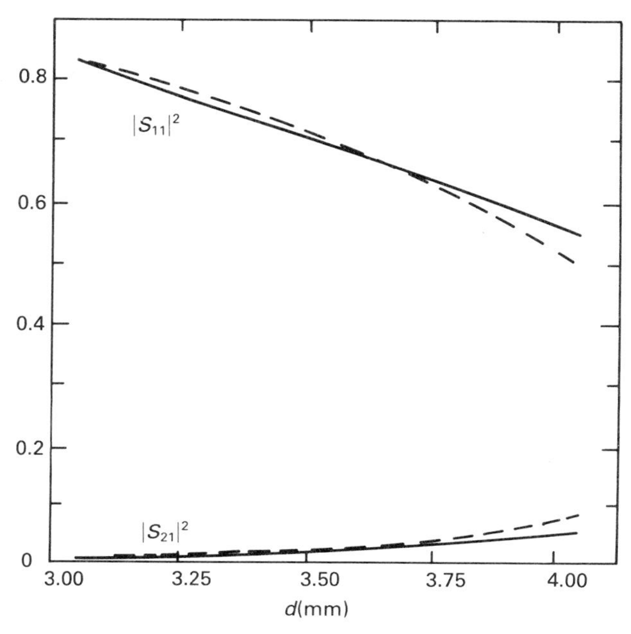

Fig. 7.22 — The theoretical curves of S-parameters of a DR. The parameters and sizes of the test model are the same as that given in Fig. 7.21. ——— experimental; ---- theoretical.

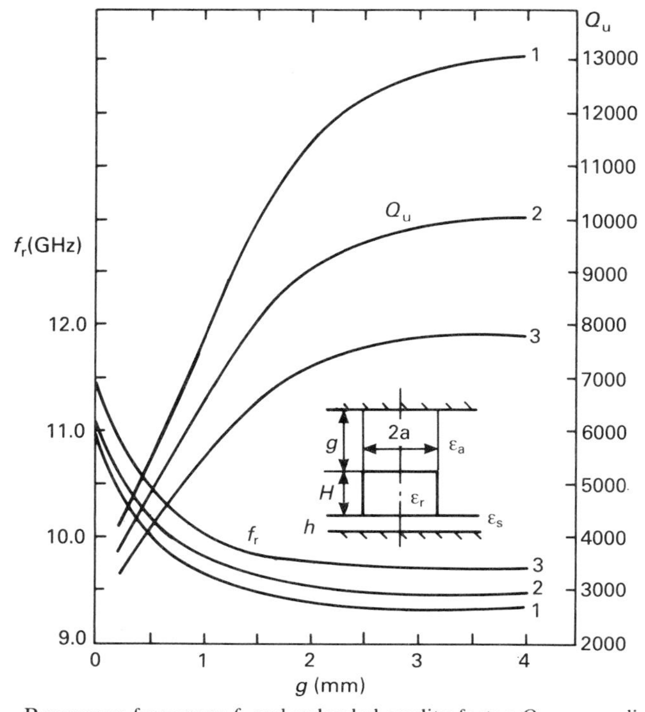

Fig. 7.23 — Resonance frequency f_r and unloaded quality factor Q_u versus distance g with substrate thickness h as parameter. The box material is aluminium. $\varepsilon_r = 40$, $\varepsilon_s = 2.5$, $\varepsilon_a = 1$, $a = \text{RA} = 3\,\text{mm}$, $H = 2.4\,\text{mm}$. 1: $h = 0.8\,\text{mm}$. 2: $h = 0.6\,\text{mm}$. 3: $h = 0.4\,\text{mm}$.

the top cylinder as parameter. One can see that the resonance frequency decreases with increasing dielectric constant (ε_{top}) and thickness (t) of the top material.

It is evident that the case decays to the case of uniform material DR when $t = 0$, $\varepsilon_{top} = 1$ *or* $\varepsilon_{top} = \varepsilon_r$. For these situations the accuracy of calculation can be checked by using the computer program DR10 (for uniform DR). This check verifies the accuracy of program DR10B.

Computer program DR20

The azimuthal electric field E of each region can be calculated by using program DR20. A family of curves of $E_{\theta 1}$, $E_{\theta 3}$, $E_{\theta 4}$ and $E_{\theta 6}$ as a function of radial distance r, with the axial distance z as the parameter is given in Fig. 7.25. This shows that the field reaches a maximum at about two thirds of the radius of the DR, but a little distance above the centre plane ($z = 0$).

Program DR20 can also be employed to design the position where the side wall should be placed. In the experiment [23] it was found that the quality factor was affected by the side wall which was placed near the DR. The side wall should thus be placed such that the Q is not affected. In the experiment [23] the sidewall was kept away from the position, where $E_{\theta 4} \sim 0.02\ E_{\theta 1 max}$ ($z = 0$), a high Q was maintained.

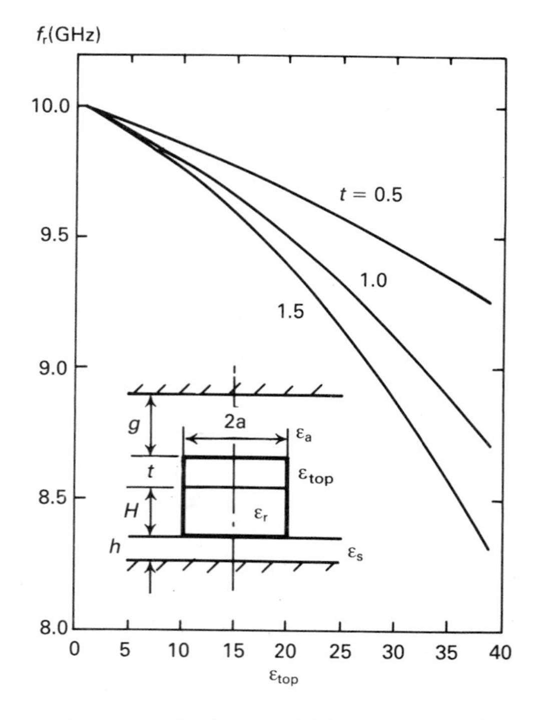

Fig. 7.24 — Resonance frequency, f_r, of a stacked DR system versus dielectric constant ε_{top} with the thickness of the top cylinder as parameter. $\varepsilon_r = 38.9$, $\varepsilon_a = 1$, $\varepsilon_s = 2.5$, $a = 3.05$ mm, $H = 2.0$ mm, $h = 0.7874$ mm, $g = 3.6126$ mm.

Figure 7.26 shows the distance r (from the centre of the DR) versus the frequency with a fixed electric field $E_{\theta 4}$ as parameter. For a frequency of 10 GHz for example one can read off $r = 6.65$ mm for $E_{\theta 4} = 0.02$. This means that the side wall should be 6.65 mm away from the centre of the DR so as to give optimum Q.

Computer program DR40

It is well known that the same resonance frequency can be achieved by DRs of a different height and radius. For a wanted frequency the radius a and height H of a DR which is placed on a substrate in an enclosed box can be calculated by using program DR40.

Computer program DR100

The external quality factor Q_e, coupling coefficient β_c and some other important parameters of shielded DR-microstrip system can be calculated by using program DR100.

Figure 7.27 shows the dimensions (a,H) of a DR and the external quality factor Q_e and stored energy in each region as functions of the shape factor $2a/H = D/H$, with a fixed DR resonance frequency and with the relative permittivity as parameter. It

should be noted that the same resonance frequency can be achieved through the use of DRs with different shape ratios.

The following explains and shows how to use the figure. Choose a shape ratio and draw a vertical line on the abscissa so that this line intersects with the curves a and H. The appropriate values for a and H can then be read off on the ordinate. For example, if we choose $2a/H = 2.25$ with $\varepsilon_r = 40$ (curve 3) then we obtain $H = 2.67$ mm $a = 3$ mm and a resonance frequency of $fr = 9$ GHz. This same resonance frequency can be obtained if we choose an aspect ratio $2a/H = 1.75$ with $\varepsilon_r = 35$ (curve 2). The DR dimensions are then $a = 2.98$ mm and $H = 3.4$ mm.

From Fig. 7.27 one can see that the higher the relative permittivity of the DR the higher Q_e, that is the weaker the coupling between the DR and microstrip. It is also seen that Q_e starts from a lower value at the left end of the curves, reaches the maximum at the right end, but the variation is relatively small.

Figure 7.27 shows also the computed curves of the percentage stored energy in each region of a given DR system versus shape ratio for a fixed resonance frequency and with the relative permittivity ε_r of the DR as parameter. From the figure one can conclude that about 98% of the energy is stored in the DR itself, and the higher the relative permittivity of the DR the more energy is stored in the DR and the less energies are stored in the other regions. It is also seen that W_6, the stored energy in region 6 where the microstrip is located, changes very slowly over the entire range of shape ratios from 1.5 to 2.5. This shows the insensitivity of Q_e to changes in shape ratio $2a/H$.

The computed curves of Fig. 7.28 were also obtained with the aid of computer program DR100. The figure displays the external quality factor Q_e, loaded quality factor Q_L, coupling coefficient β_c, mutual inductance L_m, the equivalent lumped circuit parameters R, L, C as well as the S-parameters of a DR coupled to a microstrip in an enclosure as a function of coupling distance d between the centre of the DR and the edge of the microstrip. The resonance frequency f_r is approximately

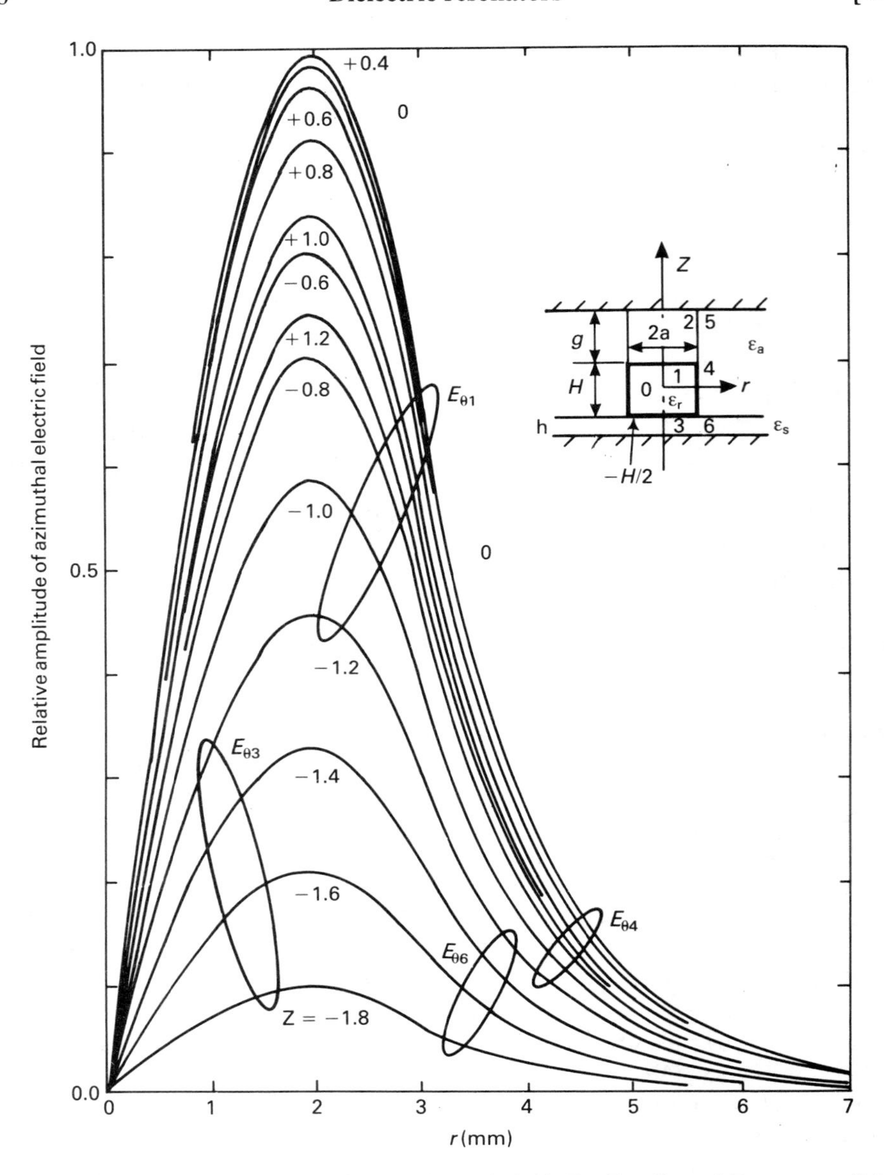

Fig. 7.25 — Relative amplitude of azimuthal electric field. $E_{\theta 1}$, $E_{\theta 3}$, $E_{\theta 4}$ and $E_{\theta 6}$ versus radial coordinate r of a DR ($TE_{01\delta}$ mode) structure. $f_r = 9.37$ GHz, $\varepsilon_r = 38.9$, $\varepsilon_a = 1$, $\varepsilon_s = 2.5$, $a = 3.05$ mm, $H = 2.4$ mm, $g = 4$ mm, $h = 0.7874$ mm. The coordinate of the point, where the field is the strongest: $r = 1.919652$ mm, $z = 0.2410672$ mm. All of the electric fields calculated here are normalized with the strongest one.

9.37 GHz, with the DR having radius $a = 3.05$ mm, height $H = 2.4$ mm and relative

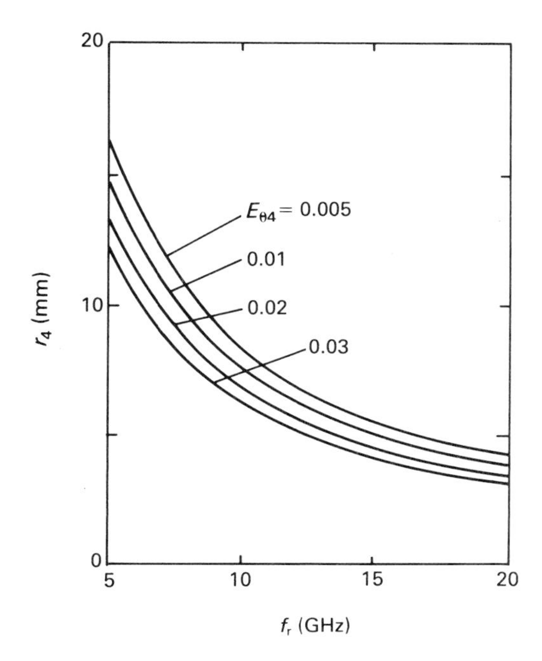

Fig. 7.26 — The distance r versus the frequency with fixed electric field $E_{\theta 4}$ as parameter. $E_{\theta 4}$ is normalized with the $E_{\theta 1 \max}$. $2a/H = 2$, $\varepsilon_r = 40$, $\varepsilon_a = 1$, $\varepsilon_s = 2.5$, $g = 0.15\lambda$, $h = 0.5$ mm, $z = 0$ (calculation was taken in the $z = 0$ plane), λ = free space wavelength.

permittivity $\varepsilon_r = 38.9$. The microstrip has a width $w = 2.25$ mm, thickness $h = 0.7874$ mm and relative permittivity $\varepsilon_s = 2.5$, impedance $Z_o = 50$ Ohms. The relative permittivity of air is $\varepsilon_a = 1$. The distance between the lid of the metal enclosure and the top of DR is $g = 4$ mm. The unloaded quality factor $Q_u = 2700$. The equivalent lumped circuit parameters of the DR itself are constant and were calculated to be the following:

$$L_r = 9.220521 \quad 10^{-8} \; H$$
$$C_r = 3.129 \quad 10^{-15} \; F$$
$$R_r = 2.010533\,\Omega$$

It can be seen from Fig. 7.28 that $|S_{11}|^2$, $|S_{11}|^2 + |S_{21}|^2$ decrease with decreasing coupling and R, L, β_c and L_m decay exponentially whilst Q_e, β_c, Q_L and $|S_{21}|^2$ rise exponentially with decreasing coupling. The condition

$$\omega_r = 1/(L_r C_r)^{0.5} = 1/(LC)^{0.5} \tag{7.81}$$

is thus always satisfied and gives the same value of resonance frequency, namely 9.37 GHz.

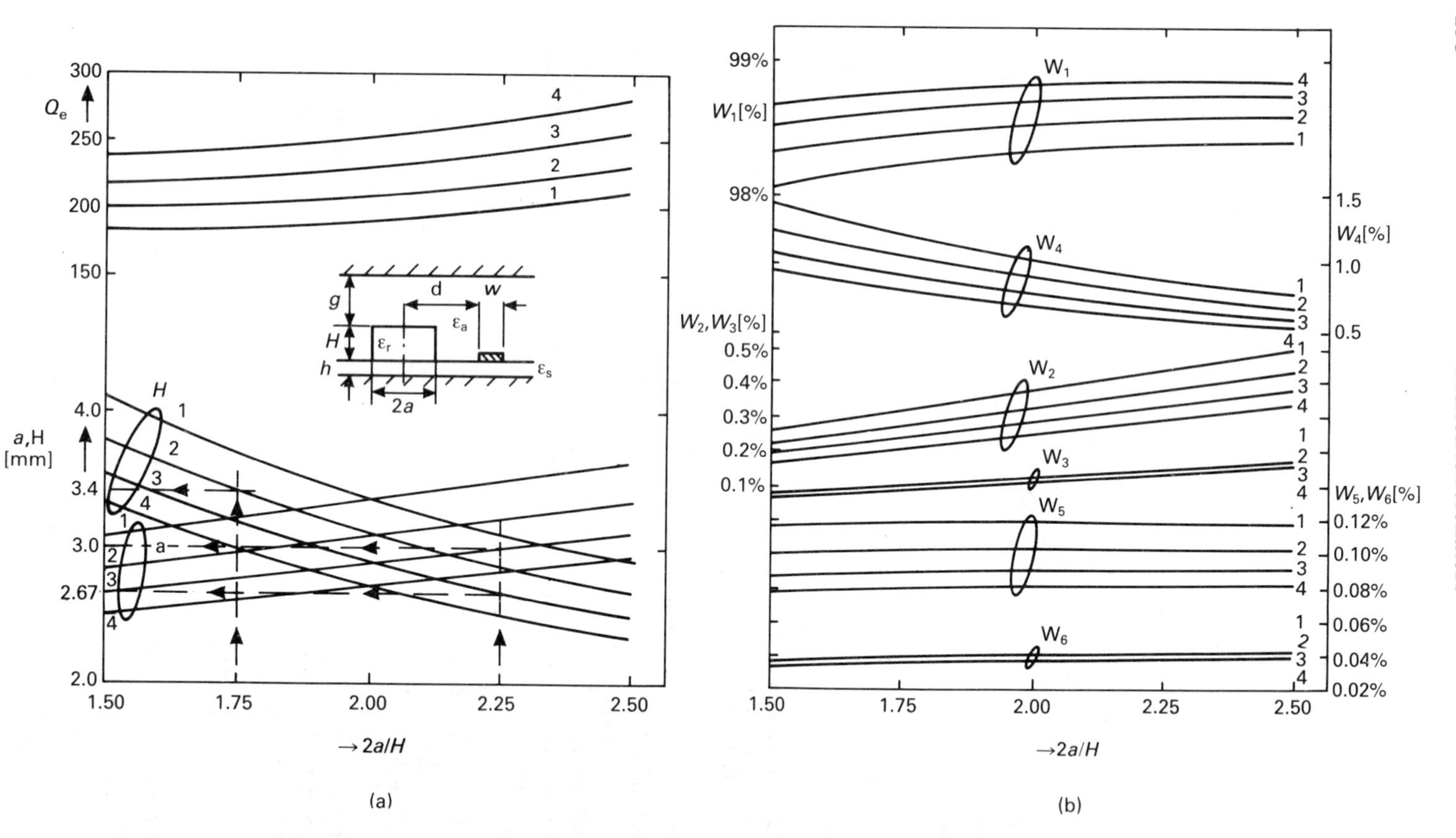

Fig. 7.27 — Computed curves showing the external quality factor Q_e and the DR dimensions a, H as a function of the shape ratio $2a/H$. The resonance frequency f_r is fixed at 9 GHz. ε_r is used as a parameter. $\varepsilon_s = 2.5$, $\varepsilon_a = 1$, $g = 3$ mm, $d = a$, $w = 2.29$ mm, $h = 0.8$ mm, $Z_0 = 50\,\Omega$. 1: $\varepsilon_r = 30$. 2: $\varepsilon_r = 35$. 3: $\varepsilon_r 40$. $\varepsilon_r = 45$.

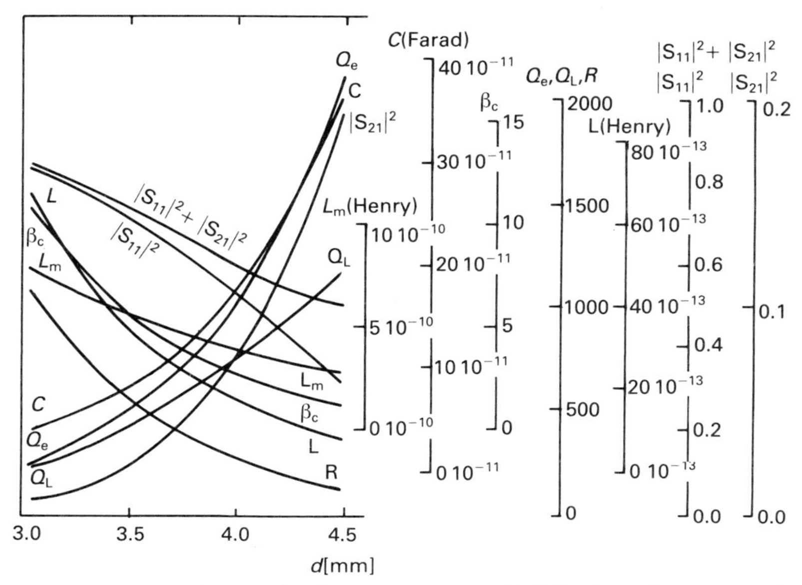

Fig. 7.28 — Computed curves for diverse parameters of a DR-microstrip structure obtained with program DR100. $f_r = 9.37$ GHz, $\varepsilon_r = 38.9$, $\varepsilon_a = 1$, $\varepsilon_s = 2.5$, $a = 3.05$ mm, $H = 2.4$ mm, $w = 2.25$ mm, $h = 0.7874$ mm, $Z_o = 50\,\Omega$, $g = 4$ mm, $Q_u = 2700$. Calculated results: $L_r = 9.220\,10^{-8}$ H, $C_r = 3.129\,10^{-15}$ F, $R_r = 2.010\,\Omega$.

Some simple Bessel function programs have been included which may prove useful in the context of DR analysis. Bessel functions can be represented in various ways. The accuracy of any numerical analysis depends on the type of series used and the number of terms in that series. The asymptotic expansions of the Bessel functions $J_0, J_1, J_2, K_0, K_1, K_2$ are given in reference [24]. Using the expansions given in eqns (7.82) to (7.84) the useful range of calculation of K_0, K_1 and K_2 can be extended to arguments of $x > 0.8$.

$$K_0(x) \simeq (\pi/2x)^{0.5} e^{-x} \left[1 - \frac{1}{8x} + \frac{9}{128x^2}\right]\left[1 - \frac{0.037}{x^3}\right] \tag{7.82}$$

$$K_1(x) \simeq (\pi/2x)^{0.5} e^{-x} \left[1 + \frac{3}{8x} - \frac{15}{128x^2}\right]\left[1 + \frac{0.037}{x^3}\right] \tag{7.83}$$

$$K_2(x) \simeq (\pi/2x)^{0.5} e^{-x} \left[1 + \frac{15}{8x} + \frac{105}{128x^2}\right]\left[1 - \frac{0.037}{x^3}\right] \tag{7.84}$$

Calculation example of DR10C

```
Programme DR10C

Choose option : Type '1' or '2'

1 - Calculate resonance frequency & unloaded quality factor of DR-
substrate-box system.

2 - Calculate resonance frequency of stacked DR-substrate-box system.

Selection ? 1

There is only TE01d mode in the DR system.

------------------------------------------

  g                 2            5
                          Ea

..............***************.............
  H          * DR           *
            *   Er   1   *              4
..............***************.............
  h                 3       Es     6
------------------------------------------

  Input the dielectric constant, Er, of DR (region 1)
Er = ? 38.9
  Input the dielectric constant, Ea, of regions 2,4,5
Ea = ? 1
  Input the dielectric constant, Es, of substrate (regions 3,6)
Es = ? 2.5
  **   Note : Er must be greater than Es & Ea    **
  Input the radius, RA, of DR in mm
RA = ? 3.05
  Input the height, H, of DR in mm
H = ? 2.4
  Input the thickness, h, of substrate in mm
h = ? .7874
  Input the distance, g, between the top of DR and upper metal shield in mm
g = ? 4


Calculation result :

Resonance frequency of DR system, Fr =  9.371498   GHz


Want to calculate the Q factor ? (Y or N)
 **  If g = ~ 0, Qu can not be calculated by the programme  **
? y
```

```
Shield material : copper, aluminum, brass or other ?

Conductivity in (SIEMENS/METER)*(E07)

If copper   then  5.8
If aluminum then  3.72
If brass    then  1.57

Enter conductivity, please.

? 3.72

Shield conductivity in SIEMENS/METER is  3.72E+07

Skin depth =  .8649365  MICRONS



Calculation result :

Resonance frequency of DR system, Fr =  9.372145   GHz


Qu (only due to shield loss) =  12624.29
```

For calculation of Qu three kinds of loss need to be taken into account: 1) shield loss on the internal surface of box including that on the metal bottom of substrate; 2) dielectric losses in DR and substrate; 3) radiation loss including that owing to field mismatch between the DR and the microstrip.
This programme only takes into account the shield loss on both the inner top surface of box and the bottom metal of substrate, and assuming that they are the same material. And it is also assumed that all side walls are far enough away from DR. The smaller the distance between DR and metal shield (upper and lower), then the shield loss become dominant, the better the accuracy of the calculation of Qu by this programme.

```
Programme DR10C

Choose option : Type '1' or '2'

1 - Calculate resonance frequency & unloaded quality factor of DR-
substrate-box system.

2 - Calculate resonance frequency of stacked DR-substrate-box system.

Selection ? 2

There is only TE01d mode in the DR system.

-----------------------------------------
                    2            5
  g
                          Ea

............***************.............
  t       * DS   Etop 7 *           8
............***************.............
          * DR          *
  H        *     Er   1 *           4
............***************.............
  h              Es   3             6
-----------------------------------------
```

```
  Input the dielectric constant, Er, of DR (region 1)
 Er = ? 38.9
  Input the dielectric constant, Ea, of regions 2,4,5,8
 Ea = ? 1
  Input the dielectric constant, Es, of substrate (regions 3,6)
 Es = ? 2.5
  **   Note : Er must be greater than Es & Ea   **
  Input the dielectric constant, Etop, of region 7
 Etop = ? 20
  Input the radius, RA, of DR in mm
 RA = ? 3.05
  Input the height, H, of DR in mm
 H = ? 2.4
  Input the thickness, h, of substrate in mm
 h = ? .7874
  Input the thickness, t, of dielectric sample (DS) in mm
 t = ? 3
  Input the distance, g, between the top of dielectric sample
(DS,which is put on the DR) and upper metal shield in mm
 g = ? 4

 **

 DEL =  .001

 Wait a minute, please.


 Input :

 Er =  38.9 ,     Ea =  1 ,    Es =  2.5
 RA =  3.05   mm
 H  =  2.4   mm
 h  =  .7874   mm
 t  =  3   mm
 g  =  4   mm

 Calculation results :

M                 Etop              Fr
 1                20               8.781221   GHz

The calculation of Fr is not very accurate if Etop > ~ 1.2 Er.

Want to calculate the Fr with some other Etops ? (Y or N)
? y
Input the number, M, of points you want to calculate.
? 9
Input Etops (M points).
? 1
? 5
? 10
? 15
? 20
? 25
? 30
? 35
? 40

Input :

Er =  38.9 ,     Ea =  1 ,    Es =  2.5
RA =  3.05   mm
H  =  2.4   mm
h  =  .7874   mm
t  =  3   mm
g  =  4   mm

Calculation results :
```

```
M           Etop        Fr
1           1           9.369739   GHz
2           5           9.294029   GHz
3           10          9.178511   GHz
4           15          9.014807   GHz
5           20          8.781221   GHz
6           25          8.484523   GHz
7           30          8.141398   GHz
8           35          7.768546   GHz
9           40          7.382671   GHz

The calculation of Fr is not very accurate if Etop > ~ 1.2 Er.

Want to calculate the Fr with some other Etops ? (Y or N)
? n
```

```
1000 PRINT" Programme DR10C "
1010 PRINT
1020 PRINT" Choose option : Type '1' or '2' "
1030 PRINT
1040 PRINT" 1 - Calculate resonance frequency & unloaded quality factor of DR-
        substrate-box system."
1050 PRINT
1060 PRINT" 2 - Calculate resonance frequency of stacked DR-substrate-box system."
1070 DIM NUM(6),DEN(6),XX(3)
1080 DIM KK(3),FCT(3),GCT(3)
1090 DIM Y(5),X(5)
1100 PRINT
1110 INPUT" Selection ";SELECT
1120 IF SELECT=1 THEN 1150
1130 IF SELECT=2 THEN 3730
1140 PRINT
1150 REM ** Programme DR10 **
1160 PRINT
1170 PRINT" There is only TE01d mode in the DR system. "
1180 PRINT
1190 PRINT"-----------------------------------------"
1200 PRINT
1210 PRINT" g                  2             5 "
1220 PRINT"                            Ea        "
1230 PRINT
1240 PRINT".............***************.............."
1250 PRINT" H           * DR          *             "
1260 PRINT"             *   Er  1   *          4 "
1270 PRINT".............***************.............."
1280 PRINT" h                  3      Es      6 "
1290 PRINT"-----------------------------------------"
1300 PRINT
1310 PRINT"  Input the dielectric constant, Er, of DR (region 1) "
1320 INPUT" Er = ";ER
1330 PRINT"  Input the dielectric constant, Ea, of regions 2,4,5 "
1340 INPUT" Ea = ";EA
1350 PRINT"  Input the dielectric constant, Es, of substrate (regions 3,6) "
1360 INPUT" Es = ";ES
1370 PRINT"  **  Note : Er must be greater than Es & Ea   ** "
1380 PRINT"  Input the radius, RA, of DR in mm "
1390 INPUT" RA = ";RA
1400 PRINT"  Input the height, H, of DR in mm "
1410 INPUT" H = ";H
1420 PRINT"  Input the thickness, h, of substrate in mm "
1430 INPUT" h = ";H1
1440 PRINT"  Input the distance, g, between the top of DR and upper metal
      shield in mm "
1450 INPUT" g = ";G
1460 IF G<.000001 THEN 1470 ELSE 1480
1470 G=.000001
1480 NQ=0
```

```
1490 XX(2)=2.9
1500 KMIN=XX(2)/SQR(ER-EA)
1510 KMAX=XX(2)/SQR(ES)
1520 KK(2)=(9*KMIN+KMAX)/10
1530 DXX=.00001
1540 DKK=.00001
1550 ITER=0
1560 PRINT
1570 XX(1)=XX(2)+DXX
1580 KK(1)=KK(2)
1590 XX(3)=XX(2)
1600 KK(3)=KK(2)+DKK
1610 FOR ITI=1 TO 3
1620 X=XX(ITI)
1630 K0=KK(ITI)
1640 K02=K0^2
1650 XIT2=X^2
1660 GOSUB 3190
1670 RRA=K02*(ER-EA)-XIT2
1680 IF RRA>0 THEN 1750
1690 WSTEPX=WSTEPX/2
1700 WSTEPK=WSTEPK/2
1710 XX(2)=XX(2)-WSTEPX
1720 KK(2)=KK(2)-WSTEPK
1730 PRINT" Start again with 1/2 smaller step. Wait a minute, please . "
1740 GOTO 1550
1750 YY=SQR(RRA)
1760 KC4A=YY
1770 GOSUB 3250
1780 FCT(ITI)=J0B+YY*K0B/X
1790 BA=SQR(K02*ER-XIT2)
1800 ALF1=SQR(XIT2-K02*ES)
1810 POW=ALF1*H1/RA
1820 IF POW>8 THEN 1870
1830 EP=EXP(POW)
1840 EI=1/EP
1850 AGU=(EP+EI)/(EP-EI)
1860 GOTO 1880
1870 AGU=1
1880 AGU=AGU*ALF1/BA
1890 FIH1=ATN(AGU)
1900 ALF2=SQR(XIT2-K02*EA)
1910 POW=ALF2*G/RA
1920 IF POW>8 THEN 1970
1930 EP=EXP(POW)
1940 EI=1/EP
1950 AGU=(EP+EI)/(EP-EI)
1960 GOTO 1980
1970 AGU=1
1980 AGU=AGU*ALF2/BA
1990 FIH2=ATN(AGU)
2000 GCT(ITI)=FIH1+FIH2-BA*H/RA
2010 NEXT ITI
2020 AL=(FCT(1)-FCT(2))/DXX
2030 AU=(GCT(1)-GCT(2))/DKK
2040 BL=(FCT(3)-FCT(2))/DXX
2050 BU=(GCT(3)-GCT(2))/DKK
2060 CL=FCT(2)-AL*XX(2)-BL*KK(2)
2070 CU=GCT(2)-AU*XX(2)-BU*KK(2)
2080 DEN0=AU*BL-AL*BU
2090 XNEW=(CL*BU-CU*BL)/DEN0
2100 KNEW=(CU*AL-CL*AU)/DEN0
2110 WSTEPX=XNEW-XX(2)
2120 WSTEPK=KNEW-KK(2)
2130 WSTEP2=WSTEPX^2+WSTEPK^2
2140 XX(2)=XNEW
2150 KK(2)=KNEW
2160 IF WSTEP2<1E-12 THEN 2220
2170 ITER=ITER+1
2180 IF ITER>10 THEN 2200
```

```
2190 GOTO 1570
2200 PRINT" Solution not found after 10 iterations "
2210 GOTO 3180
2220 K0A=KK(2)
2230 PI=3.1415926535#
2240 FIR=K0A*150/(PI*RA)
2250 EIGX=XX(2)
2260 K0A2=K0A^2
2270 EIG2=EIGX^2
2280 WRADIC=EIG2-K0A2*ES
2290 AL1A=SQR(WRADIC)
2300 WRADI=EIG2-K0A2*EA
2310 AL2A=SQR(WRADI)
2320 WRADA=K0A2*ER-EIG2
2330 BA=SQR(WRADA)
2340 AL1L1=AL1A*H1/RA
2350 AL2L2=AL2A*G/RA
2360 IF AL1L1>8 THEN 2420
2370 Z1=EXP(AL1L1)
2380 ZI1=1/Z1
2390 CT1=(Z1+ZI1)/(Z1-ZI1)
2400 SIH1=(Z1-ZI1)*.5
2410 GOTO 2430
2420 CT1=1
2430 IF AL2L2>8 THEN 2490
2440 Z2=EXP(AL2L2)
2450 ZI2=1/Z2
2460 CT2=(Z2+ZI2)/(Z2-ZI2)
2470 SIH2=(Z2-ZI2)*.5
2480 GOTO 2500
2490 CT2=1
2500 ARG1=AL1A*CT1/BA
2510 ARG2=AL2A*CT2/BA
2520 TH1=ATN(ARG1)
2530 TH2=ATN(ARG2)
2540 KC4A2=WRADA-K0A2*EA
2550 KC4A=SQR(KC4A2)
2560 GOSUB 3250
2570 X=EIGX
2580 GOSUB 3190
2590 J0B2=J0B^2
2600 TRX=J0B2-2*J0B/EIGX+1
2610 K0B2=K0B^2
2620 PRX=K0B2+2*K0B/KC4A-1
2630 SIF1=SIN(TH1*2)
2640 SIF2=SIN(TH2*2)
2650 THET=(SIF1+SIF2)*.5/(TH1+TH2)
2660 C012=RA*(COS(TH1)^2)/AL1A
2670 IF AL1L1>8 THEN 2710
2680 SECN=AL1L1/(SIH1^2)
2690 PARM1=CT1-SECN
2700 GOTO 2720
2710 PARM1=1
2720 C0PAM1=C012*PARM1
2730 C022=RA*(COS(TH2)^2)/AL2A
2740 IF AL2L2>8 THEN 2780
2750 SECN=AL2L2/(SIH2^2)
2760 PARM2=CT2-SECN
2770 GOTO 2790
2780 PARM2=1
2790 C0PAM2=C022*PARM2
2800 NUM(1)=ES*C0PAM1*TRX
2810 NUM(2)=EA*C0PAM2*TRX
2820 NUM(3)=-(WRADIC+KC4A2)*PRX*C0PAM1/K0A2
2830 NUM(4)=EA*H*(1+THET)*PRX
2840 NUM(5)=-(WRADI+KC4A2)*PRX*C0PAM2/K0A2
2850 NUM(6)=ER*H*(1+THET)*TRX
2860 DEN(1)=NUM(1)
2870 DEN(2)=NUM(2)
2880 DEN(3)=ES*PRX*C0PAM1
```

```
2890 DEN(4)=NUM(4)
2900 DEN(5)=EA*PRX*C0PAM2
2910 DEN(6)=NUM(6)
2920 DENSUM=0
2930 SURVER=-BA*RA*(SIF1+SIF2)*PRX/K0A2
2940 SURH0R=X*J0B*2*(C0PAM1+C0PAM2)/K0A2
2950 NUMSUM=SURH0R+SURVER
2960 FOR J=1 TO 6
2970 DENSUM=DENSUM+DEN(J)
2980 NUMSUM=NUMSUM+NUM(J)
2990 NEXT J
3000 VARK0A=K0A*SQR(NUMSUM/DENSUM)
3010 VARFRE=VARK0A*150/(PI*RA)
3020 PRINT
3030 PRINT
3040 FR=(FIR+VARFRE)*.5
3050 IF SELECT=1 THEN 3080 ELSE 3060
3060 Y(I)=FR
3070 GOTO 5940
3080 PRINT" Calculation result : "
3090 PRINT
3100 PRINT" Resonance frequency of DR system, Fr = "FR"  GHz "
3110 PRINT
3120 PRINT
3130 IF NQ=1 THEN RETURN
3140 PRINT" Want to calculate the Q factor ? (Y or N) "
3150 PRINT" **  If g = ~ 0, Qu can not be calculated by the programme  ** "
3160 INPUT B$
3170 IF B$="Y" OR B$="y" THEN GOSUB 3310
3180 END
3190 REM ** function J0B=J0(X)/J1(X) **
3200 XMX0=X-2.4048
3210 TEM=(.0282*XMX0-.1177)*XMX0+.2571
3220 TEM=(TEM*XMX0-.716)*XMX0+1.4282
3230 J0B=TEM*XMX0/(X-3.8317)
3240 RETURN
3250 REM ** function K0B=K0(KC4A)/K1(KC4A) **
3260 KI=1/KC4A
3270 TEM=(.00445*KI-.02679)*KI+.06539
3280 TEM=(TEM*KI-.11226)*KI+.49907
3290 K0B=1/(1+TEM*KI)
3300 RETURN
3310 REM ** Q factor **
3320 NQ=1
3330 F0=VARFRE
3340 PRINT
3350 PRINT" Shield material : copper, aluminum, brass or other ? "
3360 PRINT
3370 PRINT" Conductivity in (SIEMENS/METER)*(E07) "
3380 PRINT
3390 PRINT" If copper   then  5.8 "
3400 PRINT" If aluminum then  3.72 "
3410 PRINT" If brass    then  1.57 "
3420 PRINT
3430 PRINT" Enter conductivity, please. "
3440 PRINT
3450 INPUT CSM
3460 CSM=CSM*(1E+07)
3470 IF CSM<.1 THEN 3700
3480 SKIN=50/(SQR(F0*CSM)*PI)
3490 SKINMI=1000*SKIN
3500 PRINT
3510 PRINT" Shield conductivity in SIEMENS/METER is "CSM
3520 PRINT
3530 PRINT" Skin depth = "SKINMI" MICRONS "
3540 H1=H1-SKIN
3550 G=G-SKIN
3560 GOSUB 1550
3570 DF=VARFRE-F0
3580 IF DF/F0<.000001 THEN 3680
```

```
3590 Q=F0/DF
3600 PRINT" Qu (only due to shield loss) = "Q
3610 PRINT
3620 PRINT" For calculation of Qu three kinds of loss need  to be taken into account:

                1)      shield loss on the internal surface of box including that on
                        the metal bottom of substrate;
                2)      dielectric losses in"
3630 PRINT"             DR and substrate;
                3)      radiation  loss including  that  owing  to  field mismatch between
                        the DR and the microstrip.  This  programme  only  takes  into
                        account  the   shield  loss"
3640 PRINT"             on both the inner top surface of box  and  the  bottom  metal  of
                        substrate, and assuming that they are the same material.  And  it
                        is also assumed  that all side walls are far enough away from DR."
3650 PRINT"             The smaller the  distance  between  DR  and  metal shield (upper
                        and lower), then the shield loss   become   dominant,   the better
                        the accuracy of the calculation  of Qu by this programme.        "
3660 PRINT
3670 RETURN
3680 PRINT"             Insignificant losses in the shield, QUIT "
3690 GOTO 3180
3700 PRINT"             Conductivity too small, QUIT "
3710 GOTO 3180
3720 PRINT
3730 REM ** Programme DR10B1 **
3740 PRINT
3750 PRINT" There is only TE01d mode in the DR system. "
3760 PRINT
3770 PRINT"-----------------------------------------"
3780 PRINT"                         2             5  "
3790 PRINT"   g                                      "
3800 PRINT"                                Ea        "
3810 PRINT"                                          "
3820 PRINT"............***************.............."
3830 PRINT"   t        * DS   Etop 7 *          8  "
3840 PRINT"............***************.............."
3850 PRINT"            * DR           *             "
3860 PRINT"   H        *      Er   1 *          4  "
3870 PRINT"............***************.............."
3880 PRINT"   h                Es   3          6  "
3890 PRINT"-----------------------------------------"
3900 DIM A(16),J0(16),XSX(40)
3910 DIM Z(50),P(50),B(50)
3920 DIM H(50),C(50),D(50),W(50)
3930 PRINT
3940 PRINT"  Input the dielectric constant, Er, of DR (region 1) "
3950 INPUT" Er = ";ER
3960 PRINT"  Input the dielectric constant, Ea, of regions 2,4,5,8 "
3970 INPUT" Ea = ";EA
3980 PRINT"  Input the dielectric constant, Es, of substrate (regions 3,6) "
3990 INPUT" Es = ";ES
4000 PRINT"  **   Note : Er must be greater than Es & Ea    ** "
4010 PRINT"  Input the dielectric constant, Etop, of region 7 "
4020 INPUT" Etop = ";ETX
4030 IF ETX<1 THEN 4040 ELSE 4090
4040 PRINT
4050 PRINT"        It is meaningless to calculate the case with Etop < 1
     Input another Etop, Please. "
4060 PRINT
4070 GOTO 4020
4080 PRINT
4090 ET=.4*ER
4100 PRINT"  Input the radius, RA, of DR in mm "
4110 INPUT" RA = ";RA
4120 PRINT"  Input the height, H, of DR in mm "
4130 INPUT" H = ";H
4140 H0=H
4150 PRINT"  Input the thickness, h, of substrate in mm "
4160 INPUT" h = ";H1
```

```
4170 H10=H1
4180 PRINT"  Input the thickness, t, of dielectric sample (DS) in mm "
4190 INPUT" t = ";TT
4200 PRINT"  Input the distance, g, between the top of dielectric sample
   (DS,which is put on the DR) and upper metal shield in mm "
4210 INPUT" g = ";G
4220 G0=G
4230 IF G<.000001 THEN 4240  ELSE 4250
4240 G=.000001
4250 PI=3.1415926535#
4260 PRINT
4270 D3=G+TT+H/2
4280 D1=H1+H/2
4290 VVR=SQR(ER-EA)
4300 VVS=14.5/SQR(ES)
4310 FR=((130/VVR)+VVS)/RA
4320 PRINT
4330 LDD=.001
4340 PRINT" ** "
4350 PRINT
4360 PRINT" DEL = "LDD
4370 LDD=LDD
4380 PRINT
4390 PRINT" Wait a minute, please. "
4400 PRINT
4410 PRINT" ** "
4420 PRINT
4430 GOSUB 4640
4440 MIN0=MIN
4450 FR=FR*1.01
4460 GOSUB 4640
4470 MIN00=MIN
4480 IF MIN0<MIN00 THEN 4490 ELSE 4550
4490 FR=FR*.99
4500 GOSUB 4640
4510 PRINT" MIN = "MIN
4520 IF MIN<LDD THEN 4600
4530 FR=FR-FR*MIN*.1
4540 GOTO 4500
4550 FR=FR+FR*MIN*.1
4560 GOSUB 4640
4570 PRINT" MIN = "MIN
4580 IF MIN<LDD THEN 4600
4590 GOTO 4550
4600 FR1=FR*.972
4610 PRINT
4620 GOTO 5810
4630 PRINT
4640 REM ** MIN **
4650 PRINT
4660 GOSUB 5090
4670 PRINT
4680 KR=XXX/RA
4690 KA=YYY/RA
4700 DD=(PI*FR/150)^2
4710 II=KR^2
4720 BB=SQR(DD*ER-II)
4730 AA=SQR(II-DD*EA)
4740 LD=SQR(II-DD*ES)
4750 AK=SQR(II-DD*ET)
4760 PRINT
4770 X=AK*TT
4780 GOSUB 5040
4790 TANHX1=THX
4800 X=AA*G
4810 GOSUB 5040
4820 TANHX2=THX
4830 AA81=AA+AK*TANHX1*TANHX2
4840 AA82=AK*TANHX2+AA*TANHX1
4850 AA8=-AA81/AA82
```

```
4860 X=LD*H1
4870 GOSUB 5040
4880 TANHX3=THX
4890 RES1=AK*TANHX3*AA8
4900 RES22=BB*(LD-RES1)
4910 RES3=BB*BB*TANHX3
4920 RES4=LD*AK*AA8
4930 RES55=RES3+RES4
4940 RES66=RES22/RES55
4950 RES=ATN(RES66)+PI
4960 IF RES>2.5 THEN 4970 ELSE 4990
4970 PRINT" Quit, change input and try again. "
4980 GOTO 6900
4990 PRINT" RES = "RES
5000 BSH=BB*H
5010 PRINT" BSH = "BSH
5020 MIN=ABS(BSH-RES)/(ABS(BSH)+ABS(RES))
5030 RETURN
5040 REM ** TH(X) **
5050 THX=(EXP(X)-EXP(-X))/(EXP(X)+EXP(-X))
5060 RETURN
5070 PRINT
5080 PRINT
5090 REM ** J1(KrA)/(KrA*J0(KrA))=-K1(KaA)/(KaA*K0(KaA)), to find KrA(=XXX),
        KaA(=YYY), Kr(=KR), Ka(=KA), Bb(BB), Aa(=AA), As(=LD), Ak(=AK) **
5100 REM ** J1(KrA)=J1 **
5110 REM ** A=RA **
5120 XXX=2.7
5130 XXX=XXX
5140 IF XXX>3.3 THEN 5410
5150 XXX1=(XXX/2)^2
5160 S=1
5170 J=1
5180 T=1
5190 K=-1
5200 T=T*XXX1/J/(J+1)
5210 S=S+K*T
5220 IF T<.000001 THEN 5260
5230 K=-K
5240 J=J+1
5250 GOTO 5200
5260 J1=(XXX/2)*S
5270 REM ** J0(KrA)=J0 **
5280 A(1)=4
5290 J0(1)=1-(XXX^2)/4
5300 FOR M=1 TO 15 STEP 1
5310 A(M+1)=A(M)*((2*(M+1))^2)
5320 J0(M+1)=J0(M)+((-1)^(M+1))*(XXX^(2*(M+1)))/A(M+1)
5330 NEXT M
5340 J0=J0(16)
5350 XX=J1/(XXX*J0)
5360 REM ** K1(Y), FOR Y > 0.8 **
5370 REM ** K1(KaA)=K **
5380 MNN=(ER-EA)*(((PI*FR*RA)/150)^2)
5390 MPP=XXX^2
5400 IF MNN-MPP<0 THEN 5410 ELSE 5530
5410 PRINT
5420 IF LDD>.015 THEN 5430 ELSE 5450
5430 PRINT" The accurate result can not be obtained in 1.5 % accuracy. "
5440 GOTO 6900
5450 REM ** Change  MIN & DEL <.001 into ...<.001 + .004 * N **
5460 PRINT
5470 PRINT" ** "
5480 PRINT
5490 PRINT" Try again with lower accuracy."
5500 LDD=LDD+.004
5510 PRINT" DEL = "LDD
5520 GOTO 4370
5530 YYY=SQR(MNN-MPP)
5540 AAA=SQR(PI/(2*YYY))
```

```
5550 BBB=AAA/EXP(YYY)
5560 C=1+.375/YYY
5570 D=C-.1171875/YYY^2
5580 K=BBB*D*(1+.037/(YYY^3))
5590 REM ** K0(Y), FOR Y > 0.8 **
5600 REM ** K0(KaA)=SK **
5610 AAAA=SQR(PI/(2*YYY))
5620 BBB=AAAA/EXP(YYY)
5630 C=1-.125/YYY
5640 D=C+.0703125/YYY^2
5650 SK=BBB*D*(1-.037/(YYY^3))
5660 SSK=K/(YYY*SK)
5670 DEL=ABS(XX+SSK)/ABS(XX-SSK)
5680 IF DEL<LDD THEN 5800
5690 IF DEL>.05 THEN 5700 ELSE 5720
5700 XXX=XXX+.01
5710 GOTO 5130
5720 IF DEL>.01 THEN 5730 ELSE 5750
5730 XXX=XXX+.005
5740 GOTO 5130
5750 IF DEL>.005 THEN 5760 ELSE 5780
5760 XXX=XXX+.002
5770 GOTO 5130
5780 XXX=XXX+.001
5790 GOTO 5130
5800 RETURN
5810 REM ** Interpolation **
5820 N=2
5830 M=1
5840 PRINT
5850 FOR I=1 TO 2
5860 IF I=1 THEN 5870 ELSE 5900
5870 H=H
5880 G=G+TT
5890 GOTO 1460
5900 IF I=2 THEN 5910
5910 H=H+TT
5920 G=G
5930 GOTO 1460
5940 Y(0)=Y(1)
5950 Y(2)=Y(2)
5960 NEXT I
5970 X(0)=1
5980 X(2)=ER
5990 X(1)=.4*ER
6000 Y(1)=FR1
6010 PRINT
6020 FOR I=1 TO N
6030 H(I-1)=X(I)-X(I-1)
6040 NEXT I
6050 K=N-1
6060 FOR I=1 TO K
6070 P(I)=H(I-1)/(H(I-1)+H(I))
6080 B(I)=3*((1-P(I))*(Y(I)-Y(I-1))/H(I-1)+P(I)*(Y(I+1)-Y(I))/H(I))
6090 NEXT I
6100 P(0)=1
6110 P(N)=0
6120 B(0)=3*(Y(1)-Y(0))/H(0)
6130 B(N)=3*(Y(N)-Y(N-1))/H(N-1)
6140 FOR I=0 TO N
6150 D(I)=2
6160 NEXT I
6170 FOR I=1 TO N
6180 C(I)=1-P(I)
6190 NEXT I
6200 FOR I=0 TO N
6210 NEXT I
6220 P=N
6230 FOR I=1 TO P
6240 IF ABS(D(I))<=.000001 GOTO 6910
```

```
6250 P(I-1)=P(I-1)/D(I-1)
6260 B(I-1)=B(I-1)/D(I-1)
6270 D(I)=P(I-1)*(-C(I))+D(I)
6280 B(I)=-C(I)*B(I-1)+B(I)
6290 NEXT I
6300 B(P)=B(P)/D(P)
6310 FOR I=1 TO P
6320 B(P-I)=B(P-I)-P(P-I)*B(P-I+1)
6330 NEXT I
6340 FOR I=0 TO N
6350 NEXT I
6360 PRINT
6370 Z(1)=ETX
6380 GOTO 6480
6390 PRINT"Input the number, M, of points you want to calculate."
6400 INPUT M
6410 PRINT"Input Etops (M points)."
6420 FOR I=1 TO M
6430 INPUT Z(I)
6440 IF Z(I)<1 THEN 6450 ELSE 6470
6450 PRINT" It is meaningless to calculate the case with Etop < 1
Input another Etop, please."
6460 GOTO 6430
6470 NEXT I
6480 PRINT
6490 FOR I=1 TO M
6500 IF Z(I)<X(0) GOTO 6560
6510 FOR J=1 TO N
6520 IF Z(I)<=X(J) GOTO 6580
6530 NEXT J
6540 J=N-1
6550 GOTO 6600
6560 J=0
6570 GOTO 6600
6580 J=J-1
6590 GOTO 6600
6600 E=X(J+1)-Z(I)
6610 E1=E*E
6620 F=Z(I)-X(J)
6630 F1=F*F
6640 H1=H(J)*H(J)
6650 W(I)=(3*E1-2*E1*E/H(J))*Y(J)+(3*F1-2*F1*F/H(J))*Y(J+1)
6660 W(I)=W(I)+(H(J)*E1-E1*E)*B(J)-(H(J)*F1-F1*F)*B(J+1)
6670 W(I)=W(I)/H1
6680 NEXT I
6690 PRINT" Input : "
6700 PRINT
6710 PRINT" Er = "ER",     Ea = "EA",    Es = "ES
6720 PRINT" RA = "RA"  mm "
6730 PRINT" H  = "H0"  mm "
6740 PRINT" h  = "H10"  mm "
6750 PRINT" t  = "TT"  mm "
6760 PRINT" g  = "G0"  mm "
6770 PRINT
6780 PRINT" Calculation results : "
6790 PRINT
6800 PRINT"M";TAB(20);"Etop";TAB(40);"Fr"
6810 FOR I=1 TO M
6820 PRINT I;TAB(20);Z(I);TAB(40);W(I)"  GHz"
6830 NEXT I
6840 PRINT
6850 PRINT"The calculation of Fr is not very accurate if Etop > ~ 1.2 Er. "
6860 PRINT
6870 PRINT"Want to calculate the Fr with some other Etops ? (Y or N) "
6880 INPUT B$
6890 IF B$="Y" OR B$="y" THEN 6390 ELSE 6900
6900 END
6910 PRINT"FAIL"
6920 END
```

Calculation example of DR20

```
Programme DR20
Calculate electric field, EQ, at any point in DR-substrate-
box system.
The distance between DR and side wall of metal box is ralatively
large; And there is only TE01d mode in DR system.

-----------------------------------------
              + ZZZ
                ^
g               : 2           5
                :
                :
...........***************............
           *    :    *
H           *   0 -------*------> RRR
           * DR     1  *          4
...........***************............
h               3             6
-----------------------------------------

 Input the resonance frequency, Fr, of DR system in GHz
 ** The input of Fr must be correct **
Fr = ? 9.37
 Input the dielectric constant, Er, of DR (region 1)
Er = ? 38.9
 Input the dielectric constant, Ea, of regions 2,4,5
Ea = ? 1
 Input the dielectric constant, Es, of substrate (regions 3,6)
Es = ? 2.5
 ** Note : Er must be greater than Es & Ea **
 Input the radius, RA, of DR in mm
RA = ? 3.05
 Input the height, H, of DR in mm
H = ? 2.4
 Choosing 1.5 < 2*RA/H < 2.5 is proper for obtaining higher Q  &
suppressing TM11d mode.
 Sometimes, accurate results can not be obtained if 2*RA/H < 1.5
 Wait a minute, please.

J1(KrA)/((KrA)*(J0(KrA)) = -.536442
K1(KaA)/((KaA)*(K0(KaA)) =  .5362416
KrA =  2.925001
KaA =  2.241062
Kr  =  .9590166
Ka  =  .7347743
Bb  =  .7605295
Aa  =  .9387229
As  =  .9074319


 Input the thickness, h, of substrate in mm
h = ? .7874
 Input the distance, g, between the top of DR and upper metal shield in mm
g = ? 4

 Wait a minute, please.
```

```
The coordinate of the point where the field is the strongest :

RRR1 = 1.919652 mm ;  ZZZ1 = .2410672 mm

All of the electric fields calculated here are normalized with the strongest one
(EQ1MAX).

TYPE 0, please.

? 0

Want to calculate the electric fields :   EQ1, EQ2, EQ3, EQ4, EQ5, EQ6 ?

IF EQ1 then TYPE 1,    IF EQ2 then TYPE 2,    IF EQ3 then TYPE 3,
IF EQ4 then TYPE 4,    IF EQ5 then TYPE 5,    IF EQ6 then TYPE 6.
Want to calculate the distribution of EQ4 some distance from the DR in radial
direction ? If do then TYPE 7 .
Want to calculate the distribution of EQ1 at the top of DR ? If so then TYPE 8 .

If do not want to calculate them, then TYPE N or n.

? 1

Input the radial coordinate, RRR1, of discussed point in DR in mm

RRR1 = ? 3

Input the longitudinal coordinate, ZZZ1, of discussed point in DR        in mm

ZZZ1 = ? 1.1

The relative amplitude of EQ1 (in DR) is :

EQ1 = .5233377

Want to calculate EQ1 at another point ? (Y or N)

? n
>
? 2

Input the radial coordinate, RRR2, of discussed point in region 2 in mm

RRR2 = ? 2

Input the longitudinal coordinate, ZZZ2, of discussed point in region 2 in mm

ZZZ2 = ? 4

The relative amplitude of EQ2 is :

EQ2 = .0480944
```

```
Want to calculate EQ2 at another point ? (Y or N)

? n
>
? 3

Input the radial coordinate, RRR3, of discussed point in region 3 in mm

RRR3 = ? 2.5

Input the longitudinal coordinate, ZZZ3, of discussed point in  region 3 in mm

ZZZ3 = ? -1.6

The relative amplitude of EQ3 is :

EQ3 =  .1890077

Want to calculate EQ3 at another point ? (Y or N)

? n
>
? 4

Input the radial coordinate, RRR4, of discussed point in region 4 in mm

RRR4 = ? 7

Input the longitudinal coordinate, ZZZ4, of discussed point in region 4 in mm

ZZZ4 = ? -1.2

The relative amplitude of EQ4 is :

EQ4 =  9.715026E-03

Want to calculate EQ4 at another point ? (Y or N)

? n
>
? 5

Input the radial coordinate, RRR5, of discussed point in region 5 in mm

RRR5 = ? 4

Input the longitudial coordinate, ZZZ5, of discussed point in region 5 in mm

ZZZ5 = ? 5

The relative amplitude of EQ5 is :

EQ5 =  1.753475E-03

Want to calculate EQ5 at another point ? (Y or N)

? n
>
? 6
```

Input the radial coordinate, RRR6, of discussed point in region 6 in mm

RRR6 = ? 3.5

Input the longitudinal coordinate, ZZZ6, of discussed point in region 6 in mm

ZZZ6 = ? -1.8

The relative amplitude of EQ6 is :

EQ6 = 4.183758E-02

Want to calculate EQ6 at another point ? (Y or N)

? n
>
? 7

Input the longitudinal coordinate, ZZZ4, of discussed plane in mm

ZZZ4 = ? 0

ZZZ4 = 0 mm

RRR4 (mm)	RRR4/wave length	EQ4
5.33618000	0.16666670	0.08276136
5.49626500	0.17166670	0.07233750
5.65635000	0.17666670	0.06326051
5.81643600	0.18166670	0.05535049
5.97652100	0.18666670	0.04845252
6.13660600	0.19166670	0.04243325
6.29669200	0.19666670	0.03717750
6.45677700	0.20166670	0.03258582
6.61686300	0.20666670	0.02857207
6.77694800	0.21166670	0.02506177
6.93703300	0.21666670	0.02199027
7.09711900	0.22166670	0.01930149
7.25720400	0.22666670	0.01694674
7.41729000	0.23166670	0.01488367
7.57737500	0.23666670	0.01307545
7.73746000	0.24166670	0.01149001

TYPE 0, then goes on showing.

? 0

ZZZ4 = 0 mm

RRR4 (mm)	RRR4/wave length	EQ4
7.89754600	0.24666670	0.01009942
8.05763100	0.25166670	0.00887932
8.21771600	0.25666670	0.00780847
8.37780200	0.26166670	0.00686832
8.53788700	0.26666670	0.00604269
8.69797200	0.27166670	0.00531742
8.85805700	0.27666670	0.00468014
9.01814200	0.28166660	0.00412003
9.17822800	0.28666670	0.00362763
9.33831400	0.29166670	0.00319465
9.49840000	0.29666670	0.00281384
9.65848400	0.30166670	0.00247884
9.81856900	0.30666670	0.00218407
9.97865500	0.31166670	0.00192466
10.13874000	0.31666670	0.00169631
10.29883000	0.32166670	0.00149527

TYPE 0, then goes on showing.

? 0

ZZZ4 = 0 mm

RRR4 (mm)	RRR4/wave length	EQ4
10.45891000	0.32666670	0.00131824
10.61900000	0.33166670	0.00116233
10.77908000	0.33666670	0.00102499
10.93917000	0.34166670	0.00090400
11.09925000	0.34666670	0.00079739
11.25934000	0.35166670	0.00070343
11.41942000	0.35666670	0.00062062
11.57951000	0.36166670	0.00054762
11.73960000	0.36666670	0.00048326
11.89968000	0.37166670	0.00042651
12.05977000	0.37666660	0.00037646
12.21985000	0.38166670	0.00033232
12.37994000	0.38666670	0.00029338
12.54002000	0.39166670	0.00025903

TYPE 0, please.

? 0

>

? 8

ZZZ1 = 1.2 mm, On the top of DR

RRR1 (mm)	RRR1/RA	EQ1
0.00000000	0.00000000	0.00000000
0.10000000	0.03278689	0.06137119
0.20000000	0.06557378	0.12231940
0.30000000	0.09836066	0.18242480
0.40000000	0.13114760	0.24127410
0.50000000	0.16393440	0.29846370
0.60000000	0.19672130	0.35360250
0.70000010	0.22950820	0.40631520
0.80000010	0.26229510	0.45624500
0.90000010	0.29508200	0.50305650
1.00000000	0.32786890	0.54643760
1.10000000	0.36065580	0.58610310
1.20000000	0.39344270	0.62179550

TYPE 0, then goes on showing.

? 0

ZZZ1 = 1.2 mm, on the top of DR

RRR1 (mm)	RRR1/RA	EQ1
1.10000000	0.36065580	0.58610310
1.20000000	0.39344260	0.62179530
1.30000000	0.42622960	0.65328730
1.40000000	0.45901650	0.68038360
1.50000000	0.49180330	0.70292220
1.60000000	0.52459020	0.72077510
1.70000000	0.55737710	0.73385060
1.80000000	0.59016400	0.74209150
1.90000000	0.62295100	0.74547760
2.00000000	0.65573780	0.74402500
2.10000000	0.68852470	0.73778550

TYPE 0, then goes on showing.

? 0

```
  ZZZ1 =  1.2  mm,  on the top of DR
  RRR1 (mm)    RRR1/RA      EQ1
 2.00000000  0.65573770  0.74402510
 2.10000000  0.68852460  0.73778550
 2.20000000  0.72131150  0.72684620
 2.30000000  0.75409830  0.71132880
 2.40000000  0.78688510  0.69138880
 2.50000000  0.81967200  0.66721310
 2.60000000  0.85245880  0.63902000
 2.69999900  0.88524570  0.60705590
 2.79999900  0.91803250  0.57159380
 2.89999900  0.95081940  0.53293090
 2.99999900  0.98360620  0.49138660

 TYPE 0, please.
? 0

 Want to calculate the electric fields :                          EQ1, EQ2,
EQ3, EQ4, EQ5, EQ6  ?

 IF EQ1 then TYPE  1,     IF EQ2 then TYPE  2,     IF EQ3 then TYPE  3,
 IF EQ4 then TYPE  4,     IF EQ5 then TYPE  5,     IF EQ6 then TYPE  6.

 Want to calculate the distribution of EQ4 some distance from the DR in radial
direction ?  If do then TYPE  7 .

 Want to calculate the distribution of EQ1 at the top of DR ?  If so
 then TYPE  8 .

 If do not want to calculate them, then TYPE N or n.

? n
```

```
1000 PRINT"  Programme DR20 "
1010 PRINT"  Calculate electric field, EQ, at any point in DR-substrate-
  box system. "
1020 PRINT"  The distance between DR and side wall of metal box is ralatively
     large; And there is only TE01d mode in DR system. "
1030 DIM A(16)
1040 DIM J0(16)
1050 PRINT
1060 PRINT"----------------------------------------"
1070 PRINT"                 + ZZZ               "
1080 PRINT"                   ^                "
1090 PRINT"  g                :  2           5  "
1100 PRINT"                   :                "
1110 PRINT"                   :                "
1120 PRINT"............***************............."
1130 PRINT"            *      :      *          "
1140 PRINT"  H          *   0 -------*------> RRR   "
1150 PRINT"            * DR     1   *          4  "
1160 PRINT"............***************............."
1170 PRINT"  h                  3           6  "
1180 PRINT"----------------------------------------"
1190 PRINT
1200 PRINT"  Input the resonance frequency, Fr, of DR system in GHz "
1210 PRINT"  ** The input of Fr must be correct ** "
1220 INPUT" Fr = ";FR
1230 PRINT"  Input the dielectric constant, Er, of DR (region 1) "
1240 INPUT" Er = ";ER
1250 PRINT"  Input the dielectric constant, Ea, of regions 2,4,5 "
1260 INPUT" Ea = ";EA
1270 PRINT"  Input the dielectric constant, Es, of substrate (regions 3,6) "
1280 INPUT" Es = ";ES
1290 PRINT"  ** Note : Er must be greater than Es & Ea ** "
1300 PRINT"  Input the radius, RA, of DR in mm "
1310 INPUT" RA = ";RA
1320 PRINT"  Input the height, H, of DR in mm "
1330 INPUT" H = ";H
1340 PRINT"  Choosing 1.5 < 2*RA/H < 2.5 is proper for obtaining higher Q
  & suppressing TM11d mode. "
1350 PRINT"  Sometimes, accurate results can not be obtained if 2*RA/H < 1.5 "
1360 PRINT"  Wait a minute, please. "
1370 PI=3.1415926535#
1380 GOSUB 6220
1390 PRINT
1400 PRINT" J1(KrA)/((KrA)*(J0(KrA)) = ";XX
1410 PRINT" K1(KaA)/((KaA)*(K0(KaA)) = ";SSK
1420 KRA=XXX
1430 PRINT" KrA = "KRA
1440 KAA=YYY
1450 PRINT" KaA = "KAA
1460 KR=XXX/RA
1470 PRINT" Kr  = "KR
1480 KA=YYY/RA
1490 PRINT" Ka  = "KA
1500 DD=(PI*FR/150)^2
1510 II=KR^2
1520 IF DD*ER-II<0 THEN 1620
1530 BB=SQR(DD*ER-II)
1540 PRINT" Bb  = "BB
1550 IF II-DD*EA<0 THEN 1620
1560 AA=SQR(II-DD*EA)
1570 PRINT" Aa  = "AA
1580 IF II-DD*ES<0 THEN 1620
1590 LD=SQR(II-DD*ES)
1600 PRINT" As  = "LD
1610 GOTO 1650
```

```
1620 PRINT
1630 PRINT"  Some parameters entered are wrong, check and change them,
      please. "
1640 GOTO 5990
1650 PRINT
1660 PRINT"  Input the thickness, h, of substrate in mm "
1670 INPUT" h = ";H1
1680 PRINT"  Input the distance, g, between the top of DR and upper metal
     shield in mm "
1690 INPUT" g = ";G
1700 IF G<.000001 THEN 1710 ELSE 1720
1710 G=.000001
1720 D3=G+H/2
1730 D1=H1+H/2
1740 PRINT
1750 PRINT"  Wait a minute, please. "
1760 X=KRA
1770 GOSUB 6840
1780 X=KAA
1790 GOSUB 7070
1800 AA4=J0X/K0X
1810 GOSUB 6050
1820 GOSUB 6120
1830 GOSUB 6170
1840 AA5=AA3*AA4
1850 AA6=AA2*AA4
1860 PRINT
1870 PRINT
1880 EQ1=.05
1890 RRR1=.6*RA
1900 ZZZ1=.1*H
1910 EQ1MAX=EQ1
1920 R10=RRR1
1930 Z10=ZZZ1
1940 SS=.01/FR
1950 IF R10>RA THEN 2570 ELSE 1960
1960 IF R10<0 THEN 2570 ELSE 1970
1970 IF ABS(Z10)>H/2 THEN 2570 ELSE 1980
1980 SS=SS
1990 RRR1=R10+SS
2000 ZZZ1=Z10
2010 GOSUB 7230
2020 IF EQ1MAX-EQ1<0 THEN 2030 ELSE 2180
2030 ZZZ1=Z10+SS
2040 GOSUB 7230
2050 IF EQ1MAX-EQ1<0 THEN 1910 ELSE 2060
2060 ZZZ1=Z10-SS
2070 GOSUB 7230
2080 IF EQ1MAX-EQ1<0 THEN 1910 ELSE 2090
2090 RRR1=R10+2*SS
2100 ZZZ1=Z10
2110 GOSUB 7230
2120 IF EQ1MAX-EQ1<0 THEN 1910 ELSE 2130
2130 ZZZ1=Z10
2140 RRR1=R10+SS
2150 GOSUB 7230
2160 EQ1MAX=EQ1
2170 GOTO 2460
2180 RRR1=R10-SS
2190 GOSUB 7230
2200 IF EQ1MAX-EQ1<0 THEN 2210 ELSE 2350
2210 ZZZ1=Z10+SS
2220 GOSUB 7230
2230 IF EQ1MAX-EQ1<0 THEN 1910 ELSE 2240
2240 ZZZ1=Z10-SS
2250 GOSUB 7230
2260 IF EQ1MAX-EQ1<0 THEN 1910 ELSE 2270
```

```
2270  RRR1=R10-2*SS
2280  ZZZ1=Z10
2290  GOSUB 7230
2300  IF EQ1MAX-EQ1<0 THEN 1910 ELSE 2310
2310  RRR1=R10-SS
2320  GOSUB 7230
2330  EQ1MAX=EQ1
2340  GOTO 2460
2350  ZZZ1=Z10+SS
2360  RRR1=R10
2370  GOSUB 7230
2380  IF EQ1MAX-EQ1<0 THEN 1910 ELSE 2390
2390  ZZZ1=Z10-SS
2400  GOSUB 7230
2410  IF EQ1MAX-EQ1<0 THEN 1910 ELSE 2420
2420  ZZZ1=Z10
2430  RRR1=R10
2440  GOSUB 7230
2450  EQ1MAX=EQ1
2460  PRINT"  The coordinate of the point where the field is the strongest : "
2470  PRINT
2480  PRINT" RRR1 = "RRR1" mm ;    ZZZ1 = "ZZZ1" mm "
2490  PRINT
2500  REM **  The strongest electric field (relative value) : EQ1MAX  **
2510  PRINT"  All of the electric fields calculated here are normalized with
   the strongest one (EQ1MAX). "
2520  PRINT
2530  PRINT"  TYPE  0, please. "
2540  PRINT
2550  INPUT A$
2560  IF A$="0" THEN 3120
2570  PRINT"  The search step is too large, start again with 1/2 smaller search
   step. "
2580  PRINT
2590  SS=.5*SS
2600  GOTO 1980
2610  PRINT
2620  ZZZ1=H/2
2630  PRINT"     ZZZ1 = "ZZZ1" mm,  On the top of DR "
2640  PRINT"     RRR1 (mm)    RRR1/RA      EQ1 "
2650  FOR RRR1=0 TO .4*RA STEP .1
2660  GOSUB 7230
2670  EQ1=EQ1/EQ1MAX
2680  PRINT USING "####.########";RRR1,RRR1/RA,EQ1
2690  NEXT RRR1
2700  PRINT
2710  PRINT"  TYPE  0, then goes on showing. "
2720  PRINT
2730  INPUT A$
2740  IF A$="0" THEN 2750
2750  PRINT"     ZZZ1 = "ZZZ1" mm,  on the top of DR "
2760  PRINT"     RRR1 (mm)    RRR1/RA      EQ1 "
2770  R1=.4*RA
2780  QQQ=R1*10
2790  QQ=INT(QQQ)
2800  R1=QQ/10
2810  R1=R1-.1
2820  FOR R1=R1 TO .7*RA STEP .1
2830  RRR1=R1
2840  GOSUB 7230
2850  EQ1=EQ1/EQ1MAX
2860  PRINT USING "####.########";RRR1,RRR1/RA,EQ1
2870  NEXT R1
2880  PRINT
2890  PRINT"  TYPE  0, then goes on showing. "
2900  PRINT
```

```
2910  INPUT A$
2920  PRINT
2930  IF A$="0" THEN 2940
2940  PRINT"     ZZZ1 = "ZZZ1" mm,  on the top of DR "
2950  PRINT"     RRR1 (mm)     RRR1/RA       EQ1 "
2960  R1=.7*RA
2970  PPP=R1*10
2980  PP=INT(PPP)
2990  R1=PP/10
3000  R1=R1-.1
3010  FOR R1=R1 TO RA STEP .1
3020  RRR1=R1
3030  GOSUB 7230
3040  EQ1=EQ1/EQ1MAX
3050  PRINT USING "####.########";RRR1,RRR1/RA,EQ1
3060  NEXT R1
3070  PRINT
3080  PRINT" TYPE 0, please."
3090  PRINT
3100  INPUT A$
3110  IF A$="0" THEN 3120
3120  PRINT
3130  PRINT" Want to calculate the electric fields :                    EQ1,
EQ2, EQ3, EQ4, EQ5, EQ6  ? "
3140  PRINT
3150  PRINT" IF EQ1 then TYPE  1, IF EQ2 then TYPE  2, IF EQ3 then TYPE  3,
        IF EQ4 then TYPE  4, IF EQ5 then TYPE  5, IF EQ6 then TYPE  6."
3160  PRINT
3170  PRINT" Want to calculate the distribution of EQ4 some distance from
    the DR in radial direction ?  If do then TYPE  7 . "
3180  PRINT
3190  PRINT" Want to calculate the distribution of EQ1 at the top of DR ?
    If do then TYPE  8 . "
3200  PRINT
3210  PRINT" If do not want to calculate them, then TYPE N or n. "
3220  PRINT
3230  INPUT A$
3240  IF A$="1" THEN 3330
3250  IF A$="2" THEN 5130
3260  IF A$="3" THEN 5450
3270  IF A$="4" THEN 4250
3280  IF A$="5" THEN 3930
3290  IF A$="6" THEN 3610
3300  IF A$="7" THEN 4540
3310  IF A$="8" THEN 2610
3320  IF A$="N" OR A$="n" THEN 5990
3330  PRINT
3340  PRINT" Input the radial coordinate, RRR1, of discussed point in DR in mm
3350  PRINT
3360  INPUT" RRR1 = ";RRR1
3370  IF RRR1>RA THEN 3380 ELSE 3410
3380  PRINT
3390  PRINT" RRR1 can not be > RA, check it and try again, please. "
3400  GOTO 3330
3410  PRINT
3420  PRINT" Input the longitudinal coordinate, ZZZ1, of discussed point in DR
    in mm "
3430  PRINT
3440  INPUT" ZZZ1 = ";ZZZ1
3450  IF ABS(ZZZ1)>H/2 THEN 3460 ELSE 3490
3460  PRINT
3470  PRINT" ABS(ZZZ1) can not be > H/2, check it and try again, please. "
3480  GOTO 3410
3490  GOSUB 7230
3500  PRINT
```

```
3510  PRINT" The relative amplitude of EQ1 (in DR) is : "
3520  PRINT
3530  PRINT" EQ1 = "EQ1/EQ1MAX
3540  PRINT
3550  PRINT" Want to calculate EQ1 at another point ? (Y or N) "
3560  PRINT
3570  INPUT A$
3580  PRINT
3590  IF A$="Y" OR A$="y" THEN 3330
3600  IF A$="N" OR A$="n" THEN 3120
3610  PRINT
3620  PRINT" Input the radial coordinate, RRR6, of discussed point in region 6
      in mm "
3630  PRINT
3640  INPUT" RRR6 = ";RRR6
3650  IF RRR6<RA THEN 3660 ELSE 3690
3660  PRINT
3670  PRINT" RRR6 can not be < RA, check it and try again, please. "
3680  GOTO 3610
3690  PRINT
3700  PRINT" Input the longitudinal coordinate, ZZZ6, of discussed point in
    region 6 in mm "
3710  PRINT
3720  INPUT" ZZZ6 = ";ZZZ6
3730  IF ZZZ6>(-H/2) THEN 3750 ELSE 3740
3740  IF ZZZ6<(-H/2-H1) THEN 3750 ELSE 3780
3750  PRINT
3760  PRINT" ZZZ6 can not be > (-H/2), ZZZ6 can not be < (-H/2 -h), check
       it and try again, please. "
3770  GOTO 3690
3780  X=LD*(D1+ZZZ6)
3790  GOSUB 6020
3800  X=KA*RRR6
3810  GOSUB 7150
3820  EQ6=-(2*K1X/KA)*AA5*EXP(-LD*D1)*SHX
3830  PRINT
3840  PRINT" The relative amplitude of EQ6 is : "
3850  PRINT
3860  PRINT" EQ6 = "EQ6/EQ1MAX
3870  PRINT
3880  PRINT" Want to calculate EQ6 at another point ? (Y or N) "
3890  PRINT
3900  INPUT A$
3910  IF A$="Y" OR A$="y" THEN 3610
3920  IF A$="N" OR A$="n" THEN 3120
3930  PRINT
3940  PRINT" Input the radial coordinate, RRR5, of discussed point in region 5
      in mm "
3950  PRINT
3960  INPUT" RRR5 = ";RRR5
3970  IF RRR5<RA THEN 3980 ELSE 4010
3980  PRINT
3990  PRINT" RRR5 can not be < RA, check it and try again, please. "
4000  GOTO 3930
4010  PRINT
4020  PRINT" Input the longitudial coordinate, ZZZ5, of discussed point in
    region 5 in mm "
4030  PRINT
4040  INPUT" ZZZ5 = ";ZZZ5
4050  IF ZZZ5<H/2 THEN 4070 ELSE 4060
4060  IF ZZZ5>G+H/2 THEN 4070 ELSE 4100
4070  PRINT
4080  PRINT" ZZZ5 can not be < H/2, ZZZ5 can not be > g+H/2,
    check it and try again, please. "
```

```
4090  GOTO 4010
4100  X=AA*(D3-ZZZ5)
4110  GOSUB 6020
4120  X=KA*RRR5
4130  GOSUB 7150
4140  EQ5=-2*(K1X/KA)*AA6*EXP(-AA*D3)*SHX
4150  PRINT
4160  PRINT" The relative amplitude of EQ5 is : "
4170  PRINT
4180  PRINT" EQ5 = "EQ5/EQ1MAX
4190  PRINT
4200  PRINT" Want to calculate EQ5 at another point ? (Y or N) "
4210  PRINT
4220  INPUT A$
4230  IF A$="Y" OR A$="y" THEN 3930
4240  IF A$="N" OR A$="n" THEN 3120
4250  PRINT
4260  PRINT" Input the radial coordinate, RRR4, of discussed point in region 4
   in mm "
4270  PRINT
4280  INPUT" RRR4 = ";RRR4
4290  IF RRR4<RA THEN 4300 ELSE 4330
4300  PRINT
4310  PRINT" RRR4 can not be < RA, check it and try again, please. "
4320  GOTO 4250
4330  PRINT
4340  PRINT" Input the longitudinal coordinate, ZZZ4, of discussed point in
   region 4 in mm "
4350  PRINT
4360  INPUT" ZZZ4 = ";ZZZ4
4370  IF ABS(ZZZ4)>H/2 THEN 4380 ELSE 4410
4380  PRINT
4390  PRINT" ABS(ZZZ4) can not be > H/2, check it and try again, please. '
4400  GOTO 4330
4410  X=KA*RRR4
4420  GOSUB 7150
4430  EQ4=-(K1X/KA)*AA4*(COS(BB*ZZZ4)+AA1*SIN(BB*ZZZ4))
4440  PRINT
4450  PRINT" The relative amplitude of EQ4 is : "
4460  PRINT
4470  PRINT" EQ4 = "EQ4/EQ1MAX
4480  PRINT
4490  PRINT" Want to calculate EQ4 at another point ? (Y or N) "
4500  PRINT
4510  INPUT A$
4520  IF A$="Y" OR A$="y" THEN 4250
4530  IF A$="N" OR A$="n" THEN 3120
4540  PRINT
4550  PRINT" Input the longitudinal coordinate, ZZZ4, of discussed plane
   in mm "
4560  PRINT
4570  INPUT" ZZZ4 = ";ZZZ4
4580  IF ABS(ZZZ4)>H/2 THEN 4590 ELSE 4620
4590  PRINT
4600  PRINT" ABS(ZZZ4) can not be > H/2, check it and try again, please "
4610  GOTO 4540
4620  PRINT
4630  PRINT"        ZZZ4 = "ZZZ4" mm "
4640  PRINT"        RRR4 (mm)  RRR4/wave length     EQ4 "
4650  DS=50/FR
4660  DSS=1.5/FR
4670  FOR N=0 TO 15 STEP 1
4680  RRR4=DS+N*DSS
4690  RW=RRR4*FR/300
4700  X=KA*RRR4
```

```
4710  GOSUB 7150
4720  EQ4=-(K1X/KA)*AA4*(COS(BB*ZZZ4)+AA1*SIN(BB*ZZZ4))
4730  EQ4=EQ4/EQ1MAX
4740  PRINT USING " ######.########";RRR4,RW,EQ4
4750  NEXT N
4760  PRINT
4770  PRINT" TYPE  0, then goes on showing. "
4780  PRINT
4790  INPUT A$
4800  IF A$="0" THEN 4810
4810  PRINT"        ZZZ4 = "ZZZ4" mm "
4820  PRINT"        RRR4 (mm)  RRR4/wave length     EQ4 "
4830  FOR N=16 TO 31 STEP 1
4840  RRR4=DS+N*DSS
4850  RW=RRR4*FR/300
4860  X=KA*RRR4
4870  GOSUB 7150
4880  EQ4=-(K1X/KA)*AA4*(COS(BB*ZZZ4)+AA1*SIN(BB*ZZZ4))
4890  EQ4=EQ4/EQ1MAX
4900  PRINT USING " ######.########";RRR4,RW,EQ4
4910  NEXT N
4920  PRINT
4930  PRINT" TYPE  0, then goes on showing. "
4940  PRINT
4950  INPUT A$
4960  IF A$="0" THEN 4970
4970  PRINT"        ZZZ4 = "ZZZ4" mm "
4980  PRINT"        RRR4 (mm)  RRR4/wave length     EQ4 "
4990  FOR N=32 TO 45 STEP 1
5000  RRR4=DS+N*DSS
5010  RW=RRR4*FR/300
5020  X=KA*RRR4
5030  GOSUB 7150
5040  EQ4=-(K1X/KA)*AA4*(COS(BB*ZZZ4)+AA1*SIN(BB*ZZZ4))
5050  EQ4=EQ4/EQ1MAX
5060  PRINT USING " ######.########";RRR4,RW,EQ4
5070  NEXT N
5080  PRINT
5090  PRINT" TYPE  0, please. "
5100  PRINT
5110  INPUT A$
5120  IF A$="0" THEN 3120
5130  PRINT
5140  PRINT" Input the radial coordinate, RRR2, of discussed point in region 2
   in mm "
5150  PRINT
5160  INPUT" RRR2 = ";RRR2
5170  IF RRR2>RA THEN 5180 ELSE 5210
5180  PRINT
5190  PRINT" RRR2 can not be > RA, check it and try again, please. "
5200  GOTO 5130
5210  PRINT
5220  PRINT" Input the longitudinal coordinate, ZZZ2, of discussed point in
  region 2 in mm "
5230  PRINT
5240  INPUT" ZZZ2 = ";ZZZ2
5250  IF ZZZ2<H/2 THEN 5270 ELSE 5260
5260  IF ZZZ2>G+H/2 THEN 5270 ELSE 5300
5270  PRINT
5280  PRINT" ZZZ2 can not be < H/2, ZZZ2 can not be > g+H/2,
  check it and try again, please. "
5290  GOTO 5210
5300  X=KR*RRR2
5310  GOSUB 6930
```

```
5320 X=AA*(D3-ZZZ2)
5330 GOSUB 6020
5340 EQ2=2*(J1X/KR)*AA2*EXP(-AA*D3)*SHX
5350 PRINT
5360 PRINT" The relative amplitude of EQ2 is : "
5370 PRINT
5380 PRINT" EQ2 = "EQ2/EQ1MAX
5390 PRINT
5400 PRINT" Want to calculate EQ2 at another point ? (Y or N) "
5410 PRINT
5420 INPUT A$
5430 IF A$="Y" OR A$="y" THEN 5130
5440 IF A$="N" OR A$="n" THEN 3120
5450 PRINT
5460 PRINT" Input the radial coordinate, RRR3, of discussed point in region 3
    in mm "
5470 PRINT
5480 INPUT" RRR3 = ";RRR3
5490 PRINT
5500 IF RRR3>RA THEN 5510 ELSE 5540
5510 PRINT
5520 PRINT" RRR3 can not be > RA, check it and try again, please. "
5530 GOTO 5450
5540 PRINT
5550 PRINT" Input the longitudinal coordinate, ZZZ3, of discussed point in
  region 3 in mm "
5560 PRINT
5570 INPUT" ZZZ3 = ";ZZZ3
5580 IF ZZZ3<(-H1-H/2) THEN 5600 ELSE 5590
5590 IF ZZZ3>(-H/2) THEN 5600 ELSE 5630
5600 PRINT
5610 PRINT" ZZZ3 can not be < (-h-H/2), ZZZ3 can not be > (-H/2),
  check it and try again, please. "
5620 GOTO 5540
5630 X=KR*RRR3
5640 GOSUB 6930
5650 X=LD*(D1+ZZZ3)
5660 GOSUB 6020
5670 EQ3=2*(J1X/KR)*AA3*EXP(-LD*D1)*SHX
5680 PRINT
5690 PRINT" The relative amplitude of EQ3 is : "
5700 PRINT
5710 PRINT" EQ3 = "EQ3/EQ1MAX
5720 PRINT
5730 PRINT" Want to calculate EQ3 at another point ? (Y or N) "
5740 PRINT
5750 INPUT A$
5760 IF A$="Y" OR A$="y" THEN 5450
5770 IF A$="N" OR A$="n" THEN 3120
5780 PRINT
5790 PRINT
5800 PRINT
5810 PRINT"   The calculation shows that the input does not meet the formulae
    of DR theory :"
5820 PRINT
5830 PRINT" 1) Ka^2 = (Er-Ea)*(2*PI*Fr/c)^2 - Kr^2
5840 PRINT
5850 PRINT" 2) J1(KrA)/(KrA*J0(KrA)) = - K1(KaA)/(KaA*K0(KaA)) "
5860 PRINT
5870 PRINT" Why ?   Perhaps : "
5880 PRINT
5890 PRINT"   The input of Fr is not correct. Check and change it and try
  again, TYPE  T, please. "
```

```
5900 PRINT"  If the input of Fr is correct. TYPE U (and then TYPE 'Enter')
   for few times , please. "
5910 REM **  Change DEL < .001  into  DEL < .001 + .004 * N  **
5920 INPUT A$
5930 IF A$="T" OR A$="t" THEN 1050
5940 IF A$="U" OR A$="u" THEN 5950
5950 PRINT"  Try again with lower accuracy. "
5960 DDD=DDD+.004
5970 IF DDD>.015 THEN 5980 ELSE 6260
5980 PRINT"  The accurate result can not be obtained in 1.5 % accuracy. "
5990 END
6000 PRINT
6010 PRINT
6020 REM ** SH(X) **
6030 SHX=.5*(EXP(X)-EXP(-X))
6040 RETURN
6050 REM ** AA1 **
6060 ASD=LD*H1
6070 THASD=(EXP(ASD)-EXP(-ASD))/(EXP(ASD)+EXP(-ASD))
6080 BTHASD=BB*THASD
6090 BTHTG=BTHASD*TAN(BB*H/2)
6100 AA1=(LD-BTHTG)/(LD*TAN(BB*H/2)+BTHASD)
6110 RETURN
6120 REM ** AA2 **
6130 AADHD=AA*G
6140 SHAAD=.5*(EXP(AADHD)-EXP(-AADHD))
6150 AA2=(COS(BB*H/2)+AA1*SIN(BB*H/2))/(2*EXP(-AA*D3)*SHAAD)
6160 RETURN
6170 REM ** AA3 **
6180 ASDHDD=LD*H1
6190 SHASD=.5*(EXP(ASDHDD)-EXP(-ASDHDD))
6200 AA3=(COS(BB*H/2)-AA1*SIN(BB*H/2))/(2*EXP(-LD*D1)*SHASD)
6210 RETURN
6220 REM ** J1(KrA)/(KrA*J0(KrA))=-K1(KaA)/(KaA*K0(KaA)), to find KrA(=XXX),
       KaA(=YYY), Kr(=KR), Ka(=KA), Bb(=BB), Aa(=AA), As(=AS) **
6230 REM ** J1(KrA)=J1 **
6240 REM ** A=RA **
6250 DDD=.001
6260 DDD=DDD
6270 PRINT" DEL = "DDD
6280 XXX=2.7
6290 XXX=XXX
6300 XXX1=(XXX/2)^2
6310 S=1
6320 J=1
6330 T=1
6340 K=-1
6350 T=T*XXX1/J/(J+1)
6360 S=S+K*T
6370 IF T<.000001 THEN 6410
6380 K=-K
6390 J=J+1
6400 GOTO 6350
6410 J1=(XXX/2)*S
6420 REM ** J0(KrA)=J0 **
6430 A(1)=4
6440 J0(1)=1-(XXX^2)/4
6450 FOR M=1 TO 15 STEP 1
6460 A(M+1)=A(M)*((2*(M+1))^2)
6470 J0(M+1)=J0(M)+((-1)^(M+1))*(XXX^(2*(M+1)))/A(M+1)
6480 NEXT M
6490 J0=J0(16)
6500 XX=J1/(XXX*J0)
```

```
6510 REM ** K1(Y), FOR Y > 0.8 **
6520 REM ** K1(KaA)=K **
6530 MNN=(ER-EA)*((PI*FR*RA)/150)^2
6540 MPP=XXX^2
6550 IF MNN-MPP<0 THEN 5790
6560 YYY=SQR(MNN-MPP)
6570 AAA=SQR(PI/(2*YYY))
6580 BBB=AAA/EXP(YYY)
6590 C5=1+.375/YYY
6600 D5=C5-.1171875/YYY^2
6610 K=BBB*D5*(1+.037/(YYY^3))
6620 REM ** K0(Y), FOR Y > 0.8 **
6630 REM ** K0(KaA)=SK **
6640 AAAA=SQR(PI/(2*YYY))
6650 BBB=AAAA/EXP(YYY)
6660 C6=1-.125/YYY
6670 D6=C6+.0703125/YYY^2
6680 SK=BBB*D6*(1-.037/(YYY^3))
6690 SSK=K/(YYY*SK)
6700 DEL=ABS(XX+SSK)/ABS(XX-SSK)
6710 IF DEL<DDD THEN 6830 ELSE 6720
6720 IF DEL>.05 THEN 6730 ELSE 6750
6730 XXX=XXX+.01
6740 GOTO 6290
6750 IF DEL>.01 THEN 6760 ELSE 6780
6760 XXX=XXX+.005
6770 GOTO 6290
6780 IF DEL>.005 THEN 6790 ELSE 6810
6790 XXX=XXX+.002
6800 GOTO 6290
6810 XXX=XXX+.001
6820 GOTO 6290
6830 RETURN
6840 REM ** J0(X) **
6850 A(1)=4
6860 J0(1)=1-(X^2)/4
6870 FOR I=1 TO 15 STEP 1
6880 A(I+1)=A(I)*((2*(I+1))^2)
6890 J0(I+1)=J0(I)+((-1)^(I+1))*(X^(2*(I+1)))/A(I+1)
6900 NEXT I
6910 J0X=J0(16)
6920 RETURN
6930 REM ** J1X **
6940 X1=(X/2)^2
6950 S=1
6960 J=1
6970 T=1
6980 K=-1
6990 T=T*X1/J/(1+J)
7000 S=S+K*T
7010 IF T<.000001 THEN 7050
7020 K=-K
7030 J=J+1
7040 GOTO 6990
7050 J1X=(X/2)*S
7060 RETURN
7070 REM ** K0(X), FOR X > 0.8 **
7080 A7=SQR(PI/(2*X))
7090 B7=A7/EXP(X)
7100 C7=1-.125/X
7110 D7=C7+.0703125/X^2
7120 K=B7*D7*(1-.037/(X^3))
```

```
7130 K0X=K
7140 RETURN
7150 REM ** K1(X), FOR X > 0.8 **
7160 A8=SQR(PI/(2*X))
7170 B8=A8/EXP(X)
7180 C8=1+.375/X
7190 D8=C8-.1171875/X^2
7200 K=B8*D8*(1+.037/(X^3))
7210 K1X=K
7220 RETURN
7230 REM ** EQ1 **
7240 X=KR*RRR1
7250 GOSUB 6930
7260 EQ1=(J1X/KR)*(COS(BB*ZZZ1)+AA1*SIN(BB*ZZZ1))
7270 RETURN
```

Calculation example of DR40

```
    Programme DR40
    Calculate the radius & height of DR in a substrate-box
  system.
    There is only TE01d mode in the DR system.
    Choosing 1.5 < 2*RA/H < 2.5 is proper for obtaining
  higher Q & suppressing TM11d mode.

 -----------------------------------------


   g                       2             5

 ............***************.............
   H          * DR            *
              *             1 *          4
 ............***************.............
   h                       3             6
 -----------------------------------------


    Input the frequency wanted, F0, in GHz
  F0 = ? 9
    Input the dielectric constant, Er, of DR (region 1)
  Er = ? 40
    Input the dielectric constant, Ea, of regions 2,4,5
  Ea = ? 1
    Input the dielectric constant, Es, of substrate (regions 3,6)
  Es = ? 2.5
    ** Note : Er must be greater than Es & Ea **
    Input the proportion, P=2*RA/H, of diameter to height of DR
  2*RA/H = ? 2
    Input the thickness, h, of substrate in mm
  h = ? .8
    Input the distance, g, between the top of DR and upper metal
    shield in mm
  g = ? 4


  Calculation results :

  The resonance frequency of DR system is : Fr =  8.999849   GHz

  The radius of DR is : RA =  2.890939   mm
  The height of DR is : H  =  2.890939    mm
```

```
LIST

1000 PRINT"   Programme DR40  "
1010 PRINT"   Calculate the radius & height of DR in a substrate-box
          system. "
1020 PRINT"   There is only TE01d mode in the DR system. "
1030 PRINT"   Choosing 1.5 < 2*RA/H < 2.5 is proper for obtaining
          higher Q & suppressing TM11d mode. "
1040 DIM NUM(6),DEN(6),XX(3)
1050 DIM KK(3),FCT(3),GCT(3)
1060 PRINT
1070 PRINT"-----------------------------------------"
1080 PRINT
1090 PRINT
1100 PRINT"  g                   2              5  "
1110 PRINT
1120 PRINT".............***************.............."
1130 PRINT"  H           * DR          *              "
1140 PRINT"              *           1 *         4  "
1150 PRINT".............***************.............."
1160 PRINT"  h                   3              6  "
1170 PRINT"-----------------------------------------"
1180 PRINT
1190 PRINT"   Input the frequency wanted, F0, in GHz  "
1200 INPUT" F0 = ";F0
1210 PRINT"   Input the dielectric constant, Er, of DR (region 1) "
1220 INPUT" Er = ";ER
1230 PRINT"   Input the dielectric constant, Ea, of regions 2,4,5 "
1240 INPUT" Ea = ";EA
1250 PRINT"   Input the dielectric constant, Es, of substrate (regions 3,6) "
1260 INPUT" Es = ";ES
1270 PRINT"   ** Note : Er must be greater than Es & Ea ** "
1280 PRINT"   Input the proportion, P=2*RA/H, of diameter to height of DR "
1290 INPUT" 2*RA/H = ";P
1300 PRINT"   Input the thickness, h, of substrate in mm "
1310 INPUT" h = ";H1
1320 PRINT"   Input the distance, g, between the top of DR and upper metal
           shield in mm "
1330 INPUT" g = ";G
1340 IF G<.000001 THEN 1350 ELSE 1360
1350 G=.000001
1360 PRINT
1370 VVR=SQR(ER-EA)
1380 VVS=14.5/SQR(ES)
1390 RA=((130/VVR)+VVS)/F0
1400 RA=RA
1410 H=2*RA/P
1420 NQ=0
1430 XX(2)=2.9
1440 KMIN=XX(2)/SQR(ER-EA)
1450 KMAX=XX(2)/SQR(ES)
```

```
1460 KK(2)=(9*KMIN+KMAX)/10
1470 DXX=.00001
1480 DKK=.00001
1490 ITER=0
1500 PRINT
1510 XX(1)=XX(2)+DXX
1520 KK(1)=KK(2)
1530 XX(3)=XX(2)
1540 KK(3)=KK(2)+DKK
1550 FOR ITI=1 TO 3
1560 X=XX(ITI)
1570 K0=KK(ITI)
1580 K02=K0^2
1590 XIT2=X^2
1600 GOSUB 3190
1610 RRA=K02*(ER-EA)-XIT2
1620 IF RRA>0 THEN 1690
1630 WSTEPX=WSTEPX/2
1640 WSTEPK=WSTEPK/2
1650 XX(2)=XX(2)-WSTEPX
1660 KK(2)=KK(2)-WSTEPK
1670 PRINT" Start again with 1/2 smaller step.
           Wait a minute, please. "
1680 GOTO 1490
1690 YY=SQR(RRA)
1700 KC4A=YY
1710 GOSUB 3250
1720 FCT(ITI)=J0B+YY*K0B/X
1730 BA=SQR(K02*ER-XIT2)
1740 ALF1=SQR(XIT2-K02*ES)
1750 POW=ALF1*H1/RA
1760 IF POW>8 THEN 1810
1770 EP=EXP(POW)
1780 EI=1/EP
1790 AGU=(EP+EI)/(EP-EI)
1800 GOTO 1820
1810 AGU=1
1820 AGU=AGU*ALF1/BA
1830 FIH1=ATN(AGU)
1840 ALF2=SQR(XIT2-K02*EA)
1850 POW=ALF2*G/RA
1860 IF POW>8 THEN 1910
1870 EP=EXP(POW)
1880 EI=1/EP
1890 AGU=(EP+EI)/(EP-EI)
1900 GOTO 1920
1910 AGU=1
1920 AGU=AGU*ALF2/BA
1930 FIH2=ATN(AGU)
1940 GCT(ITI)=FIH1+FIH2-BA*H/RA
1950 NEXT ITI
1960 AL=(FCT(1)-FCT(2))/DXX
1970 AU=(GCT(1)-GCT(2))/DKK
1980 BL=(FCT(3)-FCT(2))/DXX
1990 BU=(GCT(3)-GCT(2))/DKK
2000 CL=FCT(2)-AL*XX(2)-BL*KK(2)
2010 CU=GCT(2)-AU*XX(2)-BU*KK(2)
2020 DENO=AU*BL-AL*BU
2030 XNEW=(CL*BU-CU*BL)/DENO
```

```
2040 KNEW=(CU*AL-CL*AU)/DENO
2050 WSTEPX=XNEW-XX(2)
2060 WSTEPK=KNEW-KK(2)
2070 WSTEP2=WSTEPX^2+WSTEPK^2
2080 XX(2)=XNEW
2090 KK(2)=KNEW
2100 IF WSTEP2<1E-12 THEN 2160
2110 ITER=ITER+1
2120 IF ITER>10 THEN 2140
2130 GOTO 1510
2140 PRINT" Solution not found after 10 iterations "
2150 GOTO 3180
2160 K0A=KK(2)
2170 PI=3.1415926535#
2180 FIR=K0A*150/(PI*RA)
2190 EIGX=XX(2)
2200 K0A2=K0A^2
2210 EIG2=EIGX^2
2220 WRADIC=EIG2-K0A2*ES
2230 AL1A=SQR(WRADIC)
2240 WRADI=EIG2-K0A2*EA
2250 AL2A=SQR(WRADI)
2260 WRADA=K0A2*ER-EIG2
2270 BA=SQR(WRADA)
2280 AL1L1=AL1A*H1/RA
2290 AL2L2=AL2A*G/RA
2300 IF AL1L1>8 THEN 2360
2310 Z1=EXP(AL1L1)
2320 ZI1=1/Z1
2330 CT1=(Z1+ZI1)/(Z1-ZI1)
2340 SIH1=(Z1-ZI1)*.5
2350 GOTO 2370
2360 CT1=1
2370 IF AL2L2>8 THEN 2430
2380 Z2=EXP(AL2L2)
2390 ZI2=1/Z2
2400 CT2=(Z2+ZI2)/(Z2-ZI2)
2410 SIH2=(Z2-ZI2)*.5
2420 GOTO 2440
2430 CT2=1
2440 ARG1=AL1A*CT1/BA
2450 ARG2=AL2A*CT2/BA
2460 TH1=ATN(ARG1)
2470 TH2=ATN(ARG2)
2480 KC4A2=WRADA-K0A2*EA
2490 KC4A=SQR(KC4A2)
2500 GOSUB 3250
```

```
2510 X=EIGX
2520 GOSUB 3190
2530 J0B2=J0B^2
2540 TRX=J0B2-2*J0B/EIGX+1
2550 K0B2=K0B^2
2560 PRX=K0B2+2*K0B/KC4A-1
2570 SIF1=SIN(TH1*2)
2580 SIF2=SIN(TH2*2)
2590 THET=(SIF1+SIF2)*.5/(TH1+TH2)
2600 C012=RA*(COS(TH1)^2)/AL1A
2610 IF AL1L1>8 THEN 2650
2620 SECN=AL1L1/(SIH1^2)
2630 PARM1=CT1-SECN
2640 GOTO 2660
2650 PARM1=1
2660 COPAM1=C012*PARM1
2670 C022=RA*(COS(TH2)^2)/AL2A
2680 IF AL2L2>8 THEN 2720
2690 SECN=AL2L2/(SIH2^2)
2700 PARM2=CT2-SECN
2710 GOTO 2730
2720 PARM2=1
2730 COPAM2=C022*PARM2
2740 NUM(1)=ES*COPAM1*TRX
2750 NUM(2)=EA*COPAM2*TRX
2760 NUM(3)=-(WRADIC+KC4A2)*PRX*COPAM1/K0A2
2770 NUM(4)=EA*H*(1+THET)*PRX
2780 NUM(5)=-(WRADI+KC4A2)*PRX*COPAM2/K0A2
2790 NUM(6)=ER*H*(1+THET)*TRX
2800 DEN(1)=NUM(1)
2810 DEN(2)=NUM(2)
2820 DEN(3)=ES*PRX*COPAM1
2830 DEN(4)=NUM(4)
2840 DEN(5)=EA*PRX*COPAM2
2850 DEN(6)=NUM(6)
2860 DENSUM=0
2870 SURVER=-BA*RA*(SIF1+SIF2)*PRX/K0A2
2880 SURHOR=X*J0B*2*(COPAM1+COPAM2)/K0A2
2890 NUMSUM=SURHOR+SURVER
2900 FOR J=1 TO 6
2910 DENSUM=DENSUM+DEN(J)
2920 NUMSUM=NUMSUM+NUM(J)
2930 NEXT J
2940 VARK0A=K0A*SQR(NUMSUM/DENSUM)
2950 VARFRE=VARK0A*150/(PI*RA)
2960 PRINT
2970 PRINT
2980 FR=(FIR+VARFRE)*.5
2990 PRINT
3000 IF ABS((FR-F0)/F0)>.0001 THEN 3010 ELSE 3030
```

```
3010 RA=RA+((FR-F0)/F0)*RA
3020 GOTO 1400
3030 PRINT" Input : "
3040 PRINT
3050 PRINT" Er =      "ER
3060 PRINT" Ea =      "EA
3070 PRINT" Es =      "ES
3080 PRINT" 2*RA/H = "P
3090 PRINT" h  =      "H1"  mm  "
3100 PRINT" g  =      "G "  mm  "
3110 PRINT
3120 PRINT" Calculation results : "
3130 PRINT
3140 PRINT" The resonance frequency of DR system is : Fr = "FR"  GHz "
3150 PRINT
3160 PRINT" The radius of DR is : RA = "RA"  mm "
3170 PRINT" The height of DR is : H  = "H"   mm "
3180 END
3190 REM ** function J0B=J0(X)/J1(X) **
3200 XMX0=X-2.4048
3210 TEM=(.0282*XMX0-.1177)*XMX0+.2571
3220 TEM=(TEM*XMX0-.716)*XMX0+1.4282
3230 J0B=TEM*XMX0/(X-3.8317)
3240 RETURN
3250 REM ** function K0B=K0(KC4A)/K1(KC4A) **
3260 KI=1/KC4A
3270 TEM=(.00445*KI-.02679)*KI+.06539
3280 TEM=(TEM*KI-.11226)*KI+.49907
3290 K0B=1/(1+TEM*KI)
3300 RETURN
```

Calculation example of DR100

```
    Programme DR100
    Calculate  coupling parameters of DR coupled with a
  microstrip in a metal box.
    There is only TE01d mode in the DR system.

----------------------------------------

  g                      2             5

............***************.............
  H          * DR          *
             *           1 *           4
............***************.............
  h                      3             6
----------------------------------------

   Input the resonance frequency, Fr, of DR system in GHz
   ** The input of Fr must be correct **
 Fr = ? 9.37
   Input the dielectric constant, Er, of DR (region 1)
 Er = ? 38.9
   Input the dielectric constant, Ea, of regions 2,4,5
 Ea = ? 1
   Input the dielectric constant, Es, of substrate (regions 3,6)
 Es = ? 2.5
   ** Note : Er must be greater than Es & Ea **
   Input the radius, RA, of DR in mm
 RA = ? 3.05
   Input the height, H, of DR in mm
 H = ? 2.4

   Choosing 1.5 < 2*RA/H < 2.5 is proper for obtaining higher Q
 & suppressing TM11d mode.
   Sometimes, accurate results can not be obtained if 2*RA/H < 1.5
   Wait a minute, please.
 DEL =  .001

 J1(KrA)/((KrA)*(J0(KrA)) = -.536442
 K1(KaA)/((KaA)*(K0(KaA)) =  .5362416
 KrA =  2.925
 KaA =  2.241061
 Kr =  .9590166
 Ka =  .7347742
 Bb =  .7605295
 Aa =  .9387229
 As =  .9074319
   Input the width, W, of microstrip in mm
 W = ? 2.25
```

```
  Input the thickness, h, of substrate in mm
h = ? .7874
  Input the distance, g, between the top of DR and upper metal
  shield in mm
g = ? 4
  Input the distance, d, between the centre of DR and edge of
  microstrip in mm

  **  Be sure that d >= RA
  **  In practice d < 300/(NN*Fr) -- d in mm ; Fr in GHz ,
 if want to calculate the case with a larger d, it is easy to
 change NN in line 1830  (where NN=3).

d = ? 3.05


Energy stored in region 1 in % =   98.66316
Energy stored in region 2 in % =   .4156048
Energy stored in region 3 in % =   .1816726
Energy stored in region 4 in % =   .5985928
Energy stored in region 5 in % =   .0980859
Energy stored in region 6 in % =   4.287612E-02

  Input the impedance, Zo, of microstrip coupled with DR in ohm
Zo = ? 50
  The external quality factor Qext
Qext =  251.3628


  Input the unloaded quality factor Qu
Qu = ? 2670

  The equivalent inductannce, Lr, of DR in HENRY
Lr =  9.220521E-08
  The equivalent capacitance, Cr, of DR in FARAD
Cr =  3.129E-15
  The equivalent resistance, Rr, of DR in Ohm
Rr =  2.033123

  TYPE 0, please.  Then goes on showing.
? 0

  The equivalent mutual inductance, Lm, of DR coupled with a
  microstrip in HENRY (for double loading)
Lm =  7.893463E-10
  The coupling factor Bc
Bc =  10.6221
  The loaded quality factor QL
QL =  229.7348
```

```
   TYPE  0, please. Then goes on showing.
? 0
   The equivalent inductance, L, of DR coupled with a microstrip
   in HENRY
 L =  6.7574E-12
   The equivalent capacitance, C, of DR coupled with a microstrip
   in FARAD
 C =  4.269543E-11
   The equivalent resistance, R, of DR coupled with a microstrip
   in Ohm
 R =  1062.21

 TYPE  0, please. Then goes on showing.
? 0

 Neglecting the reflection from load

 S parameter

 (ABS(S11))^2 = normalized reflection power from * DR junction *

 (ABS(S21))^2 = normalized transmission power into load

 (ABS(S11))^2 =  .8353174

 (ABS(S21))^2 =  7.403398E-03

 (ABS(S11))^2 + (ABS(S21))^2 =  .8427208

  Want to input another d, or Qu for calculation ? (Y or N)
? n
```

```
LIST

1000 PRINT"    Programme DR100 "
1010 PRINT"    Calculate  coupling parameters of DR coupled with a
           microstrip in a metal box. "
1020 PRINT"    There is only TE01d mode in the DR system. "
1030 DIM A(16)
1040 DIM J0(16)
1050 DIM XSX(40)
1060 PRINT
1070 PRINT"-----------------------------------------"
1080 PRINT
1090 PRINT
1100 PRINT"  g                    2             5  "
1110 PRINT
1120 PRINT"............**************............."
1130 PRINT"  H          * DR         *              "
1140 PRINT"             *          1 *          4  "
1150 PRINT"............**************............."
1160 PRINT"  h                    3             6  "
1170 PRINT"-----------------------------------------"
1180 PRINT
1190 PRINT"   Input the resonance frequency, Fr, of DR system in GHz "
1200 PRINT"   ** The input of Fr must be correct ** "
1210 INPUT" Fr = ";FR
1220 PRINT"   Input the dielectric constant, Er, of DR (region 1) "
1230 INPUT" Er = ";ER
1240 PRINT"   Input the dielectric constant, Ea, of regions 2,4,5 "
1250 INPUT" Ea = ";EA
1260 PRINT"   Input the dielectric constant, Es, of substrate (regions 3,6) "
1270 INPUT" Es = ";ES
1280 PRINT"   ** Note : Er must be greater than Es & Ea ** "
1290 PRINT"   Input the radius, RA, of DR in mm "
1300 INPUT" RA = ";RA
1310 PRINT"   Input the height, H, of DR in mm "
1320 INPUT" H = ";H
1330 PRINT"   Choosing 1.5 < 2*RA/H < 2.5 is proper for obtaining higher Q
          & suppressing TM11d mode. "
1340 PRINT"   Sometimes, accurate results can not be obtained if 2*RA/H < 1.5 "
1350 PRINT"   Wait a minute, please. "
1360 REM ** The effective interaction radius, RB, of DR is chosen to be 5*RA **
1370 RB=5*RA
1380 PI=3.1415926535#
1390 GOSUB 4350
1400 PRINT
1410 PRINT" J1(KrA)/((KrA)*(J0(KrA)) = ";XX
1420 PRINT" K1(KaA)/((KaA)*(K0(KaA)) = ";SSK
1430 KRA=XXX
1440 PRINT" KrA = "KRA
1450 KAA=YYY
1460 PRINT" KaA = "KAA
1470 KR=XXX/RA
1480 PRINT" Kr = "KR
1490 KA=YYY/RA
1500 PRINT" Ka = "KA
1510 DD=(PI*FR/150)^2
1520 II=KR^2
1530 IF DD*ER-II<0 THEN 1630
1540 BB=SQR(DD*ER-II)
1550 PRINT" Bb = "BB
1560 IF II-DD*EA<0 THEN 1630
```

```
1570 AA=SQR(II-DD*EA)
1580 PRINT" Aa = "AA
1590 IF II-DD*ES<0 THEN 1630
1600 LD=SQR(II-DD*ES)
1610 PRINT" As = "LD
1620 GOTO 1660
1630 PRINT
1640 PRINT"   Some parameters entered are wrong, check and change them
          please, input again. "
1650 GOTO 4120
1660 PRINT"   Input the width, W, of microstrip in mm "
1670 INPUT" W = ";W
1680 PRINT"   Input the thickness, h, of substrate in mm "
1690 INPUT" h = ";H1
1700 PRINT"   Input the distance, g, between the top of DR and upper metal
            shield in mm "
1710 INPUT" g = ";G
1720 IF G<.000001 THEN 1730 ELSE 1740
1730 G=.000001
1740 PRINT"   Input the distance, d, between the centre of DR and edge of
            microstrip in mm "
1750 PRINT
1760 PRINT"   **  Be sure that d >= RA  "
1770 PRINT"   **  In practice d < 300/(NN*Fr) -- d in mm ; Fr in GHz ,
           if want to calculate the case with a larger d, it is easy to
           change NN in line 1830  (where NN=3). "
1780 PRINT
1790 INPUT" d = ";D
1800 REM ** The effective interaction length of microstrip is
          2*SQR(((300/NN*Fr))^2)-(d^2)), which is divided
          into 2*M parts in X direction, M = 40 **
1810 D3=G+H/2
1820 D1=H1+H/2
1830 NN=3
1840 DDX=SQR(((300/(NN*FR))^2)-(D^2))*(1/40)
1850 XSX(1)=DDX/2
1860 DDWM=0
1870 FOR M=1 TO 40 STEP 1
1880 XSX(M)=XSX(1)+(M-1)*DDX
1890 ADD=SQR((XSX(M)^2)+(D+W/2)^2)
1900 X=KA*ADD
1910 GOSUB 5360
1920 DDWM=DDWM+K1X*DDX/ADD
1930 NEXT M
1940 ABC=DDWM^2
1950 X=KRA
1960 GOSUB 4970
1970 X=KAA
1980 GOSUB 5280
1990 AA4=J0X/K0X
2000 AAA4=AA4^2
2010 GOSUB 4150
2020 AAA1=AA1^2
2030 GOSUB 4240
```

```
2040 AAA2=AA2^2
2050 GOSUB 4290
2060 AAA3=AA3^2
2070 AAA5=AAA3*AAA4
2080 AAA6=AAA2*AAA4
2090 I=1
2100 X=KRA
2110 GOSUB 5060
2120 J1X=JIX
2130 GOSUB 4970
2140 J0X=J0(16)
2150 I=2
2160 GOSUB 5060
2170 J2X=JIX
2180 ZWZ=((RA/KR)^2)*((J1X^2)-J0X*J2X)
2190 SBH=(SIN(BB*H))/(BB*H)
2200 SBBH=(H*ER/2)*(1+SBH+AAA1*(1-SBH))
2210 X=2*AA*G
2220 GOSUB 5560
2230 SHXA=SHX
2240 SBBI=(EA*AAA2/AA)*EXP(-2*AA*D3)*(SHXA-2*AA*G)
2250 X=2*LD*H1
2260 GOSUB 5560
2270 SHXS=SHX
2280 SBBJ=(ES*AAA3/LD)*EXP(-2*LD*D1)*(SHXS-2*LD*H1)
2290 SBBK=(EA*AAA4*H/(KA^2))*(1+SBH+AAA1*(1-SBH))
2300 SBBL=(4*EA*AAA6/(KA^2))*EXP(-2*AA*D3)*((1/(2*AA))*SHXA-G)
2310 SBBM=(4*ES*AAA5/(KA^2))*EXP(-2*LD*D1)*((1/(2*LD))*SHXS-H1)
2320 X=KA*RB
2330 GOSUB 5360
2340 K1X=K
2350 GOSUB 5280
2360 K0X=K
2370 BMM=(RB^2)*((K1X^2)-(K0X^2)-2*(K0X*K1X)/X)
2380 X=KA*RA
2390 GOSUB 5360
2400 K1X=K
2410 GOSUB 5280
2420 K0X=K
2430 BMN=(RA^2)*((K1X^2)-(K0X^2)-2*(K0X*K1X)/X)
2440 ZZ1=ZWZ*SBBH
2450 ZZ2=ZWZ*SBBI
2460 ZZ3=ZWZ*SBBJ
2470 BMM1=.5*(BMM-BMN)
2480 ZZ4=SBBK*BMM1
2490 ZZ5=SBBL*BMM1
2500 ZZ6=SBBM*BMM1
```

```
2510 ZZ=ZZ1+ZZ2+ZZ3+ZZ4+ZZ5+ZZ6
2520 PRINT
2530 W1=ZZ1/ZZ
2540 PRINT" Energy stored in region 1 in % = "W1*100
2550 W2=ZZ2/ZZ
2560 PRINT" Energy stored in region 2 in % = "W2*100
2570 W3=ZZ3/ZZ
2580 PRINT" Energy stored in region 3 in % = "W3*100
2590 W4=ZZ4/ZZ
2600 PRINT" Energy stored in region 4 in % = "W4*100
2610 W5=ZZ5/ZZ
2620 PRINT" Energy stored in region 5 in % = "W5*100
2630 W6=ZZ6/ZZ
2640 PRINT" Energy stored in region 6 in % = "W6*100
2650 PRINT
2660 PRINT"   Input the impedance, Zo, of microstrip coupled with DR in ohm "
2670 INPUT" Zo = ";Z0
2680 X=LD*H1
2690 GOSUB 5560
2700 FFAK=SHX^2
2710 LJ1=ZZ/ABC
2720 E0=8.854187818#*(1E-15)
2730 LJ2=Z0*(PI^2)*FR*(1E+09)*E0*(KA^2)*EXP(2*LD*D1)
2740 LJ3=4*AAA5*((D+W/2)^2)*FFAK
2750 QEXT=LJ1*LJ2/LJ3
2760 PRINT"   The external quality factor Qext "
2770 PRINT" Qext = "QEXT
2780 IF QEXT>2000 THEN 2790 ELSE 2850
2790 PRINT
2800 PRINT" In practice case, the coupling between DR and microstrip is strong.
            But here QEXT > 2000, the coupling is too weak. In this case the
            radiation loss, dielectric losses and metal loss have to be taken "
2810 PRINT" into account. But this is not the purpose of this programme. "
2820 PRINT" If you want to go on calculation, then TYPE  0 "
2830 INPUT A$
2840 IF A$="0" THEN 2850
2850 I=2
2860 X=KRA
2870 GOSUB 5060
2880 J2X=JIX
2890 GGPP=J2X/(KR^2)
2900 GUO=(ER*PI*(RA^2)/BB)*SIN(BB*H/2)
2910 X=AA*G
2920 GOSUB 5530
2930 CHX1=CHX
2940 GUP=(EA*4*PI*(RA^2)*AA2/AA)*(CHX1-1)*EXP(-AA*D3)
2950 X=LD*H1
2960 GOSUB 5530
2970 CHX2=CHX
2980 GUQ=(ES*4*PI*(RA^2)*AA3/LD)*(CHX2-1)*EXP(-LD*D1)
2990 GUR=(EA*4*PI*AA4/BB)*SIN(BB*H/2)
3000 AA6=AA2*AA4
```

```
3010 GUS=(EA*4*PI*AA6/AA)*(CHX1-1)*EXP(-AA*D3)
3020 AA5=AA3*AA4
3030 GUT=(ES*4*PI*AA5/LD)*(CHX2-1)*EXP(-LD*D1)
3040 X=KA*RB
3050 GOSUB 5440
3060 K2XB=K2X
3070 G1=((KA*RB)^2)*K2XB
3080 X=KA*RA
3090 GOSUB 5440
3100 K2XA=K2X
3110 G2=((KA*RA)^2)*K2XA
3120 G12=(G1-G2)/(KA^4)
3130 GOO1=GGPP*GUO
3140 GOO2=GGPP*GUP
3150 GOO3=GGPP*GUQ
3160 GOO4=G12*GUR
3170 GOO5=G12*GUS
3180 GOO6=G12*GUT
3190 MMM=GOO1+GOO2+GOO3+GOO4+GOO5+GOO6
3200 PRINT
3210 REM ** The percentage of magnetic moment in region 1 = (GOO1/MMM)*100 **
3220 REM ** The percentage of magnetic moment in region 2 = (GOO2/MMM)*100 **
3230 REM ** The percentage of magnetic moment in region 3 = (GOO3/MMM)*100 **
3240 REM ** The percentage of magnetic moment in region 4 = (GOO4/MMM)*100 **
3250 REM ** The percentage of magnetic moment in region 5 = (GOO5/MMM)*100 **
3260 REM ** The percentage of magnetic moment in region 6 = (GOO6/MMM)*100 **
3270 PRINT
3280 PRINT"   Input the unloaded quality factor Qu "
3290 INPUT" Qu = ";QU
3300 LR=(ZZ/(MMM^2))*PI*(RA^4)/((FR^2)*(1E+18)*E0)
3310 PRINT
3320 PRINT"   The equivalent inductannce, Lr, of DR in HENRY "
3330 PRINT" Lr = "LR
3340 CR=1/(4*(PI^2)*(FR^2)*(1E+18)*LR)
3350 PRINT"   The equivalent capacitance, Cr, of DR in FARAD "
3360 PRINT" Cr = "CR
3370 RR=(2*PI*FR*LR/QU)*(1E+09)
3380 PRINT"   The equivalent resistance, Rr, of DR in Ohm "
3390 PRINT" Rr = "RR
3400 PRINT
3410 PRINT"   TYPE 0, please.  Then goes on showing. "
3420 INPUT A$
3430 IF A$="0" THEN 3440
3440 LM=SQR(Z0*LR/(PI*FR*QEXT*(1E+09)))
3450 PRINT
3460 PRINT"   The equivalent mutual inductance, Lm, of DR coupled with a
           microstrip in HENRY (for double loading) "
3470 PRINT" Lm = "LM
3480 BC=QU/QEXT
3490 PRINT"   The coupling factor Bc "
3500 PRINT" Bc = "BC
```

```
3510 QL=QU/(1+BC)
3520 PRINT"   The loaded quality factor QL "
3530 PRINT" QL = "QL
3540 SSS=S11^2+S21^2
3550 PRINT
3560 PRINT"   TYPE  0, please. Then goes on showing. "
3570 INPUT A$
3580 IF A$="0" THEN 3590
3590 L=(LM^2)/LR
3600 PRINT"   The equivalent inductance, L, of DR coupled with a microstrip
          in HENRY "
3610 PRINT" L = "L
3620 C=(LR^2)*CR/(LM^2)
3630 PRINT"   The equivalent capacitance, C, of DR coupled with a microstrip
          in FARAD "
3640 PRINT" C = "C
3650 R=2*PI*FR*(1E+09)*(LM^2)*QU/LR
3660 PRINT"   The equivalent resistance, R, of DR coupled with a microstrip
          in Ohm "
3670 PRINT" R = "R
3680 PRINT
3690 PRINT" TYPE  0, please. Then goes on showing. "
3700 INPUT A$
3710 IF A$="0" THEN 3720
3720 PRINT
3730 PRINT" Neglecting the reflection from load "
3740 PRINT
3750 PRINT" S parameter "
3760 PRINT
3770 PRINT" (ABS(S11))^2 = normalized reflection power from * DR junction * "
3780 PRINT
3790 PRINT" (ABS(S21))^2 = normalized transmission power into load "
3800 S11=R/(R+2*Z0)
3810 PRINT
3820 PRINT" (ABS(S11))^2 = "S11^2
3830 S21=SQR((1-S11^2)/(1+2*BC))
3840 PRINT
3850 PRINT" (ABS(S21))^2 = "S21^2
3860 PRINT
3870 SSS=S11^2+S21^2
3880 PRINT" (ABS(S11))^2 + (ABS(S21))^2 = "SSS
3890 PRINT
3900 PRINT"  Want to input another d, or Qu for calculation ? (Y or N) "
3910 INPUT A$
3920 IF A$="Y" OR A$="y" THEN 1740 ELSE 4120
3930 PRINT
3940 PRINT"   The calculation shows that the input does not meet the formulae
        of DR theory : "
3950 PRINT
3960 PRINT" 1) Ka^2 = (Er-Ea)*(2*PI*Fr/c)^2 - Kr^2
3970 PRINT
3980 PRINT" 2) J1(KrA)/(KrA*J0(KrA)) = - K1(KaA)/(KaA*K0(KaA)) "
3990 PRINT
4000 PRINT" Why ?  perhaps : "
4010 PRINT
4020 PRINT"   The input of Fr is not correct. Check and change it and try
        again, TYPE  T, please. "
```

```
4030 PRINT"   If the input of Fr is correct. TYPE  U (and then TYPE 'Enter')
              for few times , please. "
4040 REM **   Change DEL < .001 into  DEL < .001 + .004 * N    **
4050 INPUT A$
4060 IF A$="T" OR A$="t" THEN 1060
4070 IF A$="U" OR A$="u" THEN 4080
4080 PRINT" Try again with lower accuracy. "
4090 DDD=DDD+.004
4100 IF DDD>.015 THEN 4110 ELSE 4390
4110 PRINT" The accurate result can not be obtained in 1.5 % accuracy. "
4120 END
4130 PRINT
4140 PRINT
4150 REM ** AA1 **
4160 X=LD*H1
4170 GOSUB 5590
4180 THASD=THX
4190 BTHASD=BB*THASD
4200 TB1=BB*H/2
4210 BTHTG=BTHASD*TAN(TB1)
4220 AA1=(LD-BTHTG)/(LD*TAN(TB1)+BTHASD)
4230 RETURN
4240 REM ** AA2 **
4250 AADHD=AA*G
4260 SHAAD=.5*(EXP(AADHD)-EXP(-AADHD))
4270 AA2=(COS(TB1)+AA1*SIN(TB1))/(2*EXP(-AA*D3)*SHAAD)
4280 RETURN
4290 REM ** AA3 **
4300 ASDHDD=LD*H1
4310 SHASD=.5*(EXP(ASDHDD)-EXP(-ASDHDD))
4320 AA3=(COS(TB1)-AA1*SIN(TB1))/(2*EXP(-LD*D1)*SHASD)
4330 RETURN
4340 PRINT
4350 REM ** J1(KrA)/(KrA*J0(KrA))=-K1(KaA)/(KaA*K0(KaA)), to find KrA(=XXX),
              KaA(=YYY), Kr(=KR), Ka(=KA), Bb(=BB), Aa(=AA), As(=AS) **
4360 REM ** J1(KrA)=J1 **
4370 REM ** A=RA **
4380 DDD=.001
4390 DDD=DDD
4400 PRINT" DEL = "DDD
4410 XXX=2.7
4420 XXX=XXX
4430 XXX1=(XXX/2)^2
4440 S=1
4450 J=1
4460 T=1
4470 K=-1
4480 T=T*XXX1/J/(J+1)
4490 S=S+K*T
4500 IF T<.000001 THEN 4540
4510 K=-K
4520 J=J+1
4530 GOTO 4480
4540 J1=(XXX/2)*S
4550 REM ** J0(KrA)=J0 **
```

```
4560 A(1)=4
4570 J0(1)=1-(XXX^2)/4
4580 FOR M=1 TO 15 STEP 1
4590 A(M+1)=A(M)*((2*(M+1))^2)
4600 J0(M+1)=J0(M)+((-1)^(M+1))*(XXX^(2*(M+1)))/A(M+1)
4610 NEXT M
4620 J0=J0(16)
4630 XX=J1/(XXX*J0)
4640 REM ** K1(Y), FOR Y > 0.8 **
4650 REM ** K1(KaA)=K **
4660 MNN=(ER-EA)*((PI*FR*RA)/150)^2
4670 MPP=XXX^2
4680 IF MNN-MPP<0 THEN 3930
4690 YYY=SQR(MNN-MPP)
4700 AAA=SQR(PI/(2*YYY))
4710 BBB=AAA/EXP(YYY)
4720 C5=1+.375/YYY
4730 D5=C5-.1171875/YYY^2
4740 K=BBB*D5*(1+.037/(YYY^3))
4750 REM ** K0(Y), FOR Y > 0.8 **
4760 REM ** K0(KaA)=SK **
4770 AAAA=SQR(PI/(2*YYY))
4780 BBB=AAAA/EXP(YYY)
4790 C6=1-.125/YYY
4800 D6=C6+.0703125/YYY^2
4810 SK=BBB*D6*(1-.037/(YYY^3))
4820 SSK=K/(YYY*SK)
4830 DEL=ABS(XX+SSK)/ABS(XX-SSK)
4840 IF DEL<DDD  THEN 4960 ELSE 4850
4850 IF DEL>.05 THEN 4860 ELSE 4880
4860 XXX=XXX+.01
4870 GOTO 4420
4880 IF DEL>.01 THEN 4890 ELSE 4910
4890 XXX=XXX+.005
4900 GOTO 4420
4910 IF DEL>.005 THEN 4920 ELSE 4940
4920 XXX=XXX+.002
4930 GOTO 4420
4940 XXX=XXX+.001
4950 GOTO 4420
4960 RETURN
4970 REM ** J0(X) **
4960 A(1)=4
4990 J0(1)=1-(X^2)/4
5000 FOR I=1 TO 15 STEP 1
5010 A(I+1)=A(I)*((2*(I+1))^2)
5020 J0(I+1)=J0(I)+((-1)^(I+1))*(X^(2*(I+1)))/A(I+1)
5030 NEXT I
5040 J0X=J0(16)
5050 RETURN
5060 REM ** JI(X), I=1,2,3... **
5070 IF I=0 THEN 5090 ELSE 5080
5080 IF I<>INT(I) THEN 5090 ELSE 5110
```

```
5090 PRINT" The "I" is incorrect , check it please "
5100 GOTO 4120
5110 X1=(X/2)^2
5120 S=1
5130 J=1
5140 T=1
5150 L=1
5160 K=-1
5170 T=T*X1/J/(I+J)
5180 S=S+K*T
5190 IF T<.000001 THEN 5230
5200 K=-K
5210 J=J+1
5220 GOTO 5170
5230 FOR K=1 TO I
5240 L=L*K
5250 NEXT K
5260 JIX=(X/2)^I/L*S
5270 RETURN
5280 REM ** K0(X), FOR X > 0.8 **
5290 A7=SQR(PI/(2*X))
5300 B7=A7/EXP(X)
5310 C7=1-.125/X
5320 D7=C7+.0703125/X^2
5330 K=B7*D7*(1-.037/(X^3))
5340 K0X=K
5350 RETURN
5360 REM ** K1(X), FOR X > 0.8 **
5370 A8=SQR(PI/(2*X))
5380 B8=A8/EXP(X)
5390 C8=1+.375/X
5400 D8=C8-.1171875/X^2
5410 K=B8*D8*(1+.037/(X^3))
5420 K1X=K
5430 RETURN
5440 REM ** K2(X) , FOR X > 0.8 **
5450 IF X=0 THEN 5460 ELSE 5480
5460 PRINT" X = 0 is incorrect, check it please "
5470 GOTO 4120
5480 V=SQR(PI/(2*X))
5490 U=V/EXP(X)
5500 UC=1+(1.875/X)+(.8203125/(X^2))
5510 K2X=U*UC*(1-.036/(X^3))
5520 RETURN
5530 REM ** CH(X) **
5540 CHX=.5*(EXP(X)+EXP(-X))
5550 RETURN
5560 REM ** SH(X) **
5570 SHX=.5*(EXP(X)-EXP(-X))
5580 RETURN
5590 REM ** TH(X) **
5600 THX=(EXP(X)-EXP(-X))/(EXP(X)+EXP(-X))
5610 RETURN
```

Calculation example of DR10A

```
Programme DR10A
Calculate resonance frequency of DR-substrate-box system.
There is only TE01d mode in the DR system.
-------------------------
g          2       5
........*********........
H       * DR     *
        *    1 *       4
........*********........
h          3       6
-------------------------

DR10A is based on the formula of DR theory 'directly'.

  Input the dielectric constant, Er, of DR (region 1)
Er = ? 38.9
  Input the dielectric constant, Ea, of regions 2,4,5
Ea = ? 1
  Input the dielectric constant, Es, of substrate (regions 3,6)
Es = ? 2.5
  **  Note : Er must be greater than Es & Ea  **
  Input the thickness, h, of substrate in mm
h = ? .7874
  Input the distance, g, between the top of DR and upper metal
  shield in mm
g = ? 4
  Input the radius, RA, of DR in mm
RA = ? 3.05
  Input the height, H, of DR in mm
H = ? 2.4


Calculation result :

Fr =  9.378473   GHz
```

```
LIST

1000  PRINT" Programme DR10A "
1010  PRINT" Calculate resonance frequency of DR-substrate-box system."
1020  PRINT" There is only TE01d mode in the DR system. "
1030  REM ** An empirical coefficient of 0.972 is introduced into DR10A
             to improve the calculation accuracy.  **
1040  DIM A(16)
1050  DIM J0(16)
1060  PI=3.1415926535#
1070  PRINT"-------------------------"
1080  PRINT" g             2       5  "
1090  PRINT"........*********........"
1100  PRINT" H       * DR    *        "
1110  PRINT"         *     1 *     4  "
1120  PRINT"........*********........"
1130  PRINT" h             3       6  "
1140  PRINT"-------------------------"
1150  PRINT
1160  PRINT"   Input the dielectric constant, Er, of DR (region 1) "
1170  INPUT" Er = ";ER
1180  PRINT"   Input the dielectric constant, Ea, of regions 2,4,5 "
1190  INPUT" Ea = ";EA
1200  PRINT"   Input the dielectric constant, Es, of substrate (regions 3,6) "
1210  INPUT" Es = ";ES
1220  PRINT"   ** Note : Er must be greater than Es & Ea  ** "
1230  PRINT"   Input the thickness, h, of substrate in mm "
1240  INPUT" h = ";H1
1250  PRINT"   Input the distance, g, between the top of DR and upper metal
           shield in mm "
1260  INPUT" g = ";G
1270  PRINT"   Input the radius, RA, of DR in mm "
1280  INPUT" RA = ";RA
1290  PRINT"   Input the height, H, of DR in mm "
1300  INPUT" H = ";H
1310  IF 2*RA/H<1.5 THEN 1320 ELSE 1430
1320  PRINT
1330  PRINT" ** "
1340  PRINT
1350  PRINT" Choosing 1.5 < 2*RA/H < 2.5 is proper for obtaining higher Q
           & suppressing TM11d mode "
1360  PRINT
1370  PRINT" <*>  But, here  :  2*RA/H = "2*RA/H"  , 2*RA/H < 1.5  <*> "
1380  PRINT
1390  PRINT" The calculation is not very accurate here "
1400  PRINT
1410  PRINT" ** "
1420  PRINT
1430  IF G<.000001 THEN 1440 ELSE 1450
1440  G=.000001
1450  VVR=SQR(ER-EA)
1460  VVS=14.5/SQR(ES)
1470  FR=((130/VVR)+VVS)/RA
1480  PRINT
1490  LDD=.001
1500  PRINT" ** "
1510  PRINT
1520  PRINT" DEL = "LDD
1530  LDD=LDD
1540  PRINT
1550  PRINT" Wait a minute, please. "
```

```
1560   PRINT
1570   PRINT" ** "
1580   PRINT
1590   GOSUB 2730
1600   MINO=MIN
1610   FR=FR*1.01
1620   GOSUB 2730
1630   MINOO=MIN
1640   IF MINO<MINOO THEN 1650 ELSE 1710
1650   FR=FR*.99
1660   GOSUB 2730
1670   PRINT" MIN = "MIN
1680   IF MIN<LDD THEN 1760
1690   FR=FR-FR*MIN*.1
1700   GOTO 1660
1710   FR=FR+FR*MIN*.1
1720   GOSUB 2730
1730   PRINT" MIN = "MIN
1740   IF MIN<LDD THEN 1760
1750   GOTO 1710
1760   FR=FR*.972
1770   PRINT
1780   PRINT" Input : "
1790   PRINT
1800   PRINT" Er =      "ER
1810   PRINT" Ea =      "EA
1820   PRINT" Es =      "ES
1830   PRINT" RA =      "RA"   mm "
1840   PRINT" H  =      "H "   mm "
1850   PRINT" h  =      "H1"   mm "
1860   PRINT" g  =      "G "   mm "
1870   PRINT
1880   PRINT
1890   PRINT
1900   PRINT
1910   PRINT" Calculation result : "
1920   PRINT
1930   PRINT" Fr = "FR"  GHz "
1940   PRINT
1950   END
1960   PRINT
1970   PRINT
1980   REM ** J1(KrA)/(KrA*J0(KrA))=-K1(KaA)/(KaA*K0(KaA)), to find KrA(=XXX),
              KaA(=YYY), Kr(=KR), Ka(=KA), Bb(=BB), Aa(=AA), As(=AS) **
1990   REM ** J1(KrA)=J1 **
2000   REM ** A=RA **
2010   XXX=2.7
2020   XXX=XXX
2030   IF XXX>3.3 THEN 2300
2040   XXX1=(XXX/2)^2
2050   S=1
2060   J=1
2070   T=1
2080   K=-1
2090   T=T*XXX1/J/(J+1)
2100   S=S+K*T
2110   IF T<.000001 THEN 2150
2120   K=-K
2130   J=J+1
```

```
2140  GOTO 2090
2150  J1=(XXX/2)*S
2160  REM ** J0(KrA)=J0 **
2170  A(1)=4
2180  J0(1)=1-(XXX^2)/4
2190  FOR M=1 TO 15 STEP 1
2200  A(M+1)=A(M)*((2*(M+1))^2)
2210  J0(M+1)=J0(M)+((-1)^(M+1))*(XXX^(2*(M+1)))/A(M+1)
2220  NEXT M
2230  J0=J0(16)
2240  XX=J1/(XXX*J0)
2250  REM ** K1(Y), FOR Y > 0.8 **
2260  REM ** K1(KaA)=K **
2270  MNN=(ER-EA)*(((PI*FR*RA)/150)^2)
2280  MPP=XXX^2
2290  IF MNN-MPP<0 THEN 2300 ELSE 2420
2300  PRINT
2310  IF LDD>.015 THEN 2320 ELSE 2340
2320  PRINT" The accurate result can not be obtained in 1.5 % accuracy. "
2330  GOTO 1950
2340  REM ** Change  MIN & DEL <.001 into ...<.001 + .004 * N **
2350  PRINT
2360  PRINT" ** "
2370  PRINT
2380  PRINT" Try again with lower accuracy "
2390  LDD=LDD+.004
2400  PRINT" DEL = "LDD
2410  GOTO 1530
2420  YYY=SQR(MNN-MPP)
2430  AAA=SQR(PI/(2*YYY))
2440  BBB=AAA/EXP(YYY)
2450  C=1+.375/YYY
2460  D=C-.1171875/YYY^2
2470  K=BBB*D*(1+.037/(YYY^3))
2480  REM ** K0(Y), FOR Y > 0.8 **
2490  REM ** K0(KaA)=SK **
2500  AAAA=SQR(PI/(2*YYY))
2510  BBB=AAAA/EXP(YYY)
2520  C=1-.125/YYY
2530  D=C+.0703125/YYY^2
2540  SK=BBB*D*(1-.037/(YYY^3))
2550  SSK=K/(YYY*SK)
2560  DEL=ABS(XX+SSK)/ABS(XX-SSK)
2570  IF DEL<LDD THEN 2690
2580  IF DEL>.05 THEN 2590 ELSE 2610
2590  XXX=XXX+.01
2600  GOTO 2020
2610  IF DEL>.01 THEN 2620 ELSE 2640
2620  XXX=XXX+.005
2630  GOTO 2020
2640  IF DEL>.005 THEN 2650 ELSE 2670
2650  XXX=XXX+.002
2660  GOTO 2020
2670  XXX=XXX+.001
2680  GOTO 2020
2690  RETURN
2700  REM ** CTNH(X) **
2710  CTNHX=(EXP(X)+EXP(-X))/(EXP(X)-EXP(-X))
2720  RETURN
```

```
2730   REM ** MIN **
2740   PRINT
2750   GOSUB 1980
2760   PRINT
2770   KR=XXX/RA
2780   KA=YYY/RA
2790   DD=(PI*FR/150)^2
2800   II=KR^2
2810   BB=SQR(DD*ER-II)
2820   AA=SQR(II-DD*EA)
2830   LD=SQR(II-DD*ES)
2840   PRINT
2850   X=AA*G
2860   GOSUB 2700
2870   CAG=CTNHX
2880   ZZA=AA/BB
2890   TAA=ATN(ZZA*CAG)
2900   X=LD*H1
2910   GOSUB 2700
2920   CAH=CTNHX
2930   ZZB=LD/BB
2940   TAS=ATN(ZZB*CAH)
2950   RES=TAA+TAS
2960   BSH=BB*H
2970   MIN=ABS(BSH-RES)/(ABS(BSH)+ABS(RES))
2980   RETURN
```

Calculation example of ER

```
 RUN
INPUT:

measured resonance frequency FO of DR [GHz] ?5.5
diameter of the DR [mm]                     ?8
height of the DR [mm]                       ?12
fundamental mode of DR (enter 1)            ?1

the dielectric constant of the DR is ........Er = 40.141

INPUT:

What is the plate material used ?

Type  G for Gold
      C for Copper
      S for Silver
?C

What is the diameter of the plates used [mm] ?80
What is the room temperature [C]             ?18
What is the 3dB bandwidth at FO [MHz]        ?0.8
What is the insertion loss at FO [+ dB]      ?22


the loss tangent of the DR is ...............tan d = 1.0789E-4

the Qu of the DR at 5.5 GHz is ..............Qu = 9268.22
```

```
L.
   10 REM This programme is called ER
   20 REM This programme calculates Er,tan d and Qu of a DR
   30 REM from measured results
   40 REM ***********************************************
   50 MODE128
   60 CLS
   70 PRINT
   80 PRINT"INPUT:"
   90 PRINT
  100 INPUT "measured resonance frequency FO of DR [GHz] ",FO
  110 FO=FO*1E9
  120 INPUT "diameter of the DR [mm]                    ",D
  130 D=D*1E-3
  140 INPUT "height of the DR [mm]                      ",L
  150 L=L*1E-3
  160 INPUT "fundamental mode of DR (enter 1)           ",l
  170 PRINT
  180 REM velocity of light
  190 c=3E8
  200 mu=PI*4E-7
  210 Eo=(1E-9)/(36*PI)
  220 v=SQR( ((PI*FO*D/c)^2)*(((c/(FO*2*L/l))^2)-1))
  230 DIM A(5)
  240 A(1)=.49907
  250 A(2)=-.11226
  260 A(3)=.06539
  270 A(4)=-.02679
  280 A(5)=.00445
  290 P2=0
  300 FOR I=1 TO 5
  310 P2=P2+ A(I)/(v^I)
  320 NEXT I
  330 K=1/(1+P2)
  340 u1=2.4048
  350 u2=3.8316
  360 u=u1
  370 GOSUB 1400
  380 Fu1=F
  390 u=u2
  400 GOSUB 1400
  410 Fu2=F
  420 IF ABS(Fu1-Fu2)<.0000001 THEN 570
  430 IF((Fu1>0) AND (Fu2<0)) OR ((Fu1<0) AND (Fu2>0)) THEN 460
  440 PRINT"unsuitable mathematical approach"
  450 GOTO 1360
  460 u12=(u1+u2)/2
  470 u=u12
  480 GOSUB 1400
  490 Fu12=F
  500 IF ((Fu12>0) AND (Fu1>0)) OR ((Fu12<0) AND (Fu1<0)) THEN 530
  510 IF ((Fu12<0) AND (Fu2<0)) OR ((Fu12>0) AND (Fu2>0)) THEN 550
  520 GOTO 440
  530 u1=u12
  540 GOTO 360
  550 u2=u12
  560 GOTO 360
  570 u=u1
  580 Er=((((c/(FO*PI*D))^2)*((u^2)+(v^2)))+1)*1.06
  590 PRINT"the dielectric constant of the DR is ........Er = ";(INT(Er*1000))/1
000
  600 PRINT
```

```
 610 X=u
 620 N=0
 630 GOSUB 1680
 640 WO=JNX
 650 N=1
 660 GOSUB 1680
 670 W1=JNX
 680 N=2
 690 GOSUB 1680
 700 W2=JNX
 710 FU=(W1^2)/((W1^2)-(WO*W2))
 720 X=v
 730 N=0
 740 GOSUB 1570
 750 WO=KNX
 760 N=1
 770 GOSUB 1570
 780 W1=KNX
 790 N=2
 800 GOSUB 1570
 810 W2=KNX
 820 GV=((WO*W2)-(W1^2))/(W1^2)
 830 W=FU*GV
 840 REM PRINT"                        W = ";W
 850 A=1+W/Er
 860 PRINT"INPUT:"
 870 PRINT
 880 PRINT"What is the plate material used ?"
 890 PRINT
 900 PRINT"Type  G for Gold"
 910 PRINT"      C for Copper"
 920 PRINT"      S for Silver"
 930 INPUT"",R$
 940 PRINT
 950 IF R$="G" OR R$="g" THEN 990
 960 IF R$="C" OR R$="c" THEN 1010
 970 IF R$="S" OR R$="s" THEN 1030
 980 PRINT:GOTO 900
 990 CO=4.4503782E7 : C100=3.3670033E7
1000 GOTO 1040
1010 CO=6.3371356E7 : C100=4.3859649E7
1020 GOTO 1040
1030 CO=6.6401062E7 : C100=4.6511627E7
1040 INPUT "What is the diameter of the plates used [mm] ",PD
1050 PD=PD*1E-3
1060 INPUT "What is the room temperature [C]              ",TEMP
1070 RC=((((C100-CO)/100)*TEMP)+CO)/CO
1080 X=(v*PD)/D
1090 N=0
1100 GOSUB 1570
1110 H=KNX
1120 N=1
1130 GOSUB 1570
1140 H=H*KNX
1150 H=H*X
1160 X=v
1170 N=1
1180 GOSUB 1570
1190 H=H/((KNX)^2)
1200 DT=(2*FU*H)/(mu*Eo*Er*((D*PI*FO)^2))
1210 B1=((c*1)/(2*FO*L))*(1/(30*Er*1*(PI^2)))*(1+W)
1220 B2=((1^2)/(2*PI*Er*Eo*(mu^2)*(FO^3)*(L^3)))*(1+W)
```

```
 1230 INPUT"What is the 3dB bandwidth at FO [MHz]          ",DFO
 1240 INPUT"What is the insertion loss at FO [+ dB]        ",S210
 1250 PRINT
 1260 DFO=DFO*1E6
 1270 QU=(FO/DFO)*(1/(1-(10^(-S210/20))))
 1280 Rs=SQR((PI*FO*4*PI*1E-7)/(RC*CO))
 1290 T2=(A/QU)-(B2*Rs)-DT
 1300 PRINT
 1310 PRINT"the loss tangent of the DR is ..............tan d = ";(INT(T2*10000
0000))/100000000
 1320 PRINT
 1330 PRINT"the Qu of the DR at ";FO*1E-9;" GHz is .............Qu = ";(INT((1/
T2)*100))/100
 1340 PRINT
 1350 REM "It is up to the user to improve the accuracy of the programme"
 1360 END
 1370 REM******************************
 1380 REM    S U B R O U T I N E S   *
 1390 REM******************************
 1400 REM *** F SUBROUTINE ***
 1410 GOSUB 1440
 1420 F= J + (K*(v/u))
 1430 RETURN
 1440 REM *** J SUBROUTINE ***
 1450 A(1)=1.4282
 1460 A(2)=-.7160
 1470 A(3)=.2571
 1480 A(4)=-.1177
 1490 A(5)=.0282
 1500 P1=1/(u-3.8317)
 1510 P2=0
 1520 FOR I=1 TO 5
 1530 P2=P2+( A(I)*((u-2.4048)^I) )
 1540 NEXT I
 1550 J=P1*P2
 1560 RETURN
 1570 REM *** Kn(X) SUBROUTINE ***
 1580 T1=4*N*N-1
 1590 T2=4*N*N-9
 1600 T3=4*N*N-25
 1610 T4=4*N*N-49
 1620 T5=4*N*N-81
 1630 T6=4*N*N-121
 1640 P1=1 + (T1/(8*X)) + ((T1*T2)/(2*((8*X)^2))) + ((T1*T2*T3)/(6*((8*X)^3)))
 1650 P2=(EXP(-X))*(SQR(PI/(2*X)))
 1660 KNX=P1*P2
 1670 RETURN
 1680 REM *** Jn(X) SUBROUTINE ***
 1690 P1=(X^2)/(2*(2*N+2))
 1700 P2=(X^4)/(8*(2*N+2)*(2*N+4))
 1710 P3=(X^6)/(48*(2*N+2)*(2*N+4)*(2*N+6))
 1720 P4=(X^8)/(384*(2*N+2)*(2*N+4)*(2*N+6)*(2*N+8))
 1730 P5=(X^10)/(3840*(2*N+2)*(2*N+4)*(2*N+6)*(2*N+8)*(2*N+10))
 1740 P6=(X^12)/(46080*(2*N+2)*(2*N+4)*(2*N+6)*(2*N+8)*(2*N+10)*(2*N+12))
 1750 P7=(X^14)/(645120*(2*N+2)*(2*N+4)*(2*N+6)*(2*N+8)*(2*N+10)*(2*N+12)*(2*N+1
4))
 1760 P8=(X^16)/(10321920*(2*N+2)*(2*N+4)*(2*N+6)*(2*N+8)*(2*N+10)*(2*N+12)*(2*N
+14)*(2*N+16))
 1770 F=N
 1780 GOSUB 1810
 1790 JNX=(((X/2)^N)/FACT)*(1-P1+P2-P3+P4-P5+P6-P7+P8)
 1800 RETURN
```

```
 1810 REM *** FACTORIAL SUBROUTINE ***
 1820 IF F>1 THEN 1850
 1830 FACT=1
 1840 GOTO 1910
 1850 FK=F
 1860 FACT=F
 1870 FACT=FACT*(FK-1)
 1880 FK=FK-1
 1890 IF FK=1 THEN 1910
 1900 GOTO 1870
 1910 RETURN
 1920 END
>
>
```

DR TUNING

Mechanical tuning

A well known method to tune a resonator or dielectric oscillator is mechanical tuning. Some diagrams were given in Fig. 7.6. The tuning by means of this approach is based on the field perturbation caused by a moving metal disc or dielectric screw. Figure 7.6(c) shows the tuning by means of a metal screw from the side wall. The tuning range is small and some unwanted higher DR modes can be excited if the tuning plate is large and close to the DR.

Tuning with a metal plunger as shown in Fig. 7.6(a) results in a tuning around 3% of the centre frequency. The increased conduction loss brought about by the plunger lowers the unloaded quality factor Q_u of the resonator.The closer the plunger the lower Q_u. Additionally, this decreases the output power in the case of a dielectric resonator oscillator (DRO).

An alternative method of tuning was proposed in [30] and the configuration is shown in Fig. 7.29. Instead of using a metal plunger or dielectric screw, a second DR

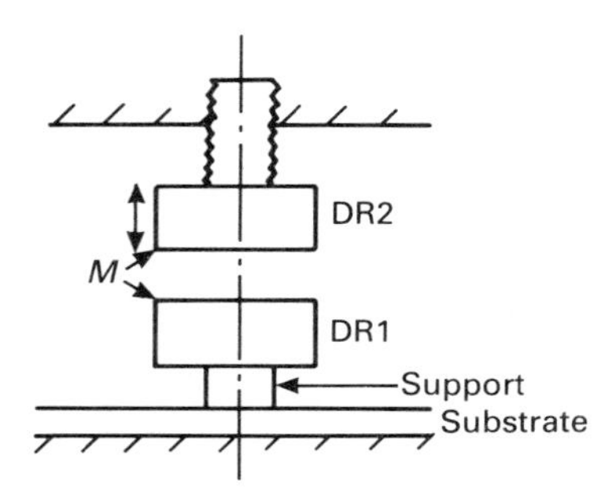

Fig. 7.29 — The basic configuration of a mechanical tuning by means of a movable 'second' DR.

is employed for frequency tuning. Moving the tuning DR closer or further away from the main DR changes the fields associated with the 'pair' of DRs and hence the resonance frequency. The main advantage to be derived from this method of tuning is the less pronounced decrease of Q_u of the fundamental mode compared to that of metal plunger tuning. Any DR oscillating structure using this method of tuning can be made to change over a much larger range without performance impairment. Typical figures are 8% frequency variation up to *X*-band [30].

Varactor tuning

The advantageous features of DRs in voltage controlled oscillators are well known and have been stated at the beginning of this chapter. One of the limiting features is the range of tuneability. Hence, having an acceptable tuning facility would open other avenues for applications and replace existing techniques. A varactor can be used to tune a DR in an oscillator. The tuning of such an arrangement depends primarily on the capacitance ratio of the varactor diode and the tightness of coupling. The tighter the coupling between the DR and varactor, the larger the tuning range. Tighter coupling, however, degrades the loaded Q and a compromise must be struck. A lower Q affects the pulling and pushing figure of the DRO, the FM noise figure and temperature stability. Some of these aspects will be discussed later.

Figure 7.30 shows an example of a varactor tuned DRO schematically. The varactor is coupled by means of a microstrip to the DR. The shift in resonance

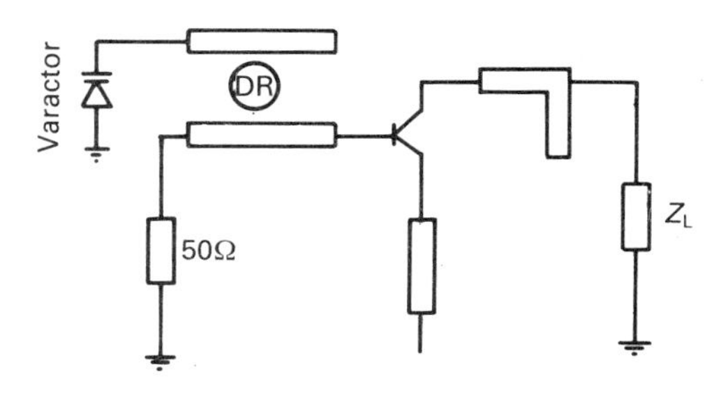

Fig. 7.30 — The varactor tuning diagram.

frequency and Q_o of the circuit can be expressed as [31]

$$\left|\frac{\Delta\omega}{\omega_o}\right| = \left|\frac{\beta_c \Delta B L Z_o}{2Q_o}\right| \tag{7.85}$$

$$Q_{o1} \simeq \frac{Q_v}{2}\left(\frac{\omega_o}{\Delta\omega}\right)$$

where

Q_o, Q_{o1} = unloaded Q of DR untuned/tuned,
β_c = coupling coefficient,
ΔBL = susceptance range of varactor,
Q_v = unloaded Q of varactor,
$\Delta\omega$ = change in resonance frequency.

The typical performance of an oscillator working at 1.69 GHz with a minimum $Q_{o1} \simeq 3500$ is as follows [31]. $P_{out} = 8$–12 dBm, tuning bandwidth = 0.55%, efficiency = 3%, frequency drift (20–50°C) = 45 kHz, FM noise = − 104 dBc/Hz.

Current controlled segmented-disc tuning

Another way of achieving tuning is by means of the perturbation of the field above the DR [32]. The schematic of this technique is shown in Fig. 7.31.

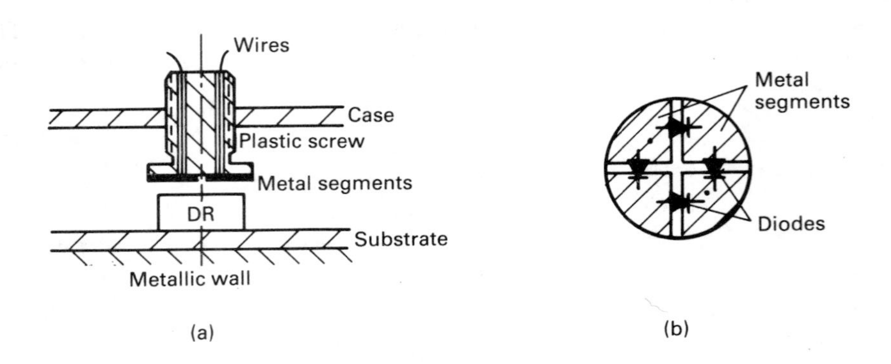

Fig. 7.31 — The diagram of a current controlled segment-disk tuning. (a) configuration of DR system; (b) disk with four conducting segments.

For a uniform disc, parallel to the upper DR surface, a continuous current is induced into the disc by virtue of the field of the DR. If the disc is now segmentized as shown in Fig. 7.31(b), then the current flow and field pattern is considerably changed. The current flow and hence tuning may be controlled by means of PIN diodes which interconnect the segments of the disc. Oscillators using this technique have produced frequency changes of 40 MHz at a centre frequency of 16 GHz with a bit rate of 20 Mb/s [32].

Magnetic tuning

The resonance frequency can also be changed by attaching a microwave ferrite to the DR and by controlling the necessary magnetic field with a current. A shielded quasi-$TE_{o1\delta}$ mode magnetically tuned DR system is shown schematically in Fig. 7.32. The field and resonance frequency are controlled by direct current coils. In order to allow for fast tuning the ferrite used must have low loss and a small scalar permeability. The physical construction of the circuit is critical in the sense that close coupling between various elements is required. A tuning bandwidth of 120 MHz at X-band was reported, including a system Q factor of 2000 [33]. The tuning power was 75 mW.

Optical tuning

The optical control of microwave devices and circuits is a relatively new area. Optical techniques offer potential advantages over other techniques in terms of speed and isolation. The optically controlled DRO was proposed in the middle of the 1980s [17,18]. The concept of an optically controlled DR system is shown in Fig. 7.33.

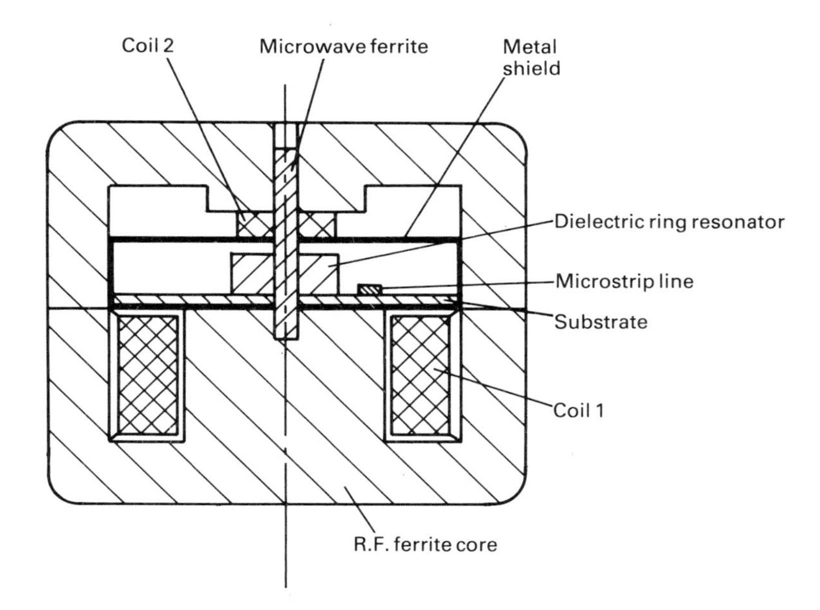

Fig. 7.32 — Shielded quasi-$TE_{o1\delta}$ mode magnetically tuned DR system with dc magnetic field circuit.

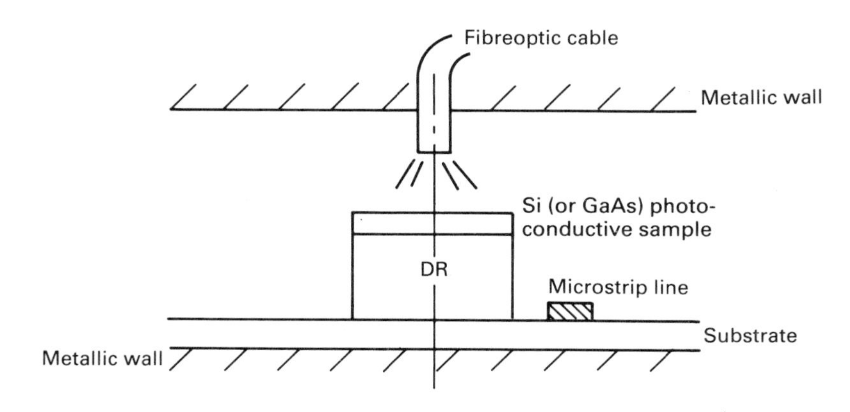

Fig. 7.33 — Optically controlled tuning concept of DR system.

When the photoconductive sample is not illuminated, it behaves like a loss-free dielectric. When it is illuminated, free carriers are generated. In the presence of the circumferential electric field of the DR (E_θ for the $TE_{01\delta}$ mode), a circumferential surface current is produced in the sample as a result of the free carriers, which is a function of the optical excitation. The surface current changes the radial magnetic field component and hence effects tuning. Further details on this tuning technique are given in [17,18].

DR measurement

The following describes a method of measuring and calculating the dielectric constant and Q factor of a DR. A network analyser is required to do this. Initially the network analyser needs to be calibrated over the frequency range of interest. If the operating frequency of the DR is not known, cannot be estimated or calculated, then it is best to calibrate the network analyser over the entire band of 10 MHz to 18.5 GHz. As can be seen from the block diagram in Fig. 7.34 and Plate 7.1, the DR is tested on a jig which comprises two rigid cables, x and y. These have loops at one end

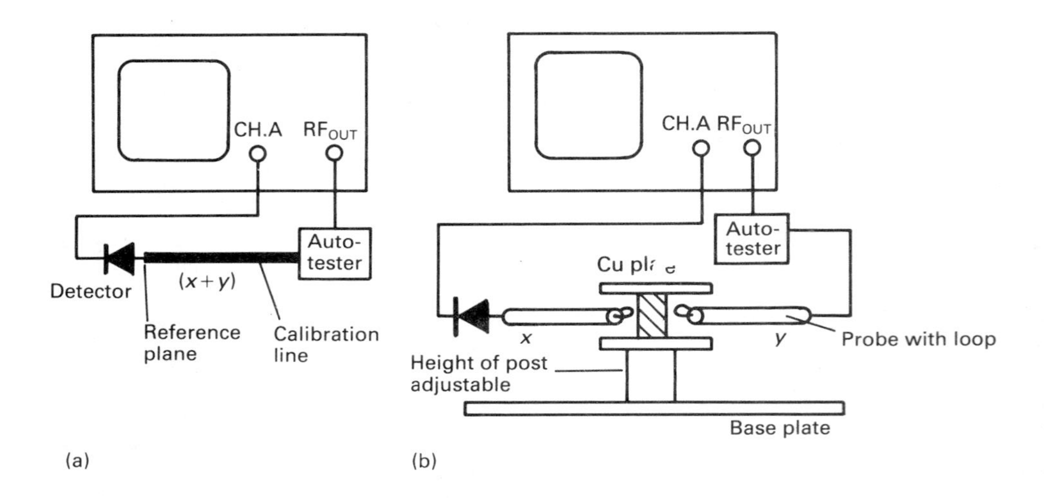

Fig. 7.34 — Wiltron network analyser. (a) calibration arrangement; (b) DR jig measuring set-up, schematic.

which are used to couple magnetically into the DR. An identical cable to that on the test jig without break is required to perform the calibration. The length of this cable will be the sum of the two individual length x and y on the test jig. Let us call this line the calibration line. This test line is then connected to the RF ouput of the network analyser bridge. The calibration is then performed as per network analyser instruction manual. This usually involves the termination of the line by an open circuit, a short circuit and a 50 Ω termination by the detector.

When the calibration is finished, the calibration line in Fig.7.34(a) is then removed. The test jig with the line and coupling loops is then connected into the measuring arrangement as per Fig. 7.34(b). A pair of suitably polished copper plates is then placed on top of the supporting post of the test jig. The DR under test is then sandwiched between these plates. If the diameter of the plates is large compared to

Plate 7.1 — Example of a DR test rig.

the DR, the DR can be considered to be in a cavity, i.e. the entire electro-magnetic field is contained within the test structure.

The two line probes x and y are then brought close to the DR in order to obtain strong coupling between the DR and the line loops. Only the channel which measures transmission on the analyser needs to be considered here. A typical display on this channel is shown in Fig. 7.35. The spikes on this display represent the modes of the DR. The mode in which one is interested in is the fundamental mode, the TE_{011} mode. It requires some experience to distinguish this mode from the others. A way to find the correct mode is to tilt the top plate of the DR structure and to observe the transmission spikes. Each spike represents energy absorption by the DR, i.e. the insertion loss (IL) at that frequency is high. When the top plate is tilted some of the spikes move. There will be one which will hardly move or change frequency, but which will change amplitude. On further tilting of the top plate the spike will then move to the left of its centre frequency. If this is the case the correct mode has been identified.

The network analyser frequency span may now be reduced so as to increase resolution around the centre frequency of the DR fundamental mode.The above procedure of tilting may now be repeated to verify the measurement. The next step is to decrease the frequency span as far as possible; this depends naturally on the make of analyser available. At the same time the coupling of the two probes to the DR should be reduced by moving the two probes away from the DR by equal amounts. The further the coupling loops are from the DR the less they disturb the magnetic field of the DR and the less its loading. The further the loops are from the DR the higher the insertion loss of the DR. The insertion loss can be read off the network analyser. The measurement should be stopped when the narrowest 3 dB bandwidth

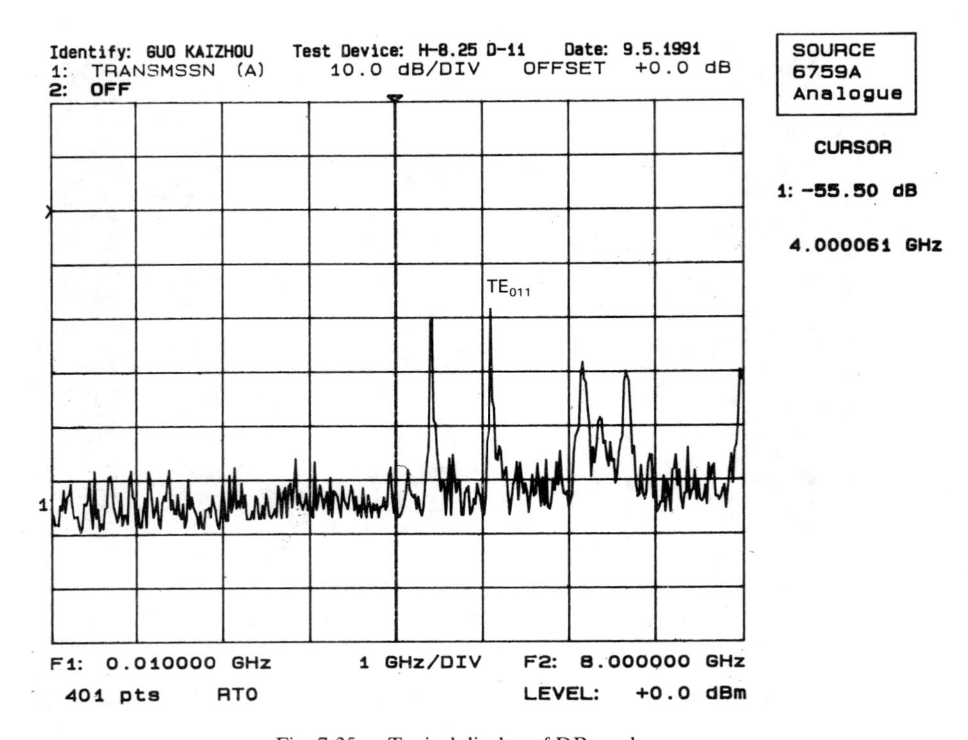

Fig. 7.35 — Typical display of DR modes.

for the highest insertion loss is obtained. As hinted earlier on, it requires some skill to identify the correct point.

An example of the final high resolution display for DR No. 2 in Table 7.4 is shown in Fig. 7.36. The centre frequency f_o may now be read off as 5.0818 GHz, the 3 dB bandwidth as 1.08 MHz and the insertion loss as 19.1 dB. These values may now be fed to a suitable computer program such as ER, in order to calculate the desired DR properties such as dielectric constant and quality factor. Finally, the DR parameters are entered into a chart to keep a record as shown in Table 7.4.

THE MEASUREMENT OF A DR-MICROSTRIP SYSTEM IN AN ENCLOSURE

The following deals with the measurement of various parameters of a DR coupled to a microstrip which is enclosed by a box. The typical resonant curve of such a system as measured with a network analyser is shown in Fig. 7.37. From the measurement we obtain a number of values which can then be used to calculate the loaded and unloaded quality factor of the system. The following equations allow the calculation of Q_u and Q_L:

$$Q_u = \frac{f_0}{f_3 - f_2} \tag{7.87}$$

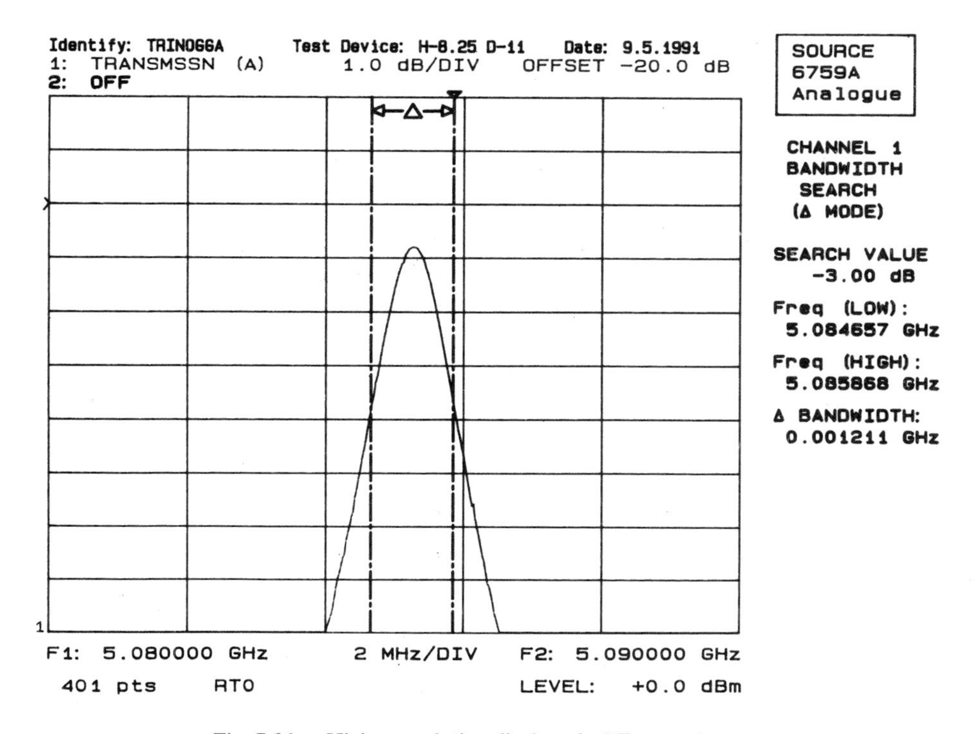

Fig. 7.36 — Higher resolution display of a TE_{011} mode.

$$Q_L = \frac{f_0}{f_4 - f_1} \tag{7.88}$$

The value for x as shown in Fig. 7.37 is calculated from the following formula [14, eqn. 10.14]:

$$x[\text{dB}] = 3 - 10\log(1 + 10^{-0.1\,L_{\text{ins}}}) \tag{7.89}$$

Figure. 7.38 shows a family of resonant curves of a DR coupled to a microstrip, enclosed in a box, for various coupling distanced d [25]. Using the following equations we can calculate the external quality factor Q_{ext} (or Q_e) and the coupling coefficient β_c. Thus

$$\frac{1}{Q_L} = \frac{1}{Q_u} + \frac{1}{Q_e} \tag{7.90}$$

$$\beta_c = \frac{Q_u}{Q_e} \tag{7.91}$$

Table 7.4 — Table for recording DR parameters

D R M E A S U R E M E N T S

date: 01 / 05 / 1991 source: ..specimen................

material: ..Ba Ti.......... ref.no: ...L.A.T.................

plate diameter ...80.....[mm] cable diameter ....1.2......[mm]

plate material Cu , Ag , Au temperature ......18........[°C]

DR mode: TE_{011} or other ....nil......

probe configuration:

****

	2a diam. [mm]	H heigh [mm]	f_o [GHz]	df 3dB [MHz]	IL loss [dB]	E_r	Q_u of DR	tan d $*10^{-4}$	Q_u of reson. struct.
No.1	8	15.03	5.3792	0.84	23.3	35.9	7587	1.318	6874
No.2	11	8.25	5.0818	1.08	19.1	37.0	10660	0.938	5292

Remarks :

Example of DR evaluation using a Wiltron Network Analyser System. The above is a typical printout for DR No.1 and 2 .

The quality factors Q_u, Q_L, Q_e, β_c and f_r against the coupling distance are shown in Fig. 7.39. There are two points which can be learned from this experiment. Firstly, one can see that the stronger the coupling the higher the resonance frequency f_r. The frequency variation in this case is small and changes from about 9.36 GHz to about 9.38 GHz. Secondly, the coupling coefficient (or coupling distance d) has only little effect on Q_u , the variation being about 7% in this work.

It should be noted that in practice there exists no perfect match at the two ends of the microstrip. Thus a standing wave will be excited in the DR-microstrip box system. Hence, the change of the DR position along the microstrip (parallel to the microstrip) will cause a change in the coupling factor. The better the match, the smaller the change in β_c.

Figure 7.40 shows the coupling coefficient β_c and the relative position of the DR with respect to the microstrip schematically. One can see, from Fig.7.40(a), that the β_c curve is symmetrical around the vertical axis. The maximum value of β_c is obtained when an appropriate part of the microstrip is covered with the DR which is shown in

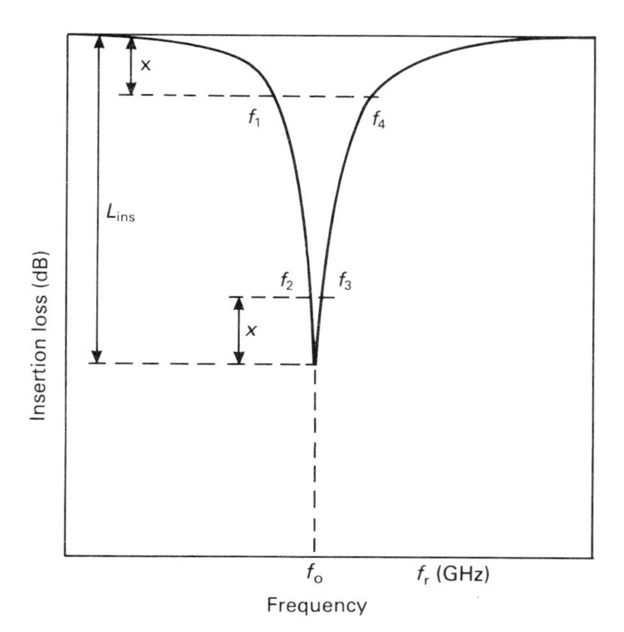

Fig. 7.37 — Resonant curve of a component.

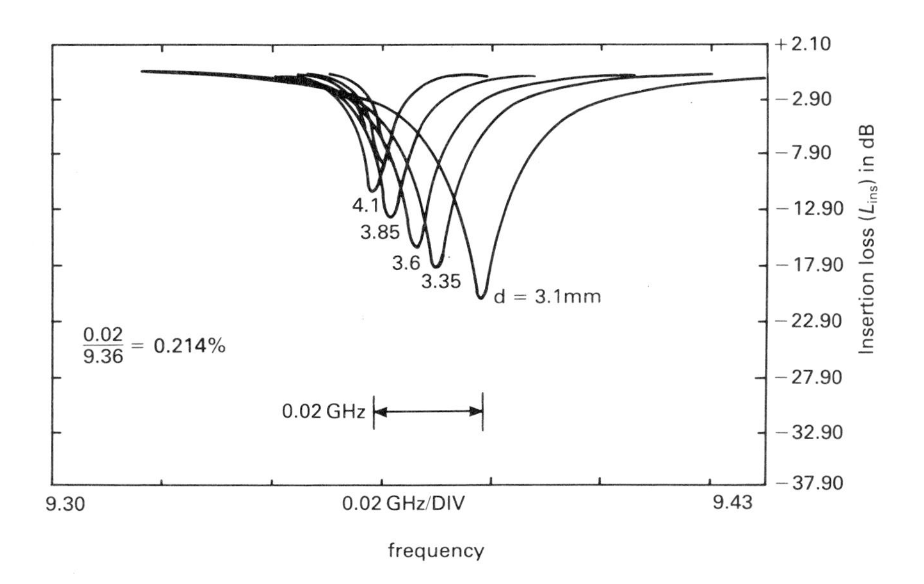

Fig. 7.38 — A family of resonant curves of a DR-substrate(microstrip)-box system with the coupling distance d as parameter.

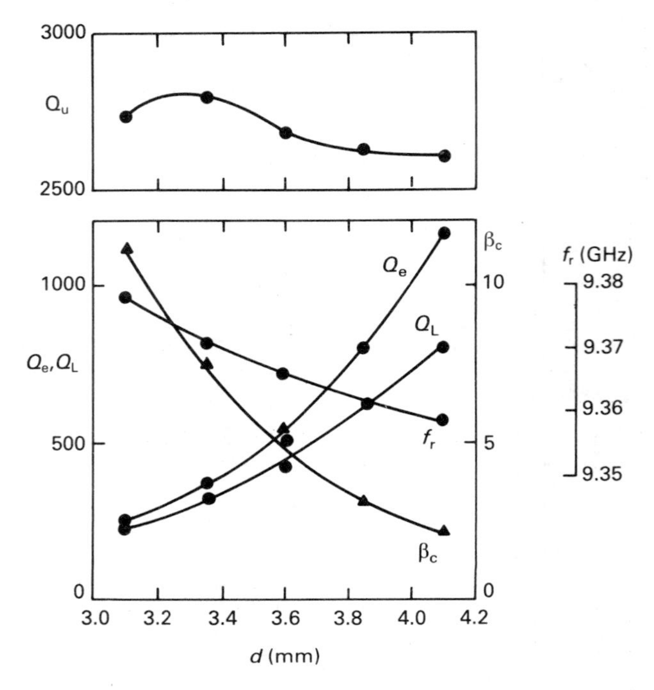

Fig. 7.39 — Q_u, Q_L, Q_e, β_c and f_r versus coupling distance. These curves are drawn from Fig. 7.38.

Fig. 7.40(c). In this position most of the magnetic lines pass around the microstrip and hence the strongest coupling is obtained. At position $d' = 0$, Fig. 7.40(d), there are no 'net' magnetic field lines around the microstrip and hence no coupling takes place between the DR and microstrip.

In practice, one mainly wants to obtain the strongest coupling between the DR and microstrip. It is easy to achieve the above aim by covering the microstrip with the DR . Since the microstrip is of finite thickness, it is impossible for the DR to lie parallel both, with respect to the lid and the ground plane. This is shown in Fig.7.41(a). As a result energy is diffused away from the DR, decreasing its Q considerably. If a spacer is placed under the DR, so that the latter is parallel as shown in Fig.7.41(b), a higher Q and β_c can be retained. Ideally, the spacer should have no loss and have the permittivity of air. In practice quartz ($\varepsilon_r = 3.8$) is frequently used.

DR AND CAVITY MODE MUTUAL INFLUENCE

In analogy with a cavity, an infinite number of modes can be excited in a DR. The microwave designer is faced with the unfortunate problem that the circuitry which he or she designs tends to couple strongly to the undesired mode, especially if this circuitry includes a DR as well. The following gives some experimental results concerning the interaction between box-modes and DR modes [25]. Figure 7.42

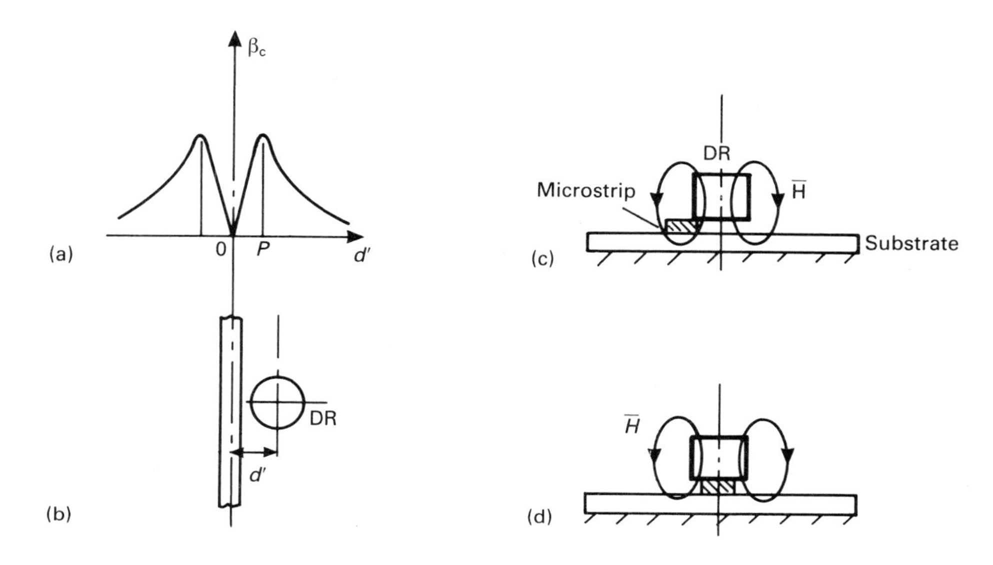

Fig. 7.40 — The coupling coefficient β_c versus relative DR position d'.

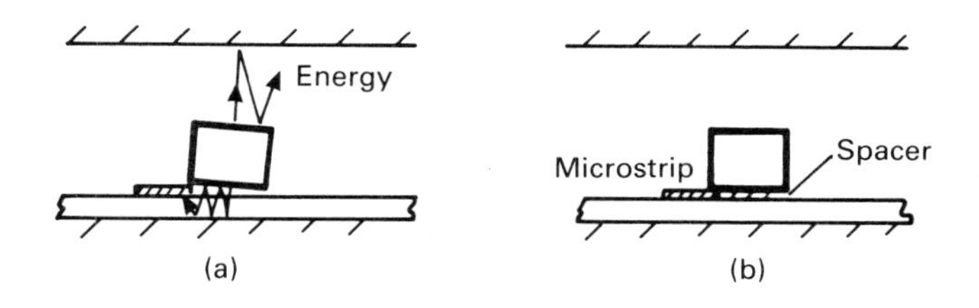

Fig. 7.41 — Using a spacer to retain a high β_c and Q.

shows a transmission plot made on a network analyser for an enclosed DR-microstrip arrangement. The resonance frequency of the $TE_{01\delta}$ mode of the DR in the enclosure is about 9.4 GHz and that of the cavity without the DR has two resonances, one at about 9.81 GHz and the other at 9.5 GHz. When the DR is now placed in the enclosure (box or cavity), mutual coupling takes place between the two previously mentioned modes. As can be seen from Fig. 7.42(b), a 'notch' is introduced in the DR $TE_{01\delta}$ mode. This means that stable coupling between a DR and microstrip cannot be achieved and any change in input power or temperature will assist mode changes and hence coupling.

Another enclosure with the same DR as used for the results in Fig. 7.42 is now used to illustrate the mode problem,but on an expanded frequency scale.A plot between 9 and 10 GHz is shown in Fig. 7.43. The height g between DR and the

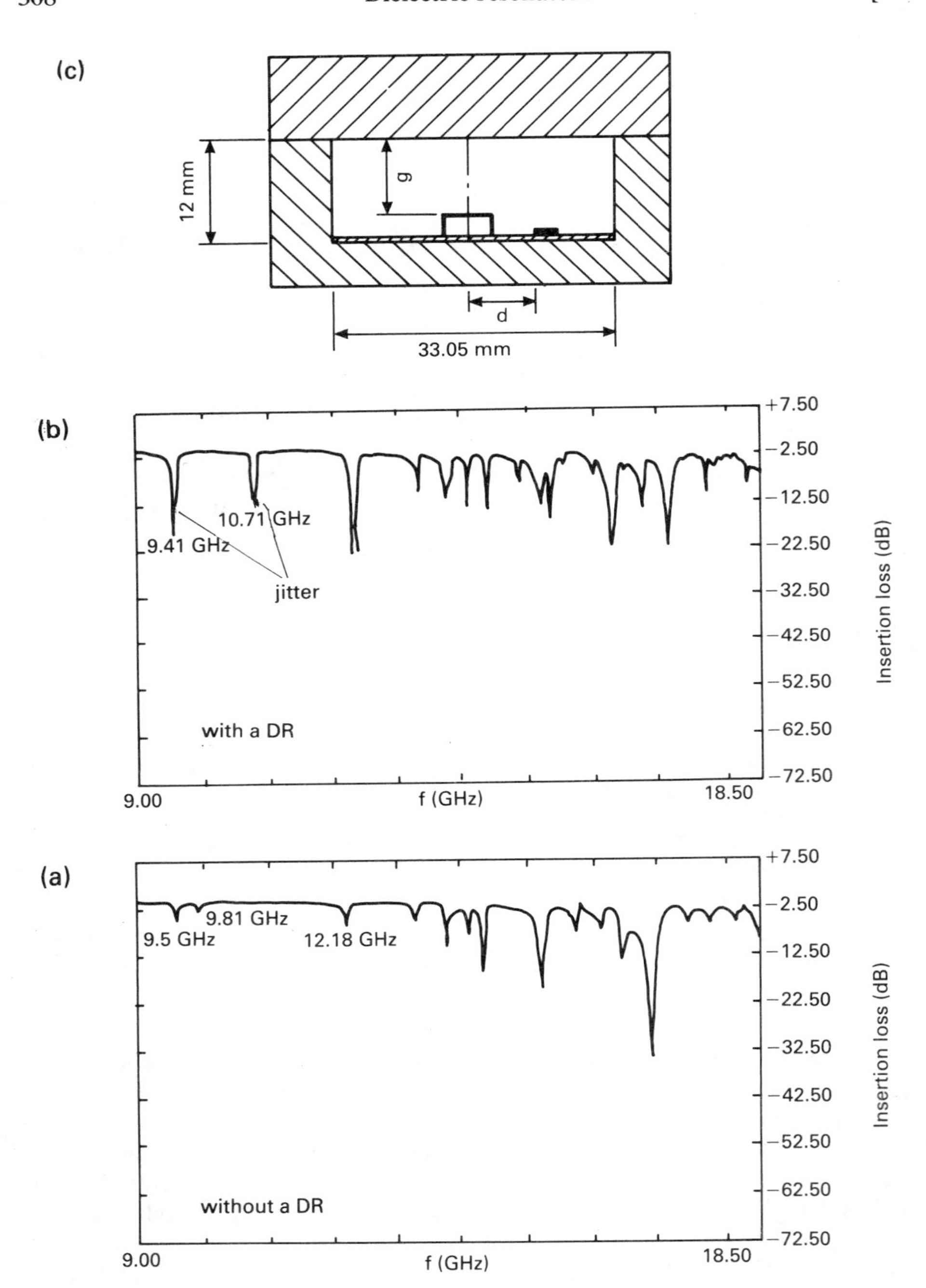

Fig. 7.42 — Insertion loss against frequency of the layout without (a) and with (b) a DR. g = 8.81 mm. (c) Test structure used in measurements. Length of metal shielding, LL = 35 mm; DR: dielectric constant $\varepsilon_r = 38.9$, height $H = 2.4$ mm, diameter $D = 6.1$ mm. Microstrip: thickness $h = 0.7874$ mm, dielectric constant $\varepsilon_s = 2.5$, width $w = 2.25$ mm.

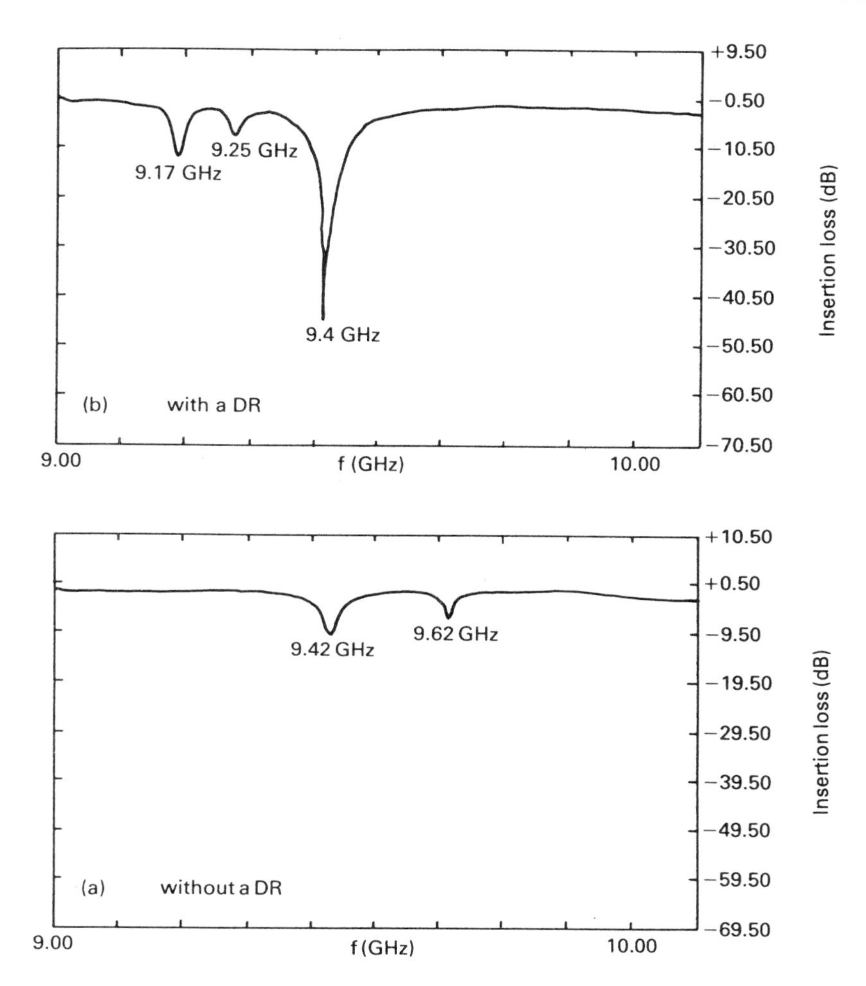

Fig. 7.43 — Insertion loss against frequency of an enclosure without and with a DR. Length of metal shielding = 35.2 mm. Width of metal shielding = 33.8 mm. Height of metal shielding = 7.9 mm. The parameters of the DR are the same as that given in Fig. 7.42.

enclosure top is smaller than in the previous experiment. There were also two cavity resonances near the DR $TE_{01\delta}$ mode. The frequency responses of the enclosure without and with a DR is given in Fig. 7.43(a,b). It is apparent that any change in enclosure dimensions owing to temperature or other causes will change the frequencies associated with the DR system.

There are two possible ways of bypassing the mode problem associated with an enclosed DR system. The first is to operate the DR at a frequency well below the lowest enclosure mode. If the enclosure has two resonant modes and they are sufficiently far apart, then DR operation can take place well between these two modes.

The experimental curves of a properly enclosed DR system are shown in Fig. 7.44. As can be seen, there are only a few modes in the entire 18.5 GHz band. It

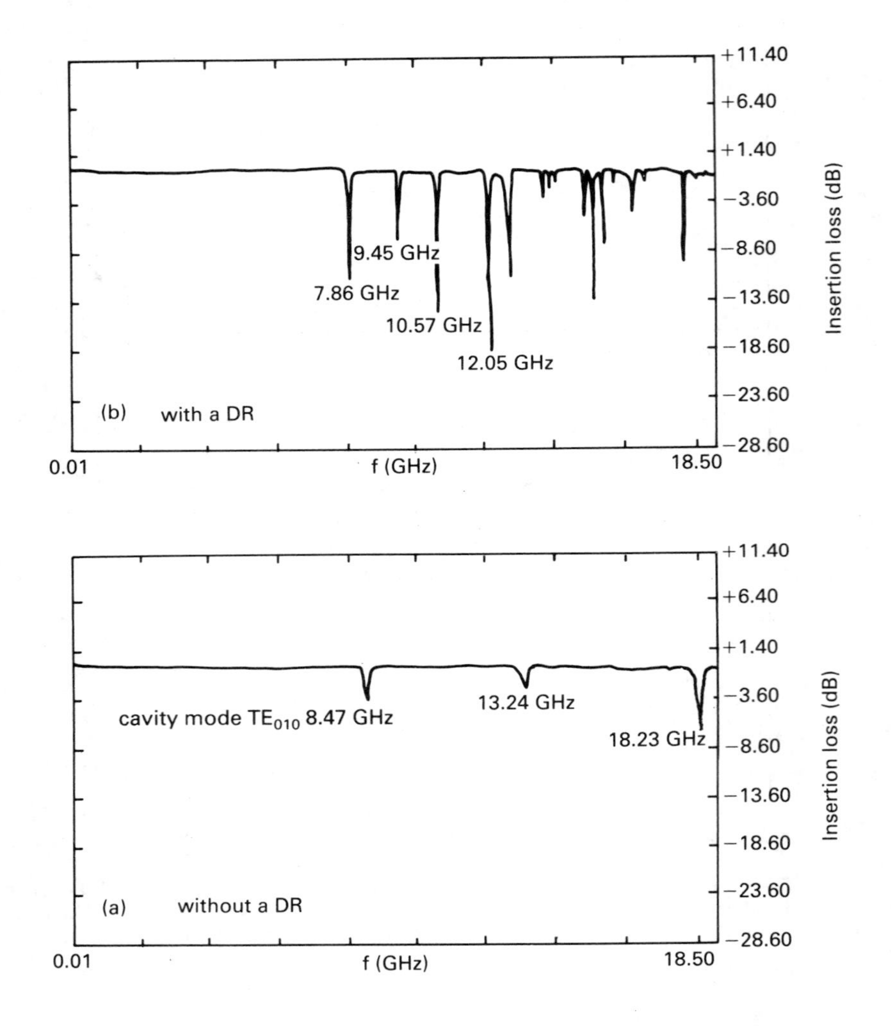

Fig. 7.44 — Insertion loss against frequency of the layout of a precise design. Length of metal shielding = 25.65 mm, Width of metal shielding = 23.4 mm, height of metal shielding = 7.8 mm.

would be reasonable to assume that a DR with a resonance frequency well within 8.47 to 13.24 GHz will be unaffected by the cavity modes. When the DR is inserted into the enclosure, however, the enclosure modes change as shown in Fig. 7.44(b) for a DR with resonance frequency of 9.45 GHz. Despite this, and in this particular case, the DR resonance still lies well between the two new modes of 7.86 GHz and 10.57 GHz.

In Chapter 8 the design of suitable enclosures or boxes for microwave circuits are discussed. Also, programs are presented which should assist box design and mode analysis.

DIELECTRIC RESONATOR OSCILLATOR TERMINOLOGY

Microwave oscillators can be designed with a variety of passive and active devices such as diodes and transistors. A dielectric resonator, suitably placed, can be used to stabilise the frequency of the oscillator. Such an oscillator which encompasses a DR for stabilization purposes is called a dielectric resonator oscillator or DRO.

A basic oscillator consists of an output port for the frequency generated and a d.c. power supply line as shown in Fig. 7.45. More ports or terminals are required if the oscillator is for example to be externally modulated.

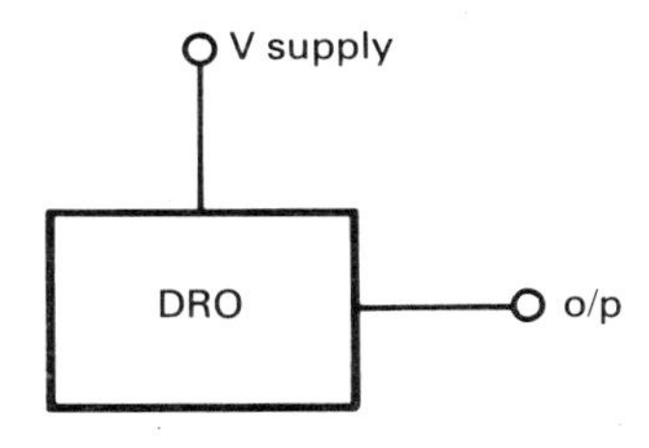

Fig. 7.45 — Basic DRO schematic.

The stability of a DRO depends on a variety of factors.This is outlined in the following together with the relevant terminology.

The following sections deal with the terminology peculiar to dielectric resonator oscillators. Basically they relate to oscillator performance variations owing to changes in supply voltage, loading and temperature.

Frequency stability

Frequency stability refers to the change in oscillator resonant frequency f_o as a function of ambient temperature changes dT, and constant supply voltage. Frequency stability is expressed in parts-per-million per degree centigrade or in short as ppm/°C. Oscillator frequency variation can be positive or negative. Frequency stability can be improved by designing the oscillator in such a way that its temperature variation is opposite to that of the DR puck. This is, however, very difficult since this must be done in the context of the complete oscillator structure, i.e. electrical and mechanical. Alternatively one can use a composite DR (stacked DR) where the positive temperature characteristics is cancelled by the negative one of the composite layers. The following is an example to illustrate frequency stability in a quantitative way.

Example

(1) Assume a centre frequency of $f_o = 12\,\text{GHz}$ at a room temperature of 20°C.
(2) Assume a temperature variation from −30°C to 70°C.
(3) From measurements in a temperature oven we have obtained
   (a) at −30°C a decrease in f_o by 2 MHz;
   (b) at +70°C an increase in f_o by 1 MHz.
(4) Total frequency change is thus $\mathrm{d}f_o = 3\,\text{MHz}$.
(5) In relation to f_o this is $\dfrac{\mathrm{d}f_o}{f_o} = \dfrac{3\,\text{MHz}}{12\,\text{GHz}} = 250\,\text{ppm}$.
(6) From (2) we obtain the temperature range as $\mathrm{d}T = 100°\text{C}$.
(7) The temperature coefficient τ at the centre frequency f_o is thus

$$\tau = \frac{\mathrm{d}f_o/f_o}{\mathrm{d}T} = \frac{250}{100} = 2.5\,\text{ppm/°C}$$

Frequency pulling

Apart from temperature effects, oscillator frequency can be affected by variations in the load which is connected to the oscillator output. This effect is known as pulling and takes place despite the high Q associated with the DR. Changes in load mean mismatch and hence changes in the VSWR. A known technique for improving the VSWR of a circuit is to incorporate an isolator between source and load or an attenuator. A third technique is to incorporate a buffer amplifier which at the same time can be used to boost the output power of the oscillator.

The following example illustrates the calculation of the frequency pulling factor, F_{pul}.

Example

(1) assume a loaded quality factor $Q_L = 1000$;
(2) assume an oscillator output power $P_o = 15\,\text{mW}$;
(3) assume a reflected power $P_r = 3\,\text{mW}$;
(4) the pulling factor F_{pul} may may then be calculated from

$$F_{\text{pul}} = \frac{\mathrm{d}f_o}{f_o} = \frac{1}{Q_L}\sqrt{\frac{P_r}{P_o}} = \frac{1}{1000}\sqrt{\frac{3\,\text{mW}}{15\,\text{mW}}} = 4.47 \times 10^{-3}$$

This shows that F_{pul} is reduced as Q_L increases or P_r decreases.

Frequency pushing

The centre frequency of a DRO changes with supply voltage variation. This is known as frequency pushing. Although this may be exploited for modulation purposes, it is usually not desirable. One way of overcoming the problem is to incorporate an efficient regulator in the DRO. The amount of pushing depends naturally on the total oscillator circuit and especially its Q. As in the previous section,no symbol appears to be assigned to frequency pushing. The symbol F_{push} is thus used here for simplicity. The pushing factor F_{push} is defined as the ratio of permissible frequency variation $\mathrm{d}f_{\mathrm{o}}$ to supply voltage variation $\mathrm{d}V_{\mathrm{s}}$ where $\mathrm{d}V_{\mathrm{s}}$ is with respect to a 1 V reference, i.e.

$$F_{\mathrm{push}} = \frac{\mathrm{d}f_{\mathrm{o}}}{\mathrm{d}V_{\mathrm{s}}} = \frac{\mathrm{d}f_{\mathrm{o}}}{1\,\mathrm{V}} \tag{7.92}$$

It will be known from the circuit into which the DRO is to be incorporated what frequency variation $\mathrm{d}f_{\mathrm{o}}$ can be tolerated without compromising circuit operation. The following is then an example of how to calculate the required supply regulation for a desired pushing factor.

Example

(1) assume an operating frequency of $f_{\mathrm{o}} = 12\,\mathrm{GHz}$;
(2) assume a supply voltage of 12 V;
(3) assume a voltage regulation is required which does not result in a frequency variation or ripple in excess of 10 kHz;
(4) from equation (7.92)

$$\mathrm{d}V_{\mathrm{s}} = \frac{\mathrm{d}f_{\mathrm{o}}(\text{required}) \times 1\,\mathrm{V}}{\mathrm{d}f_{\mathrm{o}}(\text{measured})} = \frac{10\,\mathrm{kHz} \times 1\,\mathrm{V}}{10\,\mathrm{MHz}} = 1\,\mathrm{mV}\ ;$$

(5) regulation $= \dfrac{\mathrm{d}V_{\mathrm{s}}}{V_{\mathrm{s}}} = \dfrac{1\,\mathrm{mV}}{12\,\mathrm{V}} = 0.0833 \times 10^{-3}$;
(6) regulation [%] = 0.00833.

The 12 V d.c. power supply must thus be regulated to 0.00833% in order to keep the frequency ripple to better than 10 kHz.

REFERENCES

[1] Richtmyer, R. D., Dielectric resonator, *J. Appl. Phys.*, **10**, June 1939, pp. 391–398.

[2] Okaya, A., The rutile microwave resonator, *Proc. IRE*, **48**, Nov. 1960, p. 1921.

[3] Okaya, A. and Barash, L. F., The dielectric microwave resonator, *Proc. IRE*, **50**, Oct. 1962, pp. 2081–2092.

[4] Cohn, S. B., Microwave bandpass filters containing high Q dielectric resonators, *IEEE Trans. Microwave Theory & Tech.*, **MTT-16**, April 1968, pp. 218–227.

[5] Masse, D. J., *et al.*, A new low-loss high-k temperature compensated dielectric for microwave applications, *Proc. IEEE*, **59**, Nov. 1971, pp. 1628–1629.

[6] Plourde, J. K., Linn, D. F., O'Bryan, H. M., Jr. and Thomson, J., Jr., $Ba_2Ti_9O_{20}$ as a microwave dielectric resonator, *J. Am. Ceram. Soc.*, **58**, Oct-Nov. 1975, pp. 418–420.

[7] Nishikawa, T., Ishikawa, Y. and Tamura, H., Ceramic materials for microwave applications, *Electronic Ceramics*, Spring Issue, Special Issue on Ceramic Materials for Microwave Applications, Japan, 1979.

[8] Fiedziuszko, S. J., Microwave dielectric resonators, *Microwave Journal*, Sept. 1986, p.p. 189–200.

[9] Stieglitz, M. R., Dielectric resonators: past, present and future, Microwave Journal.

[10] Purnell, M., The dielectric resonator oscillator–a new class of microwave signal source, *Microwave Journal*, Nov. 1981.

[11] Dydyk , M., Dielectric resonator adds Q to MIC filters, *Microwaves*, **16**, Dec. 1977, pp. 150–160.

[12] RESOMICS/MURATA Technical literature on ceramic components

[13] Kawthar, A. Zaki and Chunming, C., New results in dielectric loaded resonators, *IEEE Trans.*, **MTT-34**, July 1986, pp. 815–824.

[14] Kajfez, D. and Guillon, P., Dielectric resonators, 1986.

[15] Christos, T. and Pauker, V., Temperature stabilization of GaAs MESFET oscillators using dielectric resonators', *IEEE Trans. Microwave Theory & Tech.*, **MTT-31**, March 1983, pp. 312–314.

[16] Tsironis, C., Highly stable dielectric resonator FET oscillators, *IEEE Trans. Microwave Theory & Tech.*, **MTT-33**, April 1985, pp. 310–314.

[17] Herczfeld, P. R. and Daryouch, A. S., Optically tuned and FM modulated X-band dielectric resonator oscillator, *Proc. 14th European Microwave Conference*, Liege, Belgium, Sept. 1984, pp. 268–273.

[18] Herczfeld, P. R. and Daryouch, A. S. Optically controlled microwave devices and circuits, *RCA Review*, **46**, Dec. 1985, pp. 528–551.

[19] Guo, K. Z., Yang, R. S. and Chen, Z. G., Optically controlled frequency tuning of dielectric resonator, *Journal of Electronics* (China) Sept. 1991, **13** No. 5, pp. 532–537.

[20] Guillon, P., Byzery, B. and Chaubet, M. Coupling parameters between a dielectric resonator and a microstripline, *IEEE Trans.*, **MTT-33**, No. 3, March 1985, pp. 222–226.

[21] Guo, K. Z., Trinogga, L. A. and Edgar, T. H., DR coupling: improvement to an existing approach, *Electronics Letters*, IEE (UK) March 1988, **24** No. 5, pp. 285–287.

[22] Guo, K. Z. and Trinogga, L. A., 'Calculation of lumped circuit parameters (L,C,R) of a DR microstrip system, *Proc. Second Asia-Pacific Microwave Conference*, Beijing China, 26–28 October 1988, pp. 251–252.

[23] Guo, K. Z. and Trinogga, L. A., DR coupling with microstrip and computer aided analysis, *Journal of Electronics (China)* , **12**, No. 4, July 1990, pp. 385–392.

[24] Watson, G. N., *Theory of Bessel functions* (second edition), Cambridge University Press, 1962.

[25] Results obtained from research between Leeds Polytechnic and Institute of Electronics, Academia Sinica, Beijing, under the auspices of the Royal Society.

[26] Fiedziuszko, S. J. and Jelenski, A., The influence of conducting walls on resonant frequencies of the dielectric resonator, *IEEE Trans. Microwave Theory & Tech.*, **MTT-19**, Sept. 1971, p. 778.

[27] Itoh, T. and Rudokas, R., New method for computing the resonant frequencies of dielectric resonators, *IEEE Trans. Microwave Theory & Tech.*, **MTT-25**, Jan. 1977, pp. 52–54.

[28] Pospieszalski, M. W., Cylindrical dielectric resonators and their applications in TEM line microwave circuits, *IEEE Trans. Microwave Theory & Tech.*, **MTT-27**, March 1979, pp. 233–238.

[29] Jaworski, M. and Pospieszalski, M. W., An accurate solution of the cylindrical dielectric resonator problem, *IEEE Trans.*, **MTT-27**, July 1979, pp. 639–643.

[30] Wada, K., Nagata, E., and Haga, I., Wideband tunable DR VCO, *Proc. 15th European Microwave Conference*, 1985, pp. 407–412.

[31] Chan, O. Y. and Kazeminejad, S., Voltage controlled oscillator using dielectric resonator, *Electronics Letters*, June 1988, **24**, No. 13, pp. 776–777.

[32] Nesic,A., A new method for frequency modulation of dielectric resonator oscillators, *Proc. 15th European Microwave Conference*, 1985, pp. 403–406.

[33] Krupka, J., Magnetic tuning of cylindrical $H_{01\delta}$-mode dielectric resonators, *IEEE Trans.*, **MTT-37**, April 1989, pp. 743–747.

8

Mixers, circulators, isolators, cavities

Mixers have been used for many years for the conversion of higher frequency bands to a much lower frequency band which still contains the same intelligence as its high frequency counterpart. The reason for mixing, also known as frequency conversion, is that it is easier to build signal processing circuits at lower frequencies. Intermediate frequency amplifiers, limiters and detectors are examples of such signal processing networks. Although mixers have been studied extensively, there is, as in many other fields, a continuous need for better understanding and improvement. These can be in the form of extending higher frequency limitations, better reliability and reproducibility, lower conversion power loss (CPL), lower noise figure and production costs.

Mixing circuits are used in almost all microwave communication systems, radar systems and radio astronomy receivers. A mixer consists usually of a network comprising one or more non-linear devices and a means of coupling signals into and out of the mixer. The performance of a mixer can have a marked effect on the performance of a microwave system. A low power conversion loss and/or a low noise figure allow for system design trade off or allow signals of good quality to be received which would otherwise not be possible.

CLASSIFICATION OF MIXERS

Various ways can be chosen in an attempt to classifiy mixers. One could for example classify mixers in terms of the non-linear element employed, by the number or configuration in which the elements are interconnected or the kind of termination which the signal port presents to the image frequency. More specifically, if the non-linear characteristics of diodes is used in the mixing process, one distinguishes between single-ended mixers (one diode), single balanced mixers (two diodes) and double balanced mixers (four diodes). Double balanced mixers are also known as lattice mixers or ring mixers. Furthermore, mixers can be classified as harmonic mixers, image recovery and image rejection mixers. Each of these configurations will be briefly discussed later on. The presence of one or more non-linear elements is essential to the mixing process, i. e. if there is no non-linearity present then no mixing will take place. The basic circuit configurations for diode mixers are shown in Fig. 8.1.

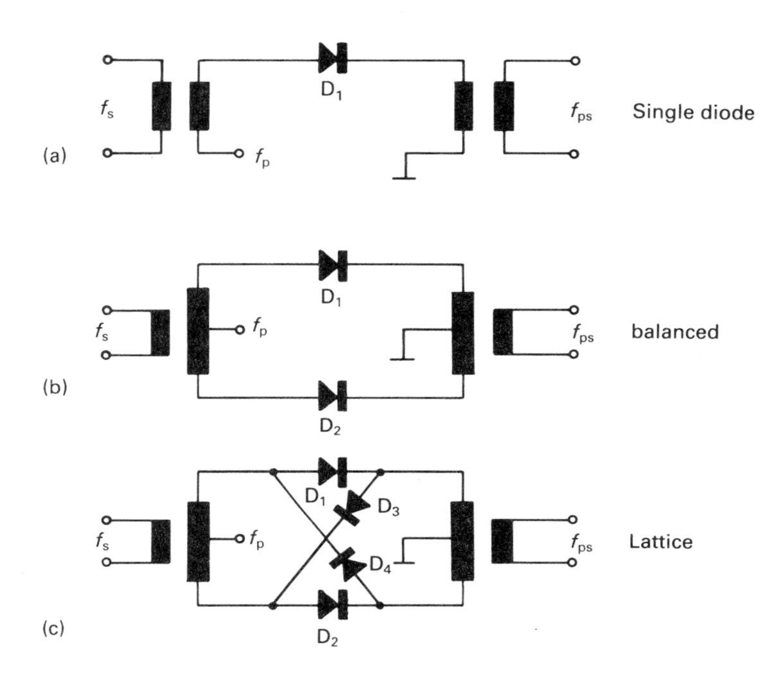

Fig. 8.1 — Diode mixer configurations.

The local oscillator, in the context of mixers, is also referred to as the pump. Thus when a signal frequency f_s and a local oscillator frequency f_p are applied to a mixer as shown in Fig. 8. 1, it produces mainly three kinds of modulation products, namely $f_p + f_s$, $f_p - f_s$ and $2f_p - f_s$. The latter is known as the image frequency and $f_p - f_s$ as the intermediate frequency (IF or f_{ps}). The image termination of the mixer can be of the same magnitude as that for the IF; it can be a short circuit or an open circuit. Accordingly, one could classify mixers as broadband, narrowband short circuit or narrowband open circuit mixers.

Single ended mixer

The single ended mixer is the simplest configuration possible and is shown diagrammatically in Fig. 8.1(a). There are two possibilities of incorporating the non-linear device into the mixer, either in shunt or in series as shown in Fig. 8.2. Since we will be only discussing diode mixers we shall refer to the non-linear device directly as diode. Filters are usually incorporated at the input and output of the mixer. The input filter stops the IF entering the input port whilst the output filter only allows the IF to enter subsequent IF stages. The diode is driven or pumped by the local oscillator which

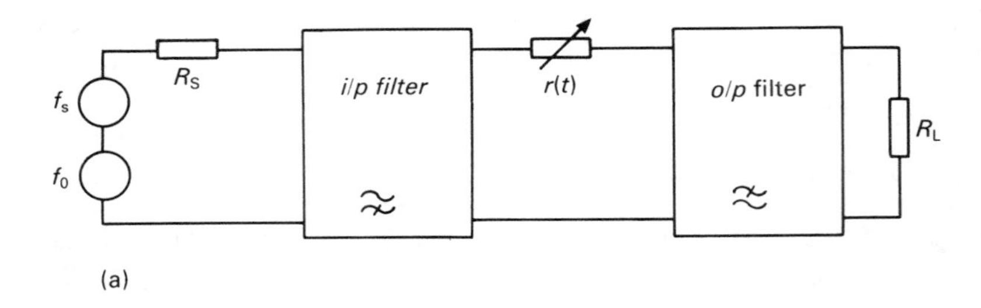

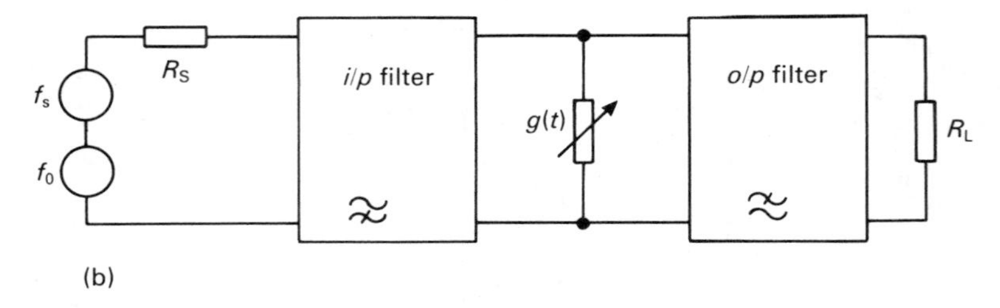

Fig. 8.2 — Single ended mixer configurations, (a) series mixer, (b) shunt mixer.

produces the time varying conductance $[g(t)]$ or resistance $[r(t)]$ as necessary for the mixing action. A more detailed description of the mixing action will be given shortly.

The advantage of simplicity brings certain drawbacks in that there exists no isolation between the input port and the local oscillator. That means that the oscillator not only drives the mixer diode, but will be able to enter the input port and radiate out of subsequent input circuitry, ultimately an aerial as far as a receiver is concerned. Although filters can circumvent this problem, their use in turn restricts the frequency range over which the mixer can be usefully employed.

Referring to Fig. 8.1(a), the mathematical aspect of mixing can be described in the following manner. If an input signal $A \cos \omega_a t$ and a constant local oscillator frequency $B \cos \omega_o t$ are applied to a mixer then this is tantamount to a multiplication. As a result, two frequencies are produced at the mixer output, the sum and difference frequencies of the applied frequencies. Thus

$$A \cos \omega_a t \times B \cos \omega_o t = \frac{AB}{2} [\cos(\omega_a + \omega_o)t + \cos(\omega_a - \omega_o)t] \tag{8. 1}$$

where $(\omega_a + \omega_o)$ is known as the sum frequency and $(\omega_a - \omega_o)$ as the difference frequency or intermediate frequency, the IF. By means of a filter one or the other frequency or frequency band can be filtered out. The frequency spectrum as it exists directly at the mixer terminals is shown in Fig. 8.3. The mirror frequency of the signal frequency with respect to the local oscillator is known as the image frequency. Depending on the mixer design and because of the closeness of this frequency to the

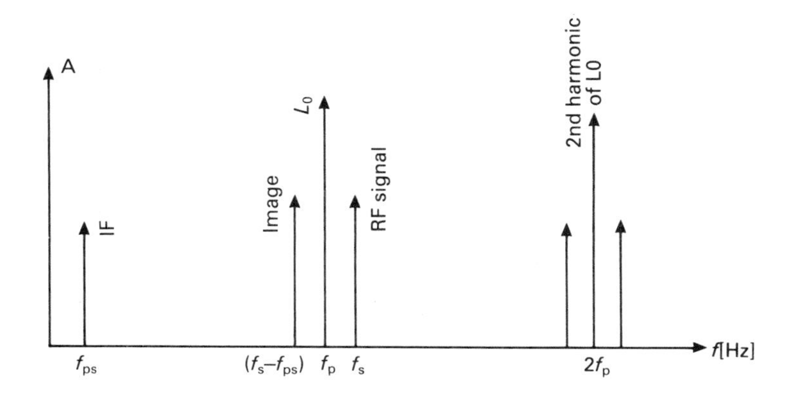

Fig. 8.3 — Frequency spectrum of a mixer.

local oscillator, the image frequency can markedly affect the performance of any mixer. Finally, another drawback of the single diode mixer is that any noise on the local oscillator will be mixed with the signal frequency and thus deteriorate the noise figure of the single ended mixer. Some of these problems can be reduced through the use of a balanced mixer.

Single balanced mixer

The single balanced mixer comprises two diodes as shown in Fig. 8.1(b). When connected they are pointing in the same direction. For a positive cycle of the oscillator at f_p both diodes D_1 and D_2 are conducting $[r(t) = 0]$ and they are cut off during the negative cycle when $r(t) = \infty$. The signal frequency f_s is thus alternately passed through the mixer depending on the pump frequency. It is thus apparent from this explanation that the diodes can be looked upon as switches controlled by the pump. Because of this the local oscillator is mathematically often represented by a square wave, which is a signum function or switching function. Because of the symmetry no oscillator voltage can develop at the signal port and, additionally, there is a tendency for any noise present on the oscillator to be reduced.

Double balanced mixer

A double balanced mixer configuration is shown in Fig. 8.1(c) and it can be seen that it employs four diodes. During the positive cycle of the local oscillator diodes D_1 and D_2 are conducting whilst D_3 and D_4 are reverse biased. During the negative cycle D_1 and D_2 are off and D_3 and D_4 are on. It can also be seen that a phase reversal of 180° takes place at the output of the mixer.

Image rejection/recovery mixer

At the beginning of this chapter and with reference to Fig. 8.3, it was pointed out that the image frequency can have a marked effect on mixer performance, that is the mixer conversion power loss (CPL). As can be seen from the same figure the image frequency is given by f_p-f_{ps} or $2f_p$-f_s. The mixer can now be designed in such a manner that the image frequency within the mixer sees a particular impedance or, in other words, is terminated in a chosen impedance. For example, the image could be terminated in an open circuit, a short circuit or a matched load. Clearly, for either case the CPL of the mixer would be different.

If the image frequency is reactively terminated in such a way that it contributes to an improved CPL of the mixer, then this is referred to as image recovery or image enhancement, and the mixer is given the appropriate name. Now consider a frequency at the signal port which has the same value as the image frequency. This frequency will mix with the local oscillator and will also produce an IF. This may or may not be desirable. Mixers which are insensitive to this second case of IF generation are called image rejection mixers (IRM). The IRM achieves its image rejection through phase cancellation and not through filtering. The frequency difference between the image and other desired inputs is thus of minor concern. In order to obtain good image rejection, however, a high coupler directivity (see Chapter 9) is required as well as low mixer VSWR, good amplitude and phase matching.

The operation of an image reject mixer is now briefly described with the aid of Fig. 8.4. The mixer unit consists of two quadrature couplers (Chapter 9) and two

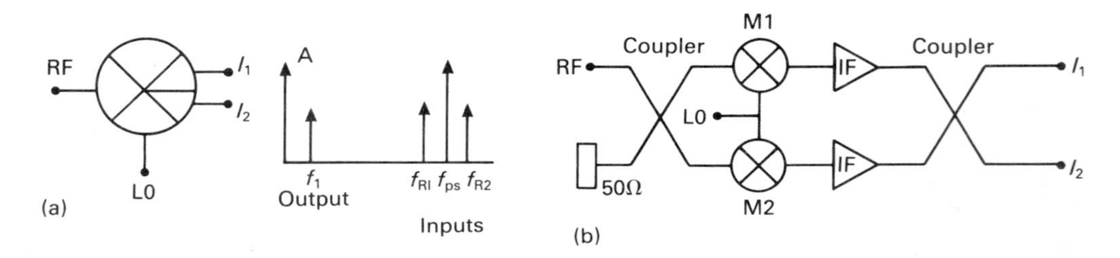

Fig. 8.4 — (a) Image reject mixer; (b) Detailed diagram.

matched (identical) mixers. The two IF-amplifiers are optional. If the frequency f_{R1} of Fig. 8.4(a) is applied to the mixer RF input, then this will be mixed (down-converted) and leave the mixer output port I_2. If f_{R2} is applied or is present it will mix and exit at port I_2. If f_{R1} is the wanted signal, then f_{R2} is its image. Theoretically none of the down-converted image frequency should exit at the desired IF-port I_1. In practical circuits, however, some of the image will enter the desired IF-port. The difference between the wanted IF and the unwanted image frequency level is defined as image rejection. If for example the mixed image frequency produces a signal level of -35 dBm at port I_1 and if the mixed wanted signal frequency produces a signal level of -10 dBm at I_1, then the image rejection is $35-10=25$ dB.

Harmonic mixer

At very high microwave frequencies, i.e. at mm-wave frequencies it is difficult or expensive to produce highly stable local oscillators for mixers. One way to overcome this problem is to use a lower frequency oscillator and taking advantage of one of the harmonics generated by the time varying conductance of the mixer diode. The IF is usually obtained from the frequency difference $f_p - f_s$, but it may also be obtained from the second harmonic $2f_p - f_s$. The efficiency of the mixer decreases, however, with the increase of the harmonic used.

GENERAL LATTICE MIXER THEORY

The lattice mixer theory has been developed over a number of years [1, 2, 3, 4]. It has been shown that low-loss mixers can be obtained by means of an externally driven diode lattice configuration. Some of the published analyses were based on the assumption that the diodes were switched between low forward and high reverse resistance (ON–OFF). Some researchers [5] based their analysis on the assumption that the diodes were current driven and thus acted as resistive switches. As a result of this, all relevant mixer relationships are expressed as functions of the oscillator current.

The local oscillator power in lattice mixers is typically 500 times larger than the signal power which permits the mixer analysis to be conducted in terms of linear network theory. The ideal criteria for mixer analysis are thus as follows:

(a) matched diode characteristics;
(b) resistive rectifiers;
(c) resistive terminations at mixer input and output port;
(d) perfect frequency selective filters;
(e) high local oscillator-to-signal power ratio

For a current driven lattice mixer as shown in Fig. 8.5, the local oscillator waveform

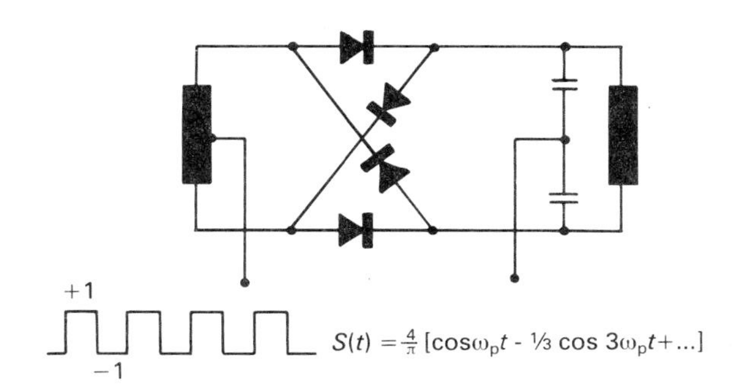

Fig. 8.5 — Switching of a lattice mixer.

may be expressed as

$$I(t) = \frac{1}{2}\, i_p \cos \omega_p t + \frac{1}{2}\, i_p \cos \omega_p t \times S(t) \tag{8. 2}$$

where i_p is the current from the local oscillator, also known as the pump. $S(t)$ is a switching function which describes the state of the outer or inner lattice diode pair. For $S(t) = +1$ the outer diode pair is conducting (ON) and

$$I(t) = \frac{1}{2}\, i_p \cos \omega_p t + \frac{1}{2}\, i_p \cos \omega_p t \times (+1) = i_p \cos \omega_p t \tag{8. 3}$$

For the same diode pair, but with $S(t) = -1$, we have $I(t) = 0$, i.e. the diode pair is reverse biased. The same reasoning applies to the inner diode pair. An alternative way of expressing the switching function is in terms of its Fourier series.

$$S(t) = \frac{4}{\pi}\left(\cos \omega_p t - \frac{1}{3} \cos 3\omega_p t + \frac{1}{5} \cos 5\omega_p t - \ldots\right) \tag{8.4}$$

Owing to this and the assumption of ideal filters within the mixer, a mathematical analysis can be more easily performed. Taking the phase difference of 180° between the diode pairs into account, the conductance $g(t)$ of the diode can be expressed as

$$g_+(t) = \frac{1 + X \cos \omega_p t + X \cos \omega_p t \times S(t)}{r_b + r_s\,(1 + X \cos \omega_p t + X \cos \omega_p t \times S(t)} \tag{8.5}$$

and

$$g_-(\mathrm{t}) = \frac{1 - X \cos \omega_p t + X \cos \omega_p t \times S(t)}{r_b + r_s\,(1 - X \cos \omega_p t + X \cos \omega_p t \times S(t)} \tag{8.6}$$

where r_b and r_s are the barrier and series resistance of the lattice diodes and where X relates the local oscillator current i_p to the diode saturation current I_s. The above equations have been used [5] to derive the general impedance and conversion loss equations for different types of mixers. Figure 8.6 indicates the constraints on the filters of the lattice mixer which make a simplified mixer analysis possible.

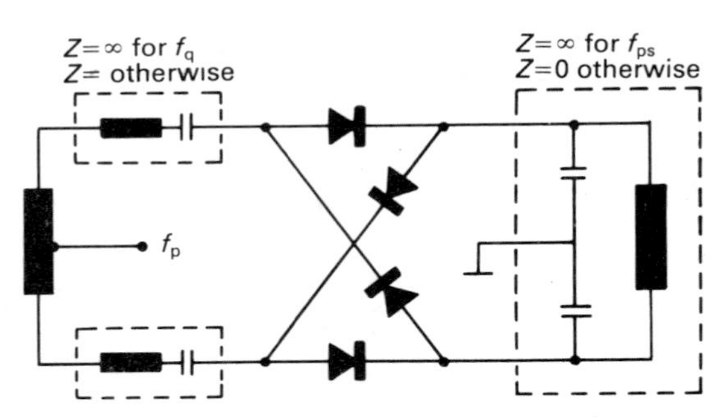

Fig. 8.6 — Filter constraints on a lattice mixer.

For mixer circuits there are normally three types of diodes in use: the point contact diode, the Schottky barrier diode and the backward diode. The static characteristic and the construction of these diodes is shown in Fig. 8.7. The spreading

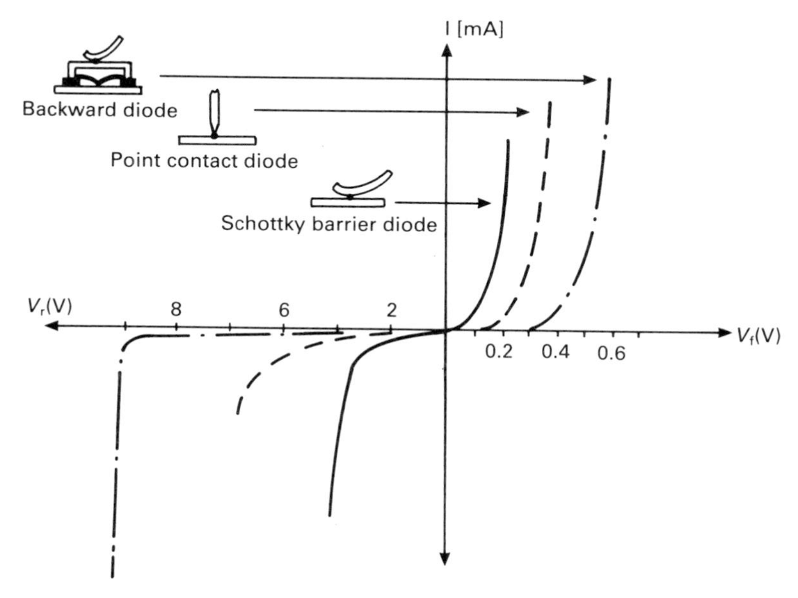

Fig. 8.7 — Static characteristics of diodes.

resistance R_s and the junction capacitance C_j of the devices have a detrimental effect on the diode conversion efficiency and the noise figure. New manufacturing techniques and better semiconductor materials have considerably reduced diode parasitics. The lowering of the R_s C_j-product resulted in higher cut-off frequencies and lower noise figures.

Using the diode equivalent circuit in Fig. 8.8, the total diode conversion loss can be expressed as $L = L_1 + L_2 + L_3$. The loss L_1 depends on the degree of match obtained at the RF and IF ports of the mixer. Since the mixer diodes are biased by the

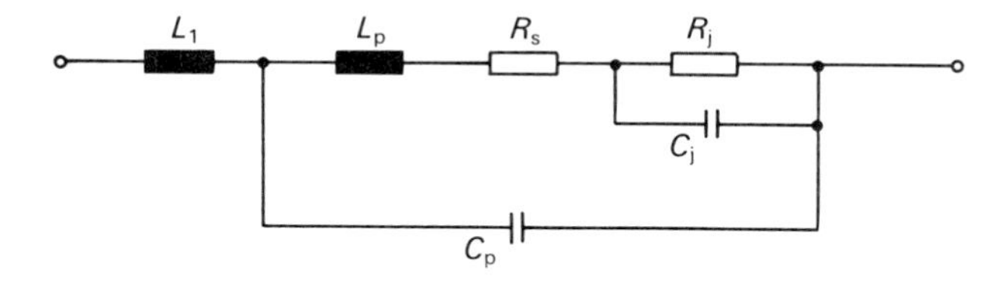

Fig. 8.8 — Equivalent circuit of a mixer diode, packaged.

local oscillator power, the optimum match settings for the oscillator and signal do occur at different tuner settings. The loss L_2 depends on the device parasitics R_s C_j. L_2 is a minimum for a certain drive. If the latter is increased beyond this value, the power dissipation in R_s increases, which, in turn, increases L_2. The third type of loss depends on the static characteristics of the diode used, i.e. the loss at the diode junction. The noise of the diode is due to the device parasitics (real part of Z) as shown in Fig. 8.8.

CONVERSION POWER LOSS MEASUREMENT

The conversion power loss (CPL) is one of the important factors of a mixer. It relates the mixer output power, which is at the intermediate frequency, to the mixer input power, which is at the radio frequency. The CPL is thus a measure of the conversion efficiency of a mixer.

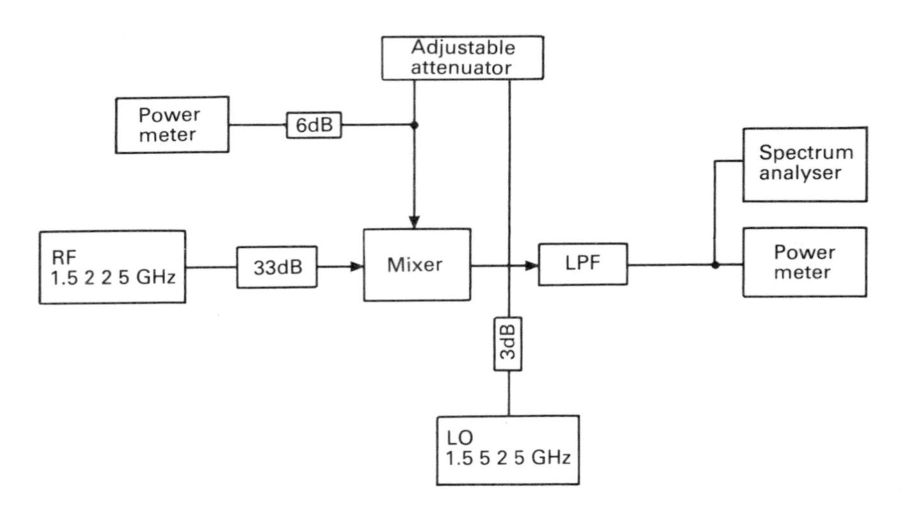

Fig. 8.9 — Arrangement for measuring CPL of a 1.5225 GHz mixer.

A typical arrangement for measuring CPL is shown in Fig. 8.9. The attenuator between the mixer input and RF or microwave signal generator reduces impedance mismatch. The same applies to the local oscillator port. The mixer output is followed by a low-pass filter which passes the IF only and which prevents any RF or oscillator breakthrough. The low-pass filter insertion loss must be known or measured beforehand. The power is then measured at the RF input port and at the IF output port and the CPL calculated using the relationship

$$\text{CPL [dB]} = 10 \log \frac{P_{\text{RF}}}{P_{\text{IF}}} \qquad (8.7)$$

A typical conversion power loss curve is shown in Fig. 8.10. From this it can be seen that if there is no oscillator drive, then CPL$\rightarrow \infty$. For a particular oscillator power the CPL will be a minimum. Thereafter the losses increase again owing to the more marked effect of losses within the diodes of the mixer, in other words the diode losses outweigh the benefits which are to be gained from a higher oscillator power.

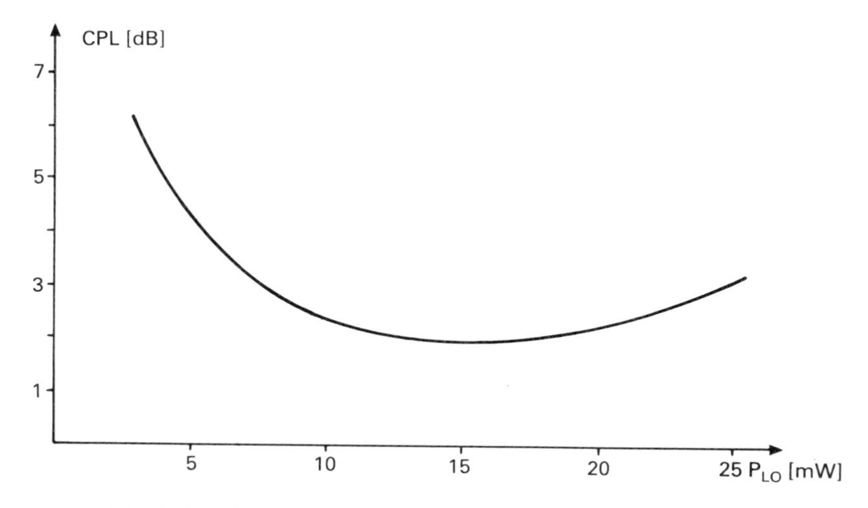

Fig. 8.10 — CPL of 1.5225 GHz mixer as a function of LO drive.

CIRCULATORS AND ISOLATORS

Circulators and isolators take advantage of the gyromagnetic properties of ferrites. The two components are non-reciprocal, i. e. they work ideally in one direction only. An isolator is a special case of the three port circulator in that one port is terminated in a resistor, usually 50 Ω. The symbols for a circulator and an isolator are shown in Fig. 8.11.

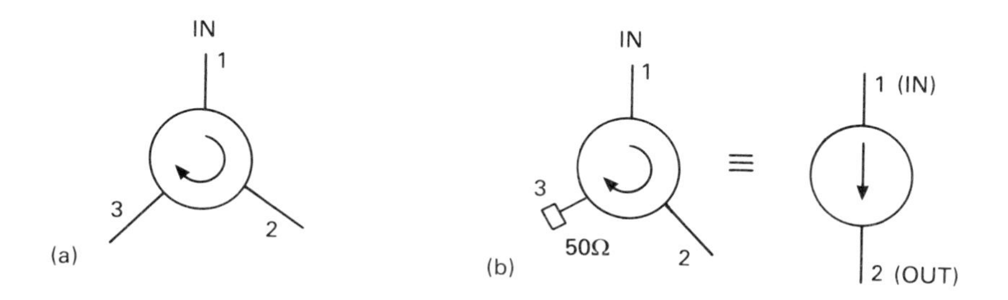

Fig. 8.11 — Symbols of (a) circulator, (b) isolator.

Referring to Fig. 8.11(a), the operation of a circulator is briefly as follows. A wave incident on port 1 couples to port 2 only. In a similar manner, a wave incident on port 2 couples to port 3 only. The applied or incident wave propagates in a clockwise direction owing to the properties of the ferrite in the component.

Referring to the symbol of an isolator, Fig. 8.11(b), a signal incident on port 1 couples to port 2. If we were to apply a signal to port 2, ideally none of it would emerge at port 1. In a good practical isolator, however, the signal emerging from port

1 would be -25 dB or more, in other words, isolation would be better than 25 dB. Applied to a microwave circuit, an isolator prevents power being reflected back to the source, i. e. it protects the source from the circuit it is supplying.

The layout of a typical isolator and circulator for microstrip applications is shown in Fig. 8.12. In the case of the isolator the third port is terminated with a 50 Ω chip

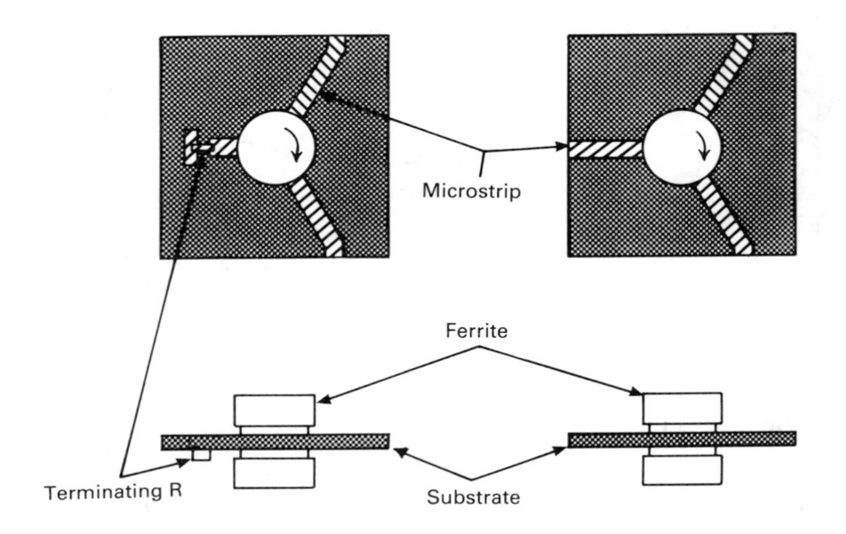

Fig. 8.12 — Layout of a typical isolator and circulator.

resistor. The normal direction of circulation is usually clockwise. The smaller the size of the component, the higher the operating frequency. The metallization on the isolator substrate is normally undertaken with a sputtering process, followed by a titanium adhesion layer and a final gold film. The required connecting pattern is produced using conventional photolithographic and etching techniques. Interconnection between the isolator or circulator to adjacent microstrips can be performed in a variety of ways which include gold ribbons, soldering or thermo-compression bonding. The component is brittle and must be handled with care owing to the materials used. This is especially true when mounting. A standard technique is clamping, as shown in Fig. 8.13. To ensure that the isolator operates as per specification, a gap of about 10 mm should be provided between the ferrite and the enclosure top and bottom. Finally, as with many microwave components, it is best to follow the manufacturer's instructions if soldering techniques are used. Otherwise it is possible that the gold film dissolves or material migration takes place, resulting in a bad joint. The performance of a typical Mullard isolator for 2.5 GHz at an ambient temperature of 23°C is stated in Table 8.1 and a measured curve is given in Fig. 8.14.

A 2.5 GHz FRONT END DESIGN

The following is an outline of how to approach the design of a simple microwave receiver input stage based on the information provided in this book. A microwave

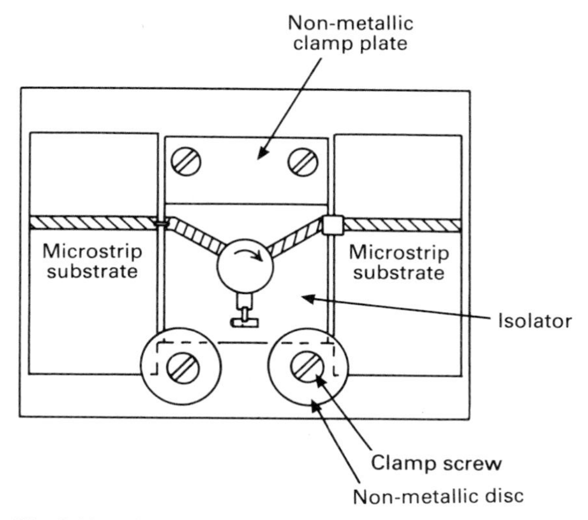

Fig. 8.13 — Clamp mounting of isolator drop insert.

Table 8.1 — Typical isolator performance

f [dB]	i/p L_{ret} [dB]	L_{ins} [dB]	Isolation [dB]	o/p L_{ret} [dB]
2.48	−25.96	−0.25	−25.73	−26.66
2.50	−26.33	−0.25	−25.84	−26.65
2.52	−26.20	−0.25	−25.74	−26.11
2.54	−25.92	−0.25	−25.54	−25.63

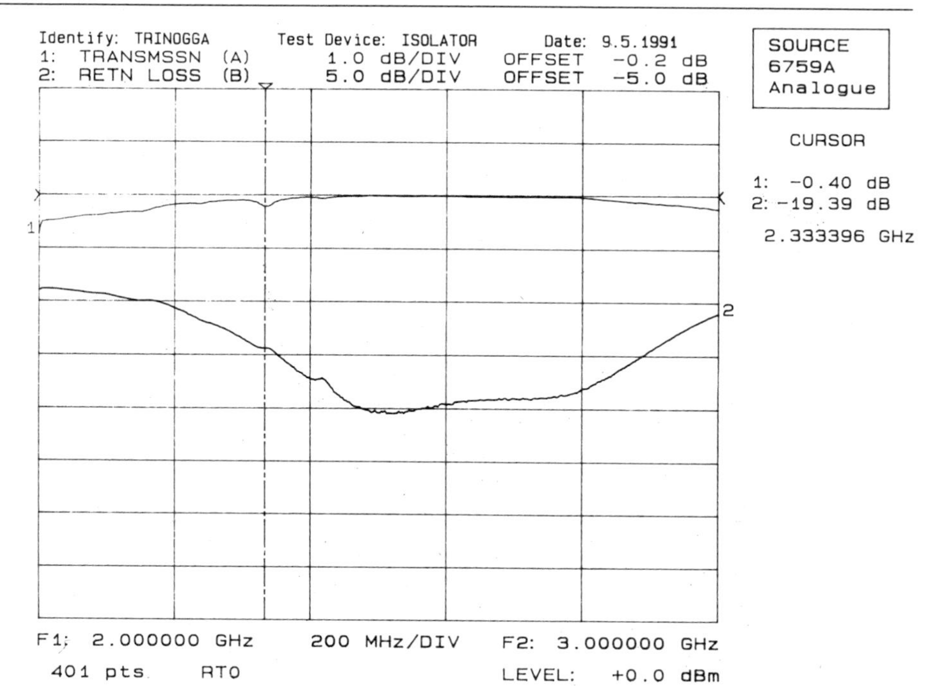

Fig. 8.14 — Frequency response and return loss of an isolator.

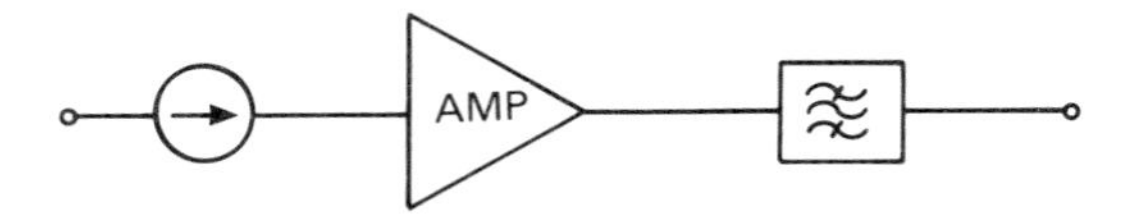

Fig. 8.15 — Block diagram of microwave front-end input stage.

input stage in the form of an amplifier with as high a gain is required with a bandwidth of about 10% of the centre frequency. The input stage is to precede a microwave receiver in order to increase its noise performance and selectivity as well as the rejection of certain out of band frequencies. The microwave receiver itself has a noise figure of 7 dB.

It is not always easy to build a microwave input stage comprising an amplifier stage alone to meet certain criteria. One way to simplify the design and at the same time improve it is through the use of filters and isolators. A possible solution is then to use an isolator, followed by the amplifier and then an output filter, all centred at 2.5 GHz. The block diagram is shown in Fig. 8.15. As was explained earlier on, isolators are commercially available in packaged and unpackaged form or drop-inserts. In order to explore the design approach a two stage amplifier and a 2.5 GHz filter were designed on a printed circuit board. This was preceded by a packaged isolator. Reasonable results were obtained and a more professional approach was then taken.

A 2.5 GHz drop insert isolator was selected and its response was measured as shown in Fig. 8.14. Based on gain/noise considerations and availability, transistor AT 41435-3 was chosen for the two stage amplifier. The S-parameters of the transistor at 2.5 GHz, $V_{ce} = 8$ V and $I_c = 15$ mA were as follows:

$S_{11} = 0.59$, 144°
$S_{12} = 0.068$, 33°
$S_{21} = 3.05$, 40°
$S_{22} = 0.41$, $-58°$

It was decided to build the amplifier on RT-Duroid and the filter on OAK microwave substrate. The parameters were as follows:

RT = Duroid	OAK laminate
$\varepsilon_r = 10.5$	$\varepsilon_r = 2.5$
$h = 0.635$ mm	$h = 0.7874$ mm
$Z_o = 50\ \Omega$	$Z_o = 50\ \Omega$
$f = 2.5$ GHz	$f = 2.5$ GHz

The reason for choosing different substrates was the physical size of the circuits to be fabricated. A filter made on RT Duroid would have been too small for the specified enclosure. This is then an example how the right choice of substrate can be used to meet special physical requirements.

Computer programs POLSTRI, KFACTOR and AMP were then used to obtain all the parameters necessary for the design and final layout. Calculations with program KFACTOR showed the transistor to be unconditionally stable, since $k = 1.175 > 1$.

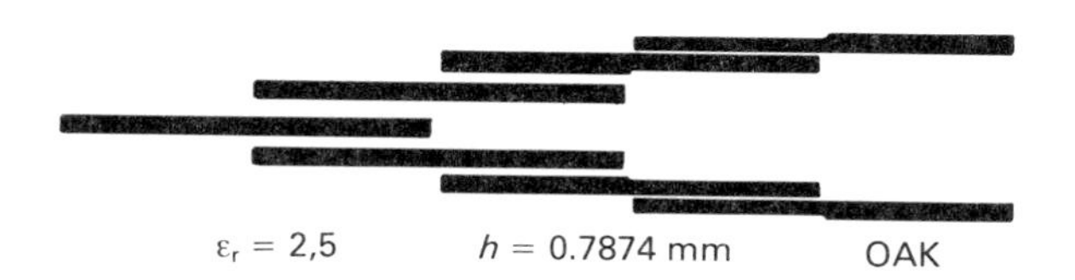

Fig. 8.16 — Folded over filter.

Measurements on the isolator/amplifier configuration showed good performance, although this could be improved upon through the use of a filter at the amplifier output. In order to accommodate the filter in the space available, a fold-over version as shown in Fig. 8.16 was used. An input connector type N was specified for the signal input and a type SMA connector for the output. The type N input socket was soldered to a rigid section of 50 Ω cable which then was soldered to the isolator drop-insert. The isolator, two-stage amplifier and bias circuit were accommodated on one side of the microwave enclosure which was milled from solid aluminium. The 2.5 GHz filter was mounted into the milled recess on the other side. The output of the amplifer was connected through a hole in the base of the box directly to the filter. The 12 V d.c. power was supplied via a feedthrough capacitor. This was connected to a simple transistor bias supply as shown in Fig. 8.17. Good quality ATC microwave capacitors are used for decoupling at the input and output lines of the amplifier stage.

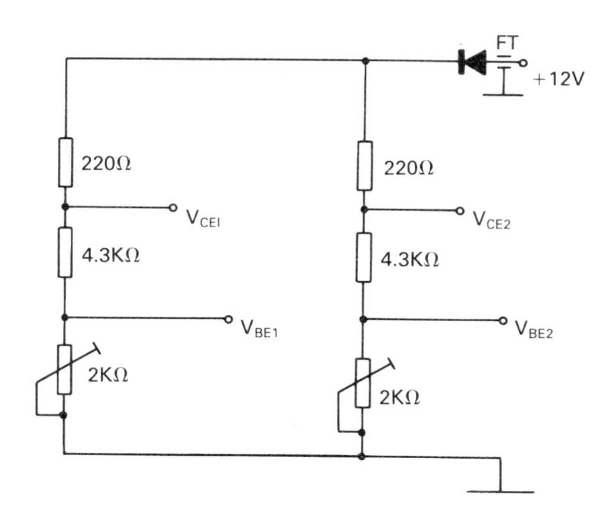

Fig. 8.17 — Example of bias suppply.

The performance of the complete circuit as measured on a Wiltron network analyser is shown in Fig. 8.18 and shows complete rejection of all out of band frequencies over the range of interest. A small amount of trimming needed to be undertaken to optimize the circuit performance. One good way of doing this is to use a small brass washer and push it along the appropriate microstrip until the desired response is obtained, and than solder it in place. Naturally, however accurately one

Identify: TRINOGGA
Test Device: 2.5 GHZ AMP
Date: 9.5.1991
1: TRANSMSSN (A)
5.0 dB/DIV
OFFSET +10.0 dB
2: RETN LOSS (B)
2.0 dB/DIV
OFFSET -20.0 dB
SOURCE
6759A
Analogue
CURSOR
1: +15.77 dB
2: -31.81 dB
2.500010 GHz
2
1
F1: 2.300000 GHz
50 MHz/DIV
F2: 2.700000 GHz
401 pts
RTO
LEVEL: -10.0 dBm

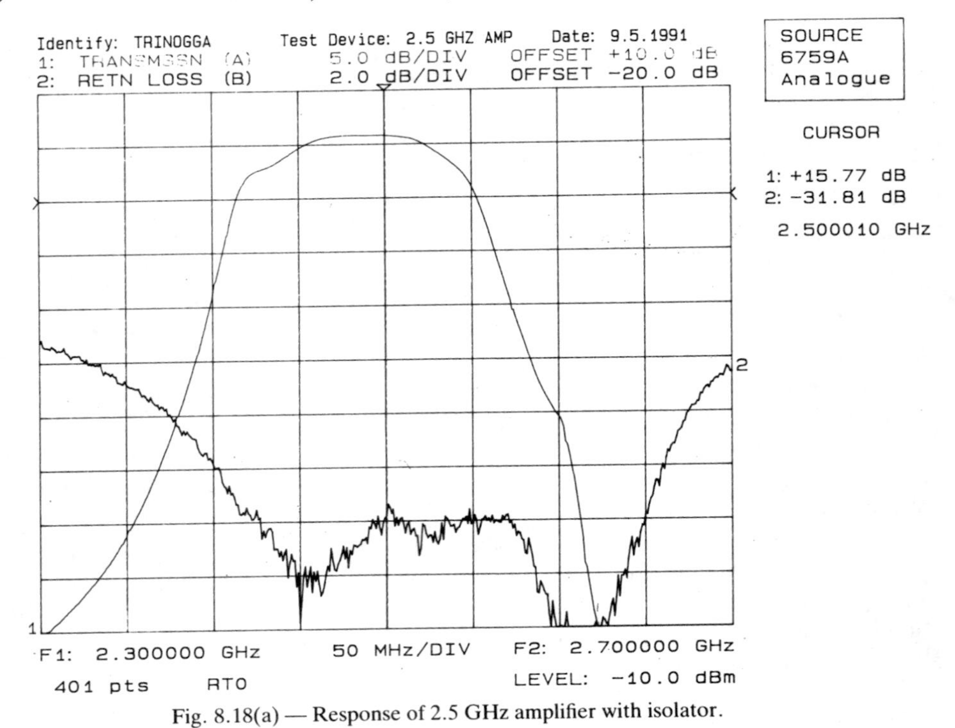

Fig. 8.18(a) — Response of 2.5 GHz amplifier with isolator.

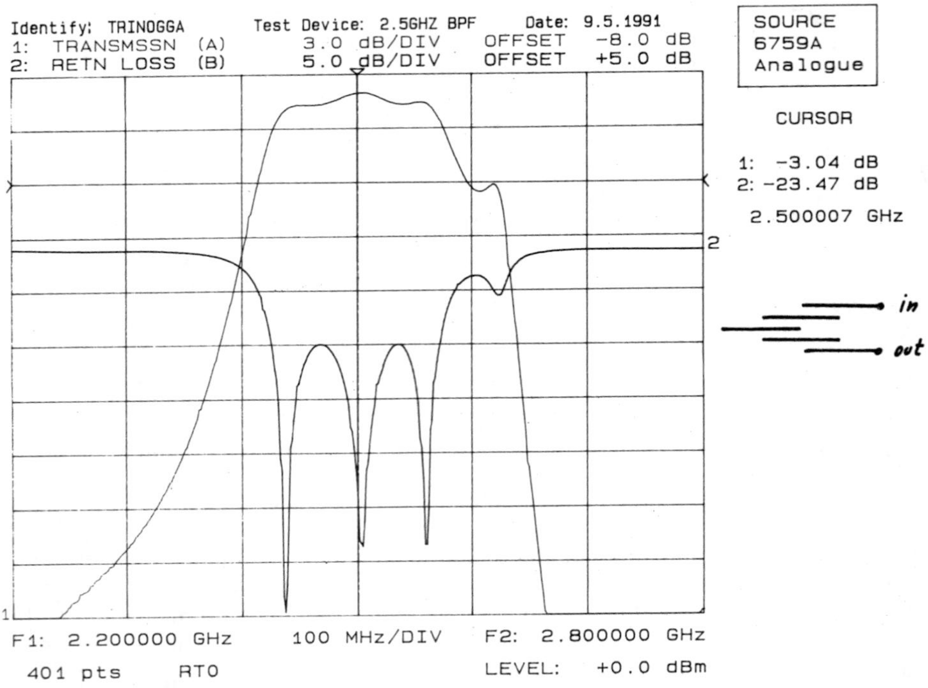

Identify: TRINOGGA
Test Device: 2.5GHZ BPF
Date: 9.5.1991
1: TRANSMSSN (A)
3.0 dB/DIV
OFFSET -8.0 dB
2: RETN LOSS (B)
5.0 dB/DIV
OFFSET +5.0 dB
SOURCE
6759A
Analogue
CURSOR
1: -3.04 dB
2: -23.47 dB
2.500007 GHz
in
out
2
1
F1: 2.200000 GHz
100 MHz/DIV
F2: 2.800000 GHz
401 pts
RTO
LEVEL: +0.0 dBm

Fig. 8.18(b) — Response of 2.5 GHz bandpass filter.

may calculate a given microwave circuit, the ultimate physical performance depends to a considerable extent on the skill of the person involved in its fabrication. The isolator/amplifier side of the front-end is shown on Plate 8.1.

MIXER SUBSYSTEM

If the front end is to form part of a subsystem, then it normally leads to a mixer unit which may or may not have an integral oscillator. A typical mixer unit based on a circular hybrid coupler is shown in Fig. 8.19. A local oscillator in the appropriate

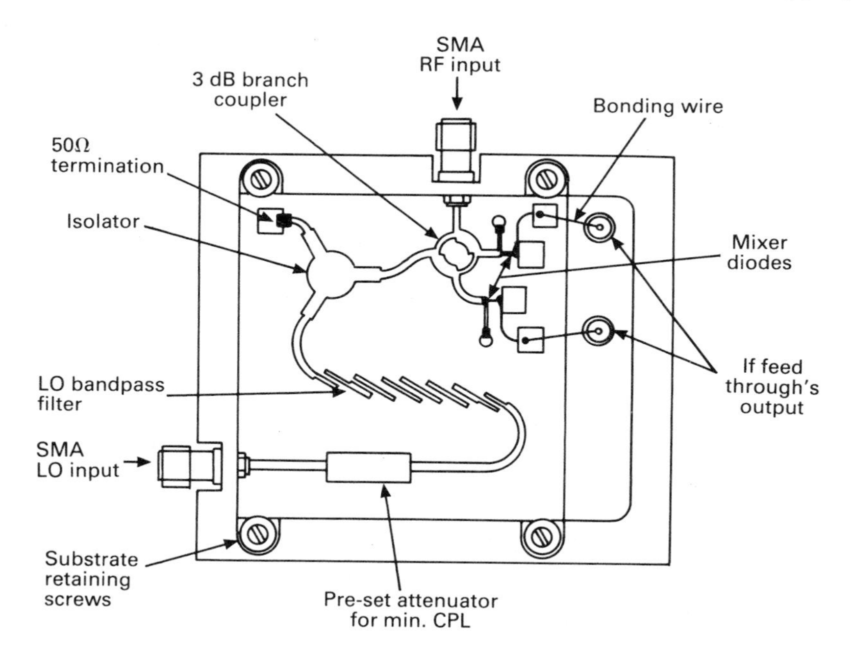

Fig. 8.19 — Mixer unit layout.

frequency range is applied via an attenuator to a bandpass filter. The purpose of the attenuator is to set the local oscillator drive to such a level as to minimize the CPL and any mismatch between filter and oscillator. The local oscillator bandpass filter improves the noise behaviour of the circuit by narrowing bandwidth and hence selectivity. The filter leads to an isolator which is made up of a circulator with a port terminated in a load of characteristic impedance of 50 Ω. The isolator will pose very little resistance to the local oscillator waveform but 20 or more decibel of attenuation to the RF. Thus it effectively stops the RF from entering the local oscillator.

The single balanced mixer is shown in more detail in Fig. 8.20. It consists of a circular 3 dB branch line coupler with a 90° phase shifter in one of its output arms. This ensures that the phase difference at the input to diodes 1 and 2 is 180°. The coupler combines the RF signal and local oscillator at a pair of mixer diodes. These

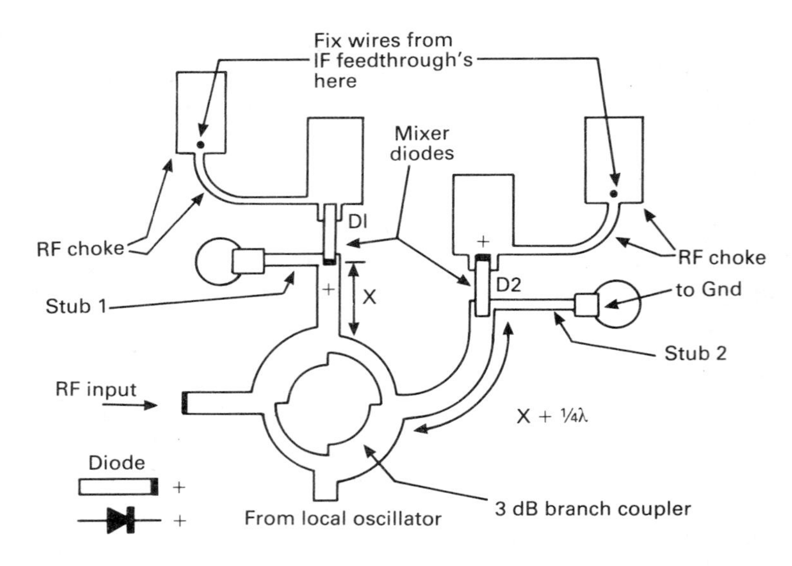

Fig. 8.20 — Mixer circuit layout.

diodes are usually matched to give best mixer performance. The RF and oscillator ports are isolated with respect to each other. The power applied to either the RF or local oscillator port is divided equally between the two output ports. The phase difference at the output ports (1, 2) of the coupler is 90°. In order to convert the 90° branch arm coupler into a circular hybrid of 180°, an additional quarter wavelength is inserted between output port 2 and mixer diode D2. If the diodes do not have an input impedance of about 50 Ω, then matching is necessary, otherwise not. The diodes are connected in opposite polarity to the branch line coupler output.

A schematic and microwave substrate layout of a 1.68 GHz mixer using a square hybrid coupler is shown in Fig. 8.21. Two hot carrier diodes are used for performing the mixing action. The diodes are biased and can be set for best mixing action. The output of the two diodes is connected to a filter matching section L_1, C_1 which filters out the difference frequency. Two RF shorts of 25 Ω are included to prevent the local oscillator from entering the IF port.

CAVITIES

Cavities are often used as a frequency stabilizing component in microwave circuits. Its operation is based on the fact that the electromagnetic field is bounded by the metal wall of the cavity. The stability is largely governed by the thermal properties of the cavity. Special materials such as Invar can be used to avoid any dimensional variations over a large temperature range. As was shown in Chapter 7, cavities are replaceable by dielectric resonators with their attendant advantages. DRs have properties very similar to those of cavities. It is thus essential to have some

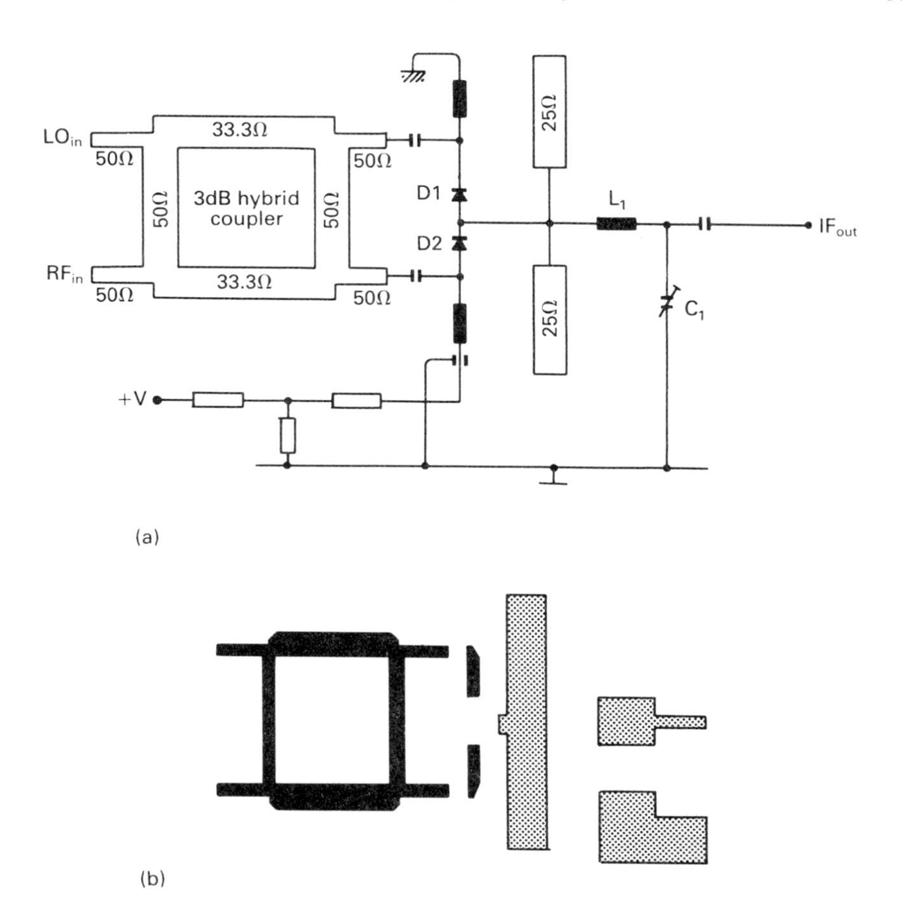

Fig. 8.21 — (a) Schematic diagram of a C-band mixer, (b) substrate layout for mixer.

understanding of those properties since they have an effect on performance when incorporated into a circuit such as an oscillator or filter. It is the quality factor Q of the cavity or DR and that of the attended load which govern overall performance.

At microwave frequencies it is impossible to employ discrete LC circuits for tuning purposes; instead one has to use distributed circuits and components. Cavities and dielectric resonators are examples of such distributed components. Cavities come in different shapes, e.g. they can be rectangular or cylindrical. From a manufacturing point of view the latter is relatively easy to make and is thus found frequently.

It is well known that a discrete component parallel LC circuit can be represented by a quarter wave transmission line, as was discussed in Chapter 4, and as is shown in Fig. 8.22. Theoretically the input impedance Z_{in} is infinite and a pure resistance at resonance. If one considers now a very large number of those quarter wave lines connected in parallel, then this will result in a cylindrical shape, which is known as a

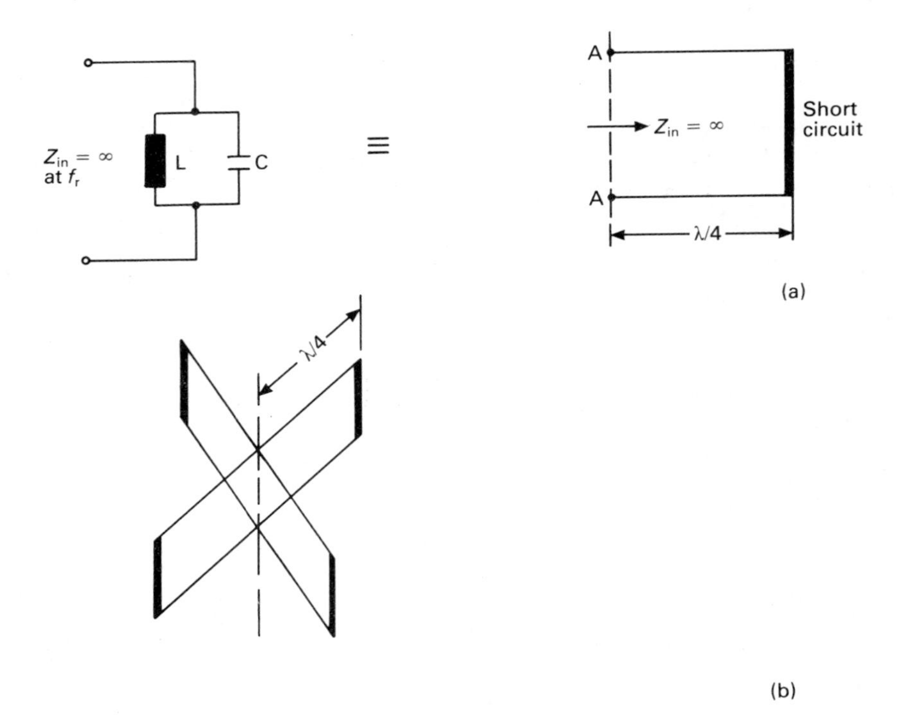

Fig. 8.22 — Evolution of a cavity.

cavity. The radius of this cavity is λ/4. Figure 8.22(b) gives an example of four parallel lines. It is obvious that in this case the line inductance is quartered and the line capacitance quadrupled. Hence the resonance frequency as given by

$$f = 1/[2\pi/(LC)^{0.5}] \tag{8.8}$$

remains the same. By the same token the radio frequency resistance is decreasing the more elements are connected in parallel. Thus, for a cavity, the high frequency resistance is very low and thus from basic circuit theory its quality factor Q is very high. In contrast to a DR, a metal cavity has no radiation loss since the electromagnetic energy is bounded by the metallic cavity walls, i.e. it cannot penetrate the wall. In the case of a DR, the electromagnetic waves leave the DR and enter the immediate air space surrounding the DR and radiation and hence losses occur. DRs with a high dielectric constant have lower losses than those with a lower constant. Losses are typically a few percent.

If a cylindrical cavity or a DR, Fig. 8.23(a), is excited, an electromagnetic pattern is set up in the component as is shown by way of examples. Figure 8.23(b) shows the electric flux lines (drawn as solid lines) parallel to the z-axis with the maximum intensity at the centre and zero intensity at the bounding walls. The magnetic flux (drawn as dashed lines) forms concentric circles around the electric flux lines with the

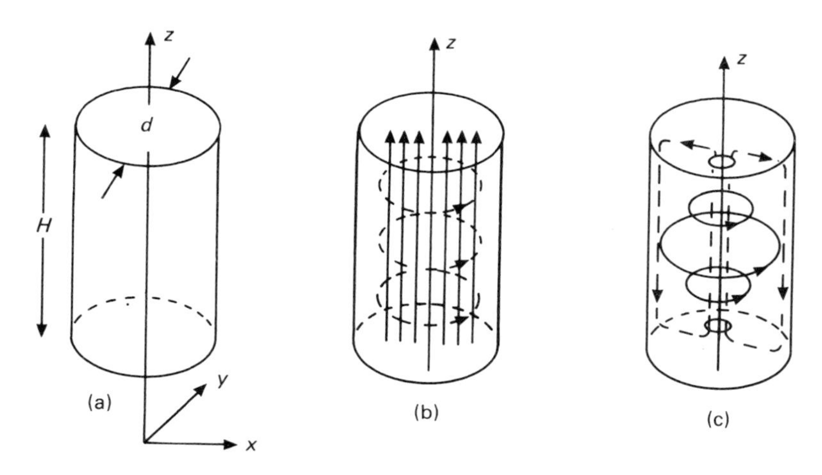

Fig. 8.23 — Examples of cavity modes. (a) cylindrical cavity, (b) TM_{010}, (c) TE_{011}.

highest intensity at the walls and the lowest at its centre. Using the z-axis as a reference, it can be seen that the magnetic field lies transverse to it and hence the notation transverse magnetic or TM. The same applies to the electric field, as shown in Fig. 8.23(c). A cavity/DR cannot only support different modes, but also a number of wave changes depending on resonator size or frequency. Subscripts are thus added, e. g. TM_{mnq} or TE_{mnq} according to the following convention:

first subscript m:	number of full wave changes around circumference
second subscript n:	number of half wave changes across diameter
third subscript q	number of half wave changes along the z-axis

The modes are thus as given in Fig. 8.23(b,c). Further details on this can be obtained from the many books on basic microwave theory.

Apart from a knowledge of which mode a DR or cavity supports, it is important to know the quality factor. The quality factor of a component is a measure of its ability to behave as a pure reactance. In the context of filters and oscillators employing a cavity or DR one has to consider three different factors, namely the unloaded quality factor Q_u, the loaded quality factor Q_L and the external quality factor Q_{ext} (Q_e) which is frequently also referred to as the quality factor of the load. These are now discussed with reference to Fig. 8.24.

Real resonators have ohmic losses and this fact may be expressed in terms of the quality factor Q. For example, the Q of a lossy inductor is

$$Q = \frac{\omega_o L}{r_s} \tag{8.9}$$

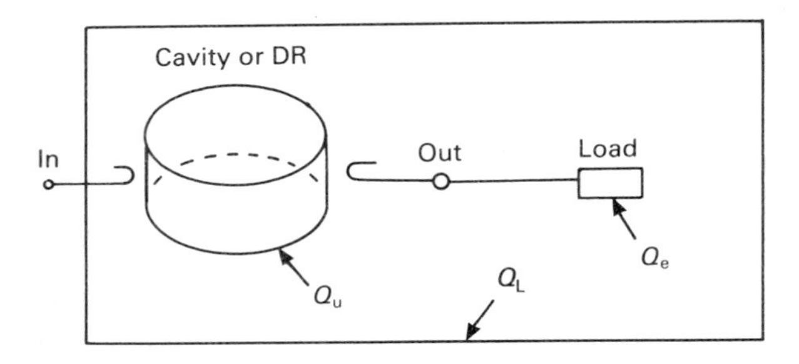

Fig. 8.24 — Definition of Q_u, Q_e and Q_L.

where $f_o = \omega_o/2\pi$ is the resonance frequency and r_s the series resistance representing the resistive losses of the inductor or the dielectric losses of the DR. Thus we have a small amount of loss associated with each cycle of energy storage. By definition the quality factor for the specific case of a DR is thus

$$Q = \omega_o \frac{1/2\ (L\ I_m^2)}{1/2\ (r_s\ I_m^2)} = Q_u = \omega_o \frac{\text{peak energy stored}}{\text{energy loss in DR}} \tag{8.10}$$

If the energy loss is entirely within the DR then the Q represents the unloaded quality factor Q_u of the DR. Clearly, if the DR is connected to a load, then the Q_u of the DR will change and will assume a new value Q_L which is known as the loaded quality factor. The larger the loading the lower Q_L. In line with the previous approach, the Q of the load itself is denominated Q_e, i.e. the Q external to the DR. The loaded quality factor Q_L should not be confused with the quality factor of the load, which is denoted by Q_e. In analogy with eqn. (8.10), the loaded quality factor Q_L is defined as

$$Q_L = \omega_o \frac{\text{energy stored}}{\text{energy loss in DR} + \text{energy loss in load}}$$

or

$$\frac{1}{Q_L} = \frac{\text{energy loss in DR}}{\omega_o \times \text{energy stored}} + \frac{\text{energy loss in load}}{\omega_o \times \text{energy stored}}$$

Hence

$$\frac{1}{Q_L} = \frac{1}{Q_u} + \frac{1}{Q_e} \tag{8. 11}$$

From this we see that Q_L is always smaller than Q_u. The loaded quality factor Q_L should not be confused with the quality factor of the load which is denoted by Q_e or Q_{ext}.

MICROWAVE HOUSING DESIGN

The housing design used for microwave circuits as well as that for integrated circuits and MMICs (packaging) is very important in that unsuitable dimensions can deteriorate the performance of carefully designed circuits or subsystems dramatically. The following gives an outline of the factors which may have to be considered when designing a housing or enclosure for a microstrip circuit.

Be it an amplifier, mixer or filter but to name a few, the circuit is usually put into a housing, cast or milled from aluminium with connectors through the walls onto the appropriate microstrip of the circuit. These connectors may be typically type N, SMA or special SMA for mm-wave applications. The main purpose of the housing is to provide mechanical strength and handling, electric shielding, and heatsinking in the case of high power applications.

A housing with a lid will form a cavity and will thus exhibit resonance. When the microstrip circuit is mounted into the housing then this may be looked upon as a dielectrically loaded cavity resonator. The problem frequently is that the lowest order cavity mode sets a limitation as to the maximum frequency for satisfactory operation for the microstrip circuit. Clearly, if for example a 1.7 GHz filter is accommodated into a housing whose fundamental cavity mode is near 1.7 GHz, or worse, at 1.7 GHz, then the filter will not work properly, if at all. There is a tendency of the lowest order cavity mode to couple to the microstrip circuit with subsequent performance deterioration.

Dielectrically loaded cavities have been treated in detail in the literature [6, 7]. Based on this, others [8, 9] have derived simple expressions which allow the quick calculation of either the box dimensions to give a specified cavity mode or to calculate and check for the cavity mode for given housing dimensions. Based on reference [8], the following provides an outline of the underlying principle. This is supplemented by a computer program and sample calculations using the program BOXMODE.

A typical microwave housing is shown in Fig. 8.25 where w is the width, l the length and h the height of the enclosure. These dimensions are the inner dimensions. The housing may then be analysed for the case where it is completely filled with a uniform dielectric, i.e. air ($\varepsilon_r = 1$) or any other dielectric with $\varepsilon_r > 1$. The TE_{101} fundamental cavity mode will exist for $h < w < l$. In analogy with waveguides there exists an electric field in the y-direction of Fig. 8.25(a) with a maximum at the centre and zero at the cavity wall. The electric field distribution is sinusoidal. The magnetic field is perpendicular to the electric field, that is parallel to the xz-plane with a maximum at he housing walls. The dielectrically loaded housing may be looked upon

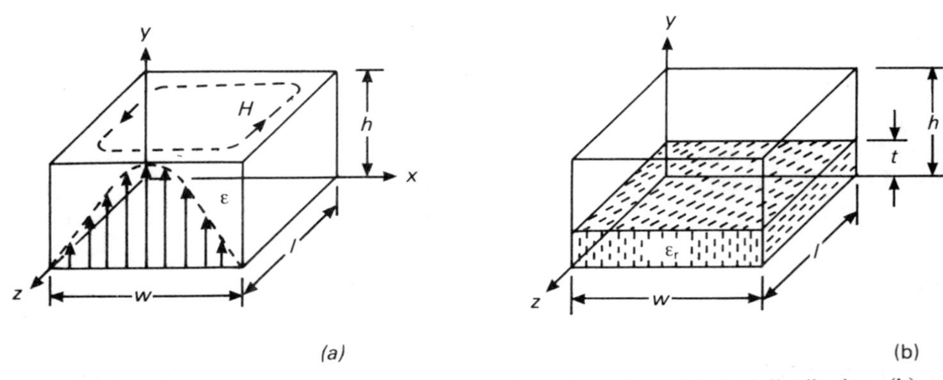

Fig. 8.25 — Typical shape of a microwave circuit enclosure/box. (a) field distribution, (b) dielectric loading by microwave substrate.

as a lumped circuit LC resonator with the resonance frequency, inductance and capacitance given by:

$$f_{101} = \frac{\frac{15}{w}\sqrt{1+\left(\frac{w}{l}\right)^2}}{\sqrt{\varepsilon_r}} \tag{8.12}$$

$$L = \pi \frac{wh}{l\left[l+\left(\frac{w}{l}\right)^2\right]} \tag{8.13}$$

$$C = 3.59\ \varepsilon_r \frac{w\ l}{h} \tag{8.14}$$

where w, l and h are in cm, f in GHz, C in pF and L in nH.

If a microstrip circuit is now placed into an identical housing (Fig. 8.25(b)) then the mathematical analysis becomes more complicated since the microstrip circuit will considerably change the previous field pattern. The previously pure TE_{101} mode becomes a quasi-TE_{101} mode. Equations (8.12) to (8.14) can, however, be used as a basis for deriving expressions for this mode.

```
RUN


Choose option 1 OR 2
(Type '1' or '2')


  1 - Calculate resonance frequency of TE101 cavity mode

  2 - Calculate dimensions of cavity


Selection ?1
```

```
Input all dimensions in mm
Ensure that HH < WW < LL

Height of cavity, HH = ?10
Width of cavity,  WW = ?20
Length of cavity, LL = ?40

Permittivity of substrate, Es =  ?2.3
Thickness of substrate, h =  ?.7874

The resonant frequency of the TE101 mode is:
 8.196 GHz


Do you wish to repeat the program   ?N
>RUN


Choose option 1 OR 2
(Type '1' or '2')


  1 - Calculate resonance frequency of TE101 cavity mode

  2 - Calculate dimensions of cavity


Selection ?2


Input all dimensions in mm

Permittivity of substrate,  Es = ?2.3
Thickness of substrate,      h = ?.7874

What is required minimum resonance frequency of the fundamental
cavity mode , TE101, in GHz ?
?8.1965

Is height (HH) of cavity fixed   ?Y

   VALUE OF HH, HH = ?10

Is width (WW) of cavity fixed    ?Y


   VALUE OF WW , WW = ?20

 Is length (LL) of cavity fixed   ?N

   MIN. VALUE OF LL, LL min = ?30
   MAX. VALUE OF LL, LL max = ?50


 Calculated results:


 Minimum resonance frequency
 OF TE101 MODE = 8.196 GHz

 THE DIMENSIONS OF THE CAVITY ARE :


     HH(mm)     WW(mm)     LL(mm)
      *****      *****      *****

     10.000     20.000     30.000
     10.000     20.000     33.000
     10.000     20.000     36.000
     10.000     20.000     39.000
```

L.

```
 10 REM This programme is called BOXMODE
 20 REM This programme is based on the paper by Edwin Johnson,"Technique
        engineers the cavity resonance in microstrip housing design",
        MSN Feb.1987, pp100-
 30 DIM H(600),W(600),L(600)
 40 PRINT:PRINT
 50 PRINT"Choose option 1 OR 2 "
 60 PRINT"(Type '1' or '2')"
 70 PRINT:PRINT
 80 PRINT"  1 - Calculate resonance frequency of TE101 cavity mode "
 90 PRINT
100 PRINT"  2 - Calculate dimensions of cavity"
110 PRINT:PRINT
120 INPUT"Selection ",SELECT
130 IF SELECT=1 THEN 1340
140 IF SELECT=2 THEN 160
150 GOTO 1550
160 PRINT
170 PRINT
180 HL=100
190 WL=0
200 LL=0
210 REM
220 PRINT"Input all dimensions in mm "
230 PRINT
240 INPUT"Permittivity of substrate,  Es = ",Er
250 INPUT"Thickness of substrate,      h = ",T:T=T/10
260 PRINT
270 PRINT"What is required minimum resonance frequency of the fundamental
          cavity mode , TE101, in GHz ? "
280 INPUT FR
290 PRINT
300 INPUT"Is height (HH) of cavity fixed    ",H$
310 PRINT
320 IF H$="Y" OR H$="y" THEN GOTO 360
330 INPUT "   MIN. VALUE OF HH, HH min = ",H1:H1=H1/10
340 INPUT "   MAX. VALUE OF HH, HH max = ",H2:H2=H2/10
350 GOTO 390
360 INPUT "   VALUE OF HH, HH = ",H:H=H/10
370 H1=H
380 H2=H
390 PRINT
400 INPUT"Is width (WW) of cavity fixed     ",W$
410 PRINT
420 IF W$="Y" OR W$="y" THEN GOTO 460
430 INPUT "   MIN. VALUE OF WW, WW min = ",W1:W1=W1/10
440 INPUT "   MAX. VALUE OF WW, WW max = ",W2:W2=W2/10
450 GOTO 490
460 INPUT "   VALUE OF WW , WW = ",W:W=W/10
470 W1=W
480 W2=W
490 PRINT
500 INPUT"Is length (LL) of cavity fixed    ",L$
510 PRINT
520 IF L$="Y" OR L$="y" THEN GOTO 560
530 INPUT "   MIN. VALUE OF LL, LL min = ",L1:L1=L1/10
540 INPUT "   MAX. VALUE OF LL, LL max = ",L2:L2=L2/10
550 GOTO 590
560 INPUT "   VALUE OF LL, LL = ",L:L=L/10
570 L1=L
580 L2=L
```

```
 590 PRINT
 600 PRINT
 610 PRINT
 620 IF (H1<W2) OR (H2<W2) THEN 640
 630 GOTO 650
 640 IF (W1<L2) OR (W2<L2) THEN 710
 650 PRINT"  The condition HH < WW < LL has NOT been met"
 660 PRINT"  Try again !"
 670 PRINT
 680 PRINT
 690 PRINT
 700 GOTO 220
 710 F1MIN=FR
 720 K1=0
 730 PRINT" Calculated results: "
 740 PRINT
 750 IF ((H2-H1)/(.1*H1))<200 THEN 780
 760 HSTEP=(H2-H1)/200
 770 GOTO 790
 780 HSTEP=(.1*H1)
 790 IF((W2-W1)/(.1*W1))<200 THEN 820
 800 WSTEP=(W2-W1)/200
 810 GOTO 830
 820 WSTEP=(.1*W1)
 830 IF((L2-L1)/(.1*L1))<200 THEN 860
 840 LSTEP=(L2-L1)/200
 850 GOTO 870
 860 LSTEP=(.1*L1)
 870 FOR H=H1 TO H2 STEP (.1*H1)
 880 FOR W=W1 TO W2 STEP (.1*W1)
 890 FOR L=L1 TO L2 STEP (.1*L1)
 900 IF (H<W) AND (W<L) THEN GOTO 920
 910 GOTO 1040
 920 GOSUB 1610
 930 IF F1<FR THEN 990
 940 K1=K1+1
 950 H(K1)=H
 960 W(K1)=W
 970 L(K1)=L
 980 GOTO 1040
 990 IF F1>F1MIN THEN 1040
1000 F1MIN=F1
1010 HF1MIN=H
1020 WF1MIN=W
1030 LF1MIN=L
1040 NEXT L
1050 NEXT W
1060 NEXT H
1070 PRINT
1080 REM SOUND 17,-10,P,3
1090 IF K1>0 THEN 1190
1100 PRINT"A minimum resonance frequency of ";FR;" GHz is NOT possible with the
            above cavity dimensions"
1110 PRINT
1120 PRINT"The maximum frequency achievable is :"
1130 PRINT""; (INT(F1MIN*10000))/10000;" GHz with cavity dimensions of :"
1140 PRINT
1150 PRINT"HEIGHT HH = ";HF1MIN*10;" mm"
1160 PRINT"WIDTH  WW = ";WF1MIN*10;" mm"
1170 PRINT"LENGTH LL = ";LF1MIN*10;" mm"
1180 GOTO 1550
1190 PRINT"Minimum resonance frequency"
```

```
1200 PRINT"OF TE101 MODE = ";FR;" GHz"
1210 PRINT
1220 PRINT"THE DIMENSIONS OF THE CAVITY ARE :"
1230 PRINT:PRINT
1240 PRINT"    HH(mm)    WW(mm)    LL(mm)"
1250 PRINT"     *****     *****     *****"
1260 PRINT
1270 FOR K2=1 TO K1
1280 IF K2=1 THEN 1310
1290 IF (INT(H(K2)*100))/100=(INT(H(K2-1)*100))/100 THEN 1310
1300 PRINT
1310 PRINT (INT(H(K2)*100))/10,(INT(W(K2)*100))/10,(INT(L(K2)*100))/10
1320 NEXT K2
1330 GOTO 1550
1340 CLS
1350 PRINT
1360 PRINT
1370 PRINT"Input all dimensions in mm"
1380 PRINT"Ensure that HH < WW < LL"
1390 PRINT
1400 INPUT"Height of cavity, HH = ",H:H=H/10
1410 INPUT"Width of cavity,  WW = ",W:W=W/10
1420 INPUT"Length of cavity, LL = ",L:L=L/10
1430 IF (H<W) AND (W<L) THEN 1480
1440 PRINT
1450 PRINT"The condition HH < WW < LL has not been met"
1460 PRINT"Please try again !"
1470 GOTO 1390
1480 PRINT
1490 INPUT"Permittivity of substrate, Es =  ",Er
1500 INPUT"Thickness of substrate, h =  ",T:T=T/10
1510 GOSUB 1610
1520 PRINT
1530 PRINT"The resonant frequency of the TE101 mode is:"
1540 PRINT" ";(INT(F1*10000))/10000;" GHz"
1550 PRINT
1560 PRINT
1570 INPUT"Do you wish to repeat the program   ";R3$
1580 IF R3$="Y" OR R3$="y" THEN 40
1590 REM  SOUND 17,-12,P,2
1600 END
1610 REM *****  SUBROUTINE  *****
1620 F101=(15/W)*SQR(1+((W/L)^2))
1630 F1=F101*SQR(1-((Er-1)*(T/(Er*H))))
1640 RETURN
>
```

For either case, unloaded or loaded, the inner dimensions of the housing have an effect on the modes in the housing since the box may be either long and narrow, short and wide or square. Thus, an analysis must take this into account. The following are examples for calculating resonance frequency or housing dimensions using the program BOXMODE.

From the computer printout we see that for the box dimensions given, the fundamental TE_{101} mode is 8.1965 GHz. If we were for example to use a box with fixed height and width, then we can use option 2 to calculate the mode for a specified range of box lengths. This gives a minimum frequency of 8.1965 GHz. This may now be re-checked by using the minimum and maximum box length of 30 and 39 mm with option 1. The appropriate frequencies are 8.811 GHz and 8.2389 GHz which is higher than 8.1965 GHz. This means that the enclosure can accommodate microstrip circuits working up to about 8 GHz without circuit performance deterioration.

MICROWAVE NETWORK ANALYSER MEASUREMENTS

When a microwave circuit or subsystem has been designed and constructed its performance needs to be experimentally verified. There are various techniques and equipments available for doing this. The kind of measurement to be undertaken depends also on the specific information which is required about the system. Amongst others this information may be about the noise figure, the frequency response, losses, compression point or spectral performance. One of the most useful and powerful tools is probably a network analyser system in conjunction with a computer and hard copy printout facility.

The Wiltron 562 scalar network analyser when combined with a sweeper or synthesizer forms such a powerful tool. The choice of sweeper or synthesizer depends entirely upon the required frequency accuracy. Narrow band filter measurements for example demand synthesizer accuracy. The 562 scalar analyser is also compatible via its dedicated GPIB interface with sources from Hewlett Packard, namely the 8340 and 8350B series. The above configurations provide automatic full screen frequency annotation. A zero to 10 V sweep ramp output mode is also available to control VCOs and thereby configure a simple swept frequency scalar measurement system. In this mode of operation the frequency labelling has to be entered manually on the 562 display.

The Wiltron 562 system measures insertion loss, insertion gain or RF power with a 76 dB dynamic range over the widest possible frequency range in coax (10 MHz to 40 GHz). It also measures device match as return loss in dB or as VSWR. Separate detectors may be used on all four inputs for multipole measurements. Direct detection allows simultaneous RF power measurement at different frequencies, for example, at the RF, IF and local oscillator frequencies of mixers and converters. Measurements are assisted by a comprehensive range of precision coaxial directional bridges, detectors and terminations.

The 562 network analyser system uses d.c. detection which eliminates uncertainty from RF modulation. A detector low level calibration is made on every retrace giving a sensitivity of -60 dBm. When used with the Wiltron 6700 swept frequency synthesizer in step sweep mode, all measurement frequencies, including markers and cursors, have synthesizer accuracy. The equipment also features trace memory functions. Trace memory permits simple de-embedding and greatly simplifies amplifier measurements.

The 562 is easy to use and is menu driven. At each step the instrument provides a comprehensive display of all pertinent parameters. Ten display cursor functions are available to locate important frequencies, amplitudes, deltas or bandwidths. Nine complete system setups, including source settings, may be saved in a non-volatile memory for later recall, four memory locations may include calibration data. All can be previewed on the CRT prior to selection. Hard copy output may be obtained either via a buffered Centronics port to a inkjet printer or via a plotter connected to the system GPIB.

The scalar network analyser is shown in Plate 8.1 and a block diagram of a typical test setup in Fig. 8.26.

REFERENCES

[1] Howson, D. P., Minimum conversion loss in input match conditions in the broad band mixer, *Radio and Electronic Engineer*, **42**, No. 5, May 1972, pp. 237–242.

[2] Johnson, K. M., X-band integrated circuit mixer with reactively terminated image, *IEEE* **MTT-16**, No. 7, July 1968, pp. 388–397.

[3] Barber, M. R., Noise figure and conversion loss of the Schottky barrier mixer diode, *IEEE*, **MTT-15**, No. 111, Nov. 1967, pp. 629–635.

[4] Tucker, D. G., The input impedance of rectifier modulators, *Proc. IEE*, Jan. 1960, pp. 273–281.

[5] Kulesza, B. L. J., General theory of a lattice mixer, *Proc. IEEE*, **118**, No. 7, July 1971, pp. 864–870.

[6] Armstrong, A. F. and Cooper, P. D., Techniques for investigating spurious propagation in enclosed microstrip, *Radio and Electronic Engineer*, **48**, No. 1/2, Jan./Feb. 1978, pp. 64–72.

[7] Bedair, S. S., Predict enclosure effects on shielded microstrip, *Microwaves & RF*, July 1985, pp. 97.

[8] Johnson, E. F., Technique engineers the cavity resonance in microstrip housing design, *MSN & CT*, Feb. 1987, pp. 100–109.

[9] Gardiol, F. E., Careful MIC design prevents waveguide modes, *Microwaves*, **16**, May 1977, pp. 188–191.

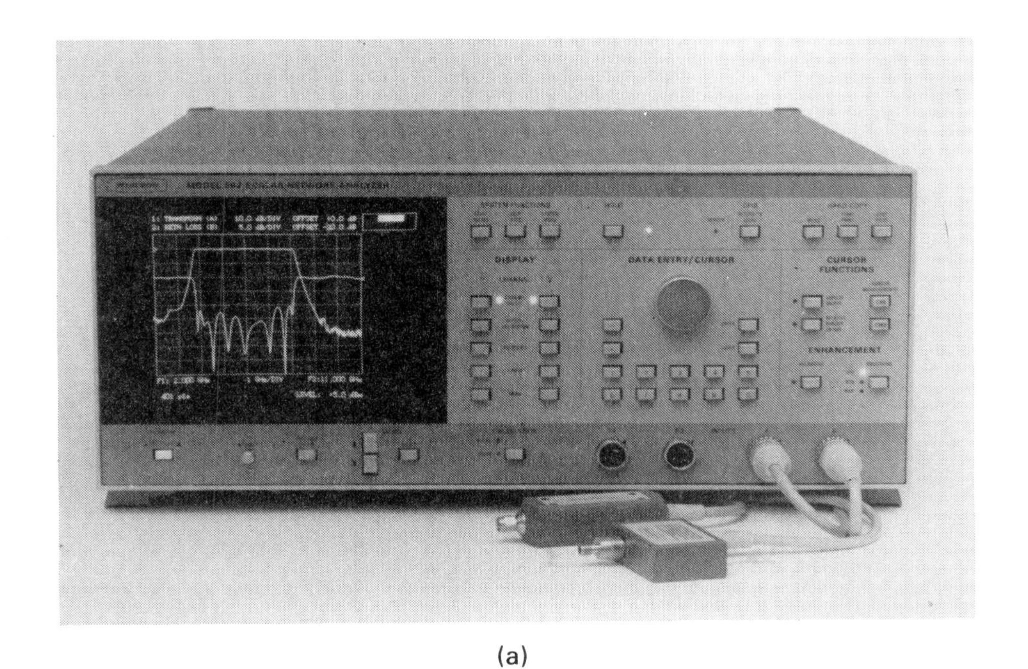

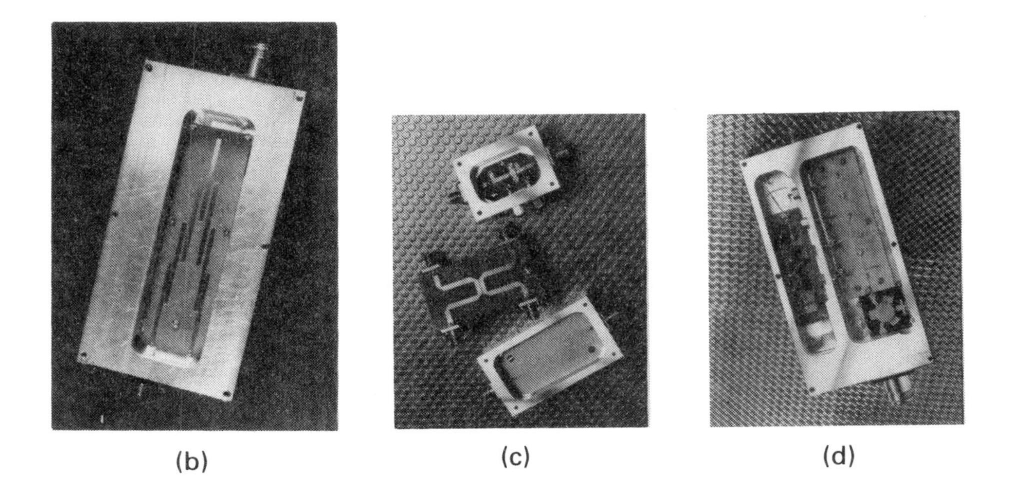

Plate 8.1 — (a) Wiltron 562 scalar network analyser. Examples of microstrip circuits: (b) filter (Chapter 6), (c) couplers (Chapter 9), (d) amplifier (Chapter 5), (e) 2.5 GHz front-end (Chapter 8).

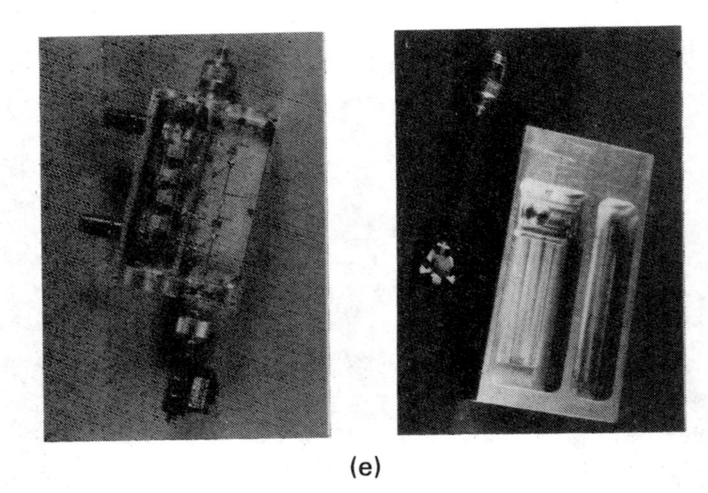

(e)

Plate 8.1 continued.

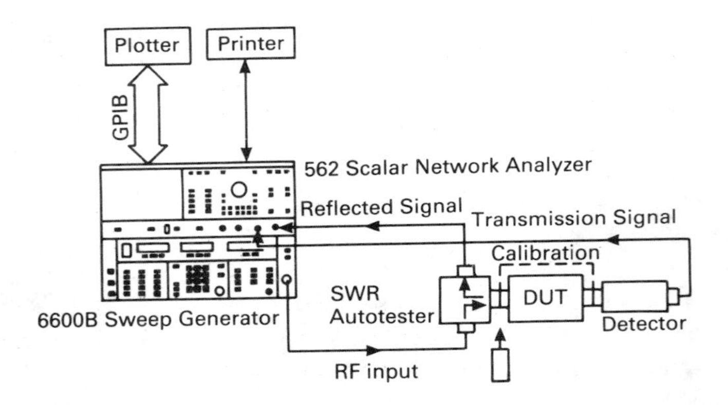

Fig. 8.26 — Typical scalar network analyser measuring set-up.

9

Microwave couplers

INTRODUCTION

A coupler is one of the fundamental components in microwaves. It finds applications in circuits such as balanced amplifiers, mixers, power dividers and combiners. Different coupler configurations and designs are available to cater for specific requirements.

In Fig. 9.1 a coupler is used as a power divider to measure, for example, the

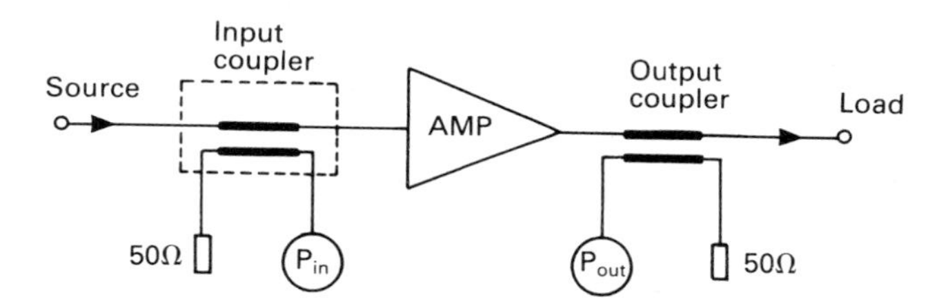

Fig. 9.1 — Example of the use of a coupler.

power gain of an amplifier. The coupler is needed since one cannot simply tap the input and output at microwave frequencies.

In the most general form a coupler is a 4-port device with two of the ports being

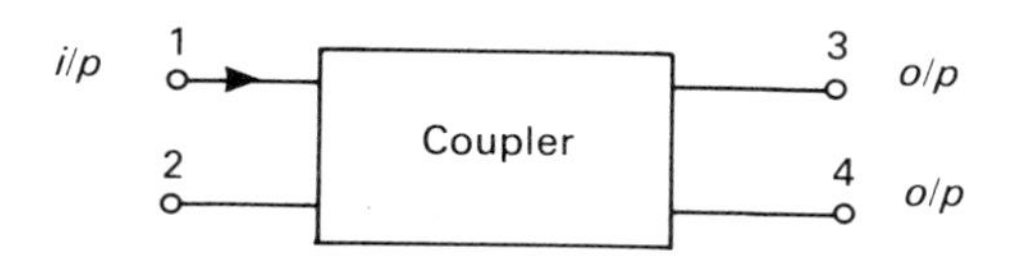

Fig. 9.2 — Block diagram of a coupler.

mutually decoupled with respect to the other two ports. Referring to Fig. 9.2, consider microwave power applied to port 1. This power will appear at port 3 and 4 but not at port 2. Port 2 is thus said to be decoupled with respect to port 1. If power was introduced to port 3, then power would appear at port 1 and 2 but not at port 4.

The question now arises as to the amount of power which appears at the output ports. There are two distinct possibilities depending on the coupler design. First of all, the applied power is equally divided between the two output ports as in the case, say, for balanced amplifiers. This case of coupler is usually referred to as hybrid coupler. On the other hand (Fig. 9.1) where the power is not equally divided at the output the coupler is referred to as a directional coupler. The hybrid coupler is a special case of the directional coupler. The directional coupler gets its name from the fact that the power available at its ports depends on the direction of power flow, i.e. forward travelling (transmission) or backward travelling (reflection).

PARALLEL LINE DIRECTIONAL COUPLER

In the parallel line directional coupler (Fig. 9.3), like in any coupler, a predetermined fraction of the power incident on the primary arm is coupled to a secondary arm and then onto another circuit. In a parallel line coupler the coupling increases as the gap between primary and secondary arm decreases. The main difficulty which arises in the construction of such a coupler is that the spacing s or coupling gap is very difficult to realize for very tight coupling. The power applied to port 1 of the primary arm passes through the coupler and appears at port 2. Some of the power entering the coupler is also transferred or 'coupled' across the coupling gap to the secondary arm of the coupler and appears at port 3. As per eqn. (9.4), virtually no power should appear at port 4 which is usually terminated in the characteristic line impedance of the coupler, e.g. $Z_o = 50\ \Omega$.

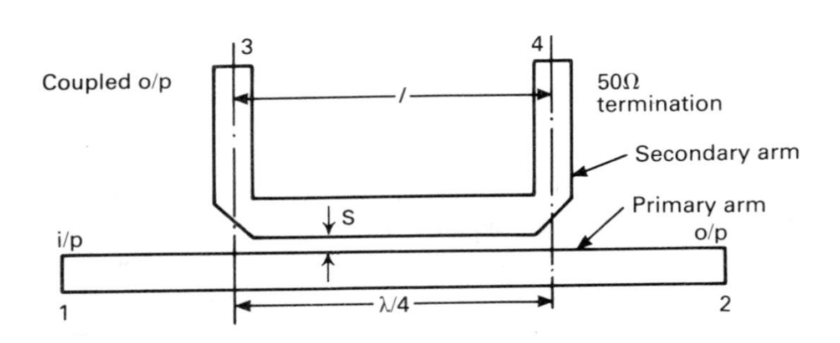

Fig. 9.3 — A typical microstrip coupler (parallel line).

The major parameters of a directional coupler are stated and defined in voltage terms as follows, where

C = coupling factor
D = directivity
I = isolation
T = transmission factor

The coupling factor (C) is the ratio of the voltage at port 3 to the voltage at port 1.

$$C = \frac{V_3}{V_1} = 20 \log \frac{V_3}{V_1} \text{ (dB)} \tag{9.1}$$

This is a measure of the amount of coupling which has occurred. A 6 dB coupler, for example, is one in which one quarter of the power at port 1 appears at port 3. Alternatively, it can be said that the voltage at port 3 is −6 dB relative to that at port 1. The transmission factor (T) is the ratio of the voltage at port 2 to that at port 1.

$$T = \frac{V_2}{V_1} = 20 \log \frac{V_2}{V_1} \text{ (dB)} \tag{9.2}$$

This is a measure of the transmission directly through the 'primary' arm of the coupler. The directivity (D) is the ratio of the voltage at port 4 to that at port 3.

$$D = \frac{V_4}{V_3} = 20 \log \frac{V_4}{V_3} \text{ (dB)} \tag{9.3}$$

The directivity is a measure of the undesirable coupling which occurs at port 4. Port 4 is normally terminated in its characteristic impedance. For an ideal coupler $D = 0$, and the power is split entirely between ports 2 and 3. The fourth property of a coupler is isolation (I) which is the ratio of the voltage at port 4 to that at port 1, i.e.

$$I = \frac{V_4}{V_1} = 20 \log \frac{V_4}{V_1} \text{ (dB)} \tag{9.4}$$

The length over which coupling takes place is called the coupling length, l.

The theory of a parallel-coupled microstrip was discussed in Chapter 2. The signal in the primary arm, travelling from port 1 to port 2, couples across to the other arm, and travels in the opposite direction appearing at port 3. The relative polarities of the voltages in the two microstrips, taken at any reference plane along the structure, will either be alike or opposite. The two field patterns set up by such voltages are referred to as even-mode and odd-mode as described previously. This fact must be taken into account when designing parallel.

DESIGN PROCEDURE

The following outlines the design procedure of a parallel line coupler based on Fig. 9.4 with the following parameters:

s = separation between the two microstrip lines in the coupled region;
w_1 = width of the microstrips in the coupled region;
l = length of the two microstrips in the coupled region;
w_2 = width of the microstrips outside the coupled region.

To calculate the above dimensions the following data (design specification) is required:

Z_o = single microstrip characteristic impedance
C = coupling factor;
ε_r = substrate permittivity;
h = thickness of the substrate;
f = mid-band operating frequency.

When synthesizing a coupler, then Z_{oe} and Z_{oo} are known since they are related to the coupling factor C and characteristic impedance Z_o (Chapter 2) through eqns (9.5) and (9.6):

$$Z_{oe} = Z_o \left\{ \frac{1 + 10^{C/20}}{1 - 10^{C/20}} \right\}^{0.5} \tag{9.5}$$

$$Z_{oo} = Z_o \left\{ \frac{1 - 10^{C/20}}{1 + 10^{C/20}} \right\}^{0.5} \tag{9.6}$$

Z_{oe} and Z_{oo} are the even and odd mode impedances for coupled microstrip. For an equivalent single microstrip, the even mode characteristic impedance Z_{ose} is given by

$$Z_{ose} = \frac{Z_{oe}}{2} \tag{9.7}$$

and the odd mode characteristic impedance is given by

$$Z_{oso} = \frac{Z_{oo}}{2} \tag{9.8}$$

The next stage is to calculate the value of the equivalent single microstrip $(w/h)_s$, where the suffix s denotes a single line. Naturally there will be a different $(w/h)_s$ value for each of the two modes. These are given by $(w/h)_{se}$ for the even mode and $(w/h)_{so}$ for the odd mode and are calculated using the single line static TEM equations given in Chapter 2.

The computer program used substitutes the value of Z_{oso} instead of Z_o into the appropriate equation and then calculates the value of $(w/h)_{so}$. It then repeats the sequence, except this time substituting a value of Z_{ose} in eqn. (9.7) instead of Z_o and calculates a value for $(w/h)_{se}$.

With these two equivalent single microstrip values it is now possible to calculate the required w/h and s/h values for the coupler. Using the following equation, we can calculate the value of s/h, thus

$$\frac{s}{h}=\frac{2}{\pi}=\cosh^{-1}\left\{\frac{\cosh\,[(\pi/2)(w/h)_{se}]+\cosh\,[(\pi/2)(w/h)_{so}]}{\cosh\,[(\pi/2)(w/h)_{so}]-\cosh\,[(\pi/2)(w/h)_{se}]}\right\}^{-2} \tag{9.9}$$

The program substitutes values for $(w/h)_{se}$ and $(w/h)_{so}$ into eqn. (9.10) to obtain an initial value of s/h. Substituting this initial value of s/h into the following equations produces the initial value for w/h.

$$(w/h)_{se}=(2/\pi)\cosh^{-1}\left(\frac{2d-g+1}{g+1}\right) \tag{9.10}$$

where

$$g=\cosh\left[\left(\frac{\pi}{2}\right)\left(\frac{s}{h}\right)\right] \tag{9.11}$$

and

$$d=\cosh\left(\frac{\pi w}{h}+\frac{\pi s}{2h}\right) \tag{9.12}$$

Equation (9.10) can be rearranged to give

$$d=\frac{g-1+\{(g+1)\cosh[(\pi/2)(w/h)_{se}]\}}{2} \tag{9.13}$$

The program substitutes values into eqn. (9.13) to obtain a value for d. This value is then entered into eqn. (9.14). Equation (9.14) is eqn. (9.12) solved for w/h.

$$\frac{w}{h}=\frac{\cosh^{-1} d}{\pi}-\frac{s}{2h} \tag{9.14}$$

The values of d and s/h are entered into eqn. (9.14) and an initial value for w/h is obtained. Using the following equations [1] and the s/h and w/h values just calculated gives the w/h shape ratio for a single line in the odd mode.

For $\varepsilon_r <= 6$

$$(w/h)_{so} = \frac{2}{\pi}\cosh^{-1}\left[\frac{2d-g-1}{g-1}\right] + \left[\frac{4}{\pi(1+\varepsilon_r/2)}\right]\cosh^{-1}\left[1+\frac{2w/h}{s/h}\right] \tag{9.15}$$

For $\varepsilon_r > 6$

$$(w/h)_{so} = \frac{2}{\pi}\cosh^{-1}\left[\frac{2d-g-1}{g-1}\right] + \frac{1}{\pi}\cosh^{-1}\left[1+\frac{2w/h}{s/h}\right] \tag{9.16}$$

where g and d are given by eqns (9.11) and (9.12). It will be found that the two $(w/h)_{so}$ values do not match. An iterative technique is then used to change the initial s/h shape ratio, recalculate w/h and $(w/h)_{so}$ until the two $(w/h)_{so}$ values are identical. This results in the final values of s/h and w/h for the coupler.

The third step is now to calculate the coupling length, l. The physical length of the microstrip depends upon the wavelength, which in turn is a function of the substrate permittivity and signal frequency. It is known that the maximum degree of coupling occurs when the length is equal to $\lambda_{gm}/4$, where λ_{gm} is the mid-band wavelength.

For greater accuracy the average of the even mode and odd mode wavelengths have to be calculated. For the even mode and odd mode wavelengths the program uses the formulae:

$$\lambda_{ge} = \frac{300\ Z_{oe}}{f\ Z_{o1e}} \tag{9.17}$$

$$\lambda_{go} = \frac{300\ Z_{oo}}{f\ Z_{o1o}} \tag{9.18}$$

where Z_{o1e} and Z_{o1o} are the even mode and odd mode characteristic impedances with no substrate present as shown in Fig. 9.4. Since f, Z_{oe} and Z_{oo} are already known, only Z_{o1e} and Z_{o1o} need to be calculated.

Therefore, from Chapter 2:

$$Z_{o1e} = \frac{1}{c\ C_e} \tag{9.19}$$

$$Z_{o1o} = \frac{1}{c\ C_o} \tag{9.20}$$

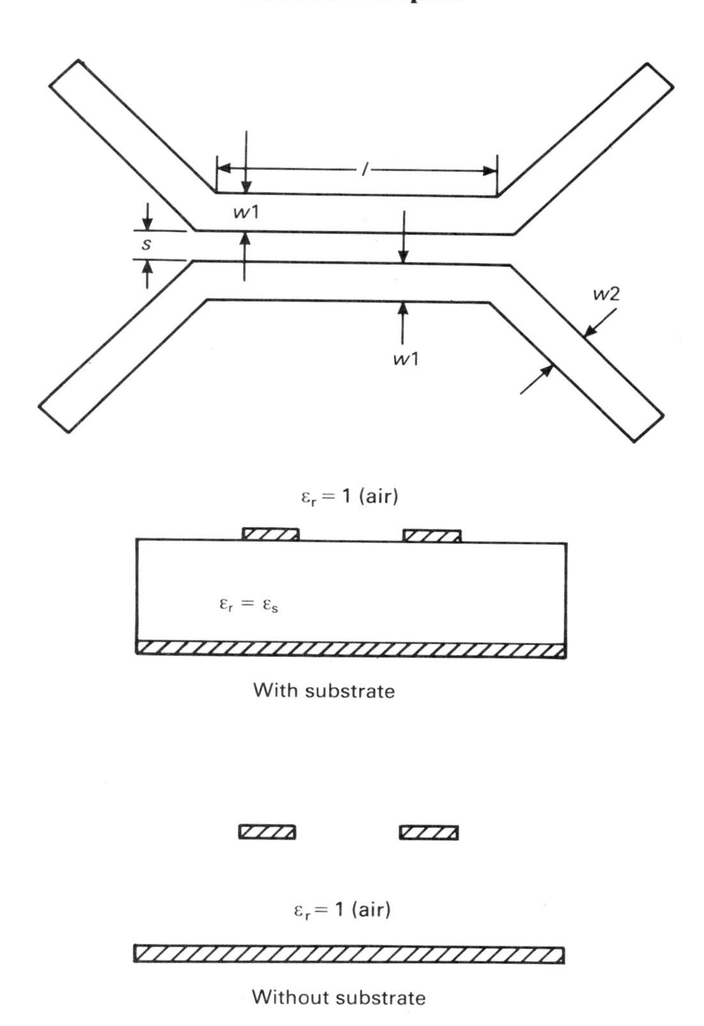

Fig. 9.4 — Parallel coupled microstrips with and without substrate.

The wavelengths λ_{ge} and λ_{go} can now be calculated and hence the coupling length as:

$$l = \frac{\lambda_{go} + \lambda_{ge}}{8} = (2n - 1)\frac{\lambda_{go} + \lambda_{ge}}{8} \tag{9.21}$$

where the value of n may be chosen to give a convenient odd number of quarter wavelength and hence coupling length l. A design example using the program COUPDES is given in the following. An example of different physical layouts of parallel line couplers is given in Fig. 9.5(a). Careful layout will minimize any performance variations between the same coupler but different layout.

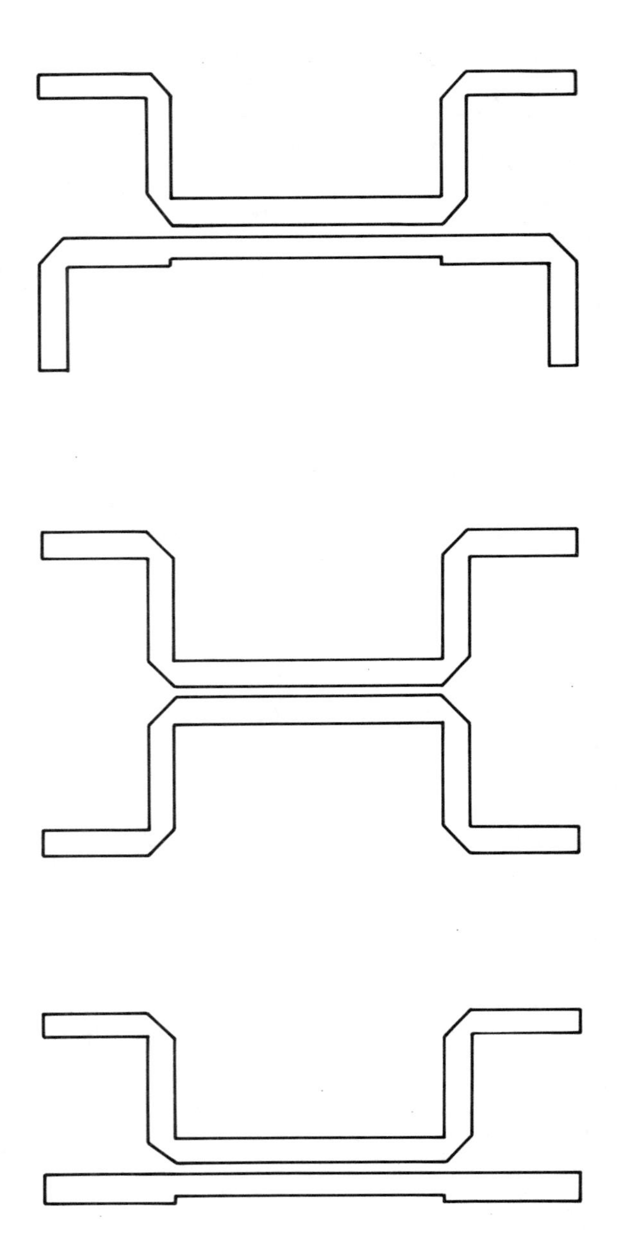

Fig. 9.5 — (a) Three different configurations of parallel line couplers.

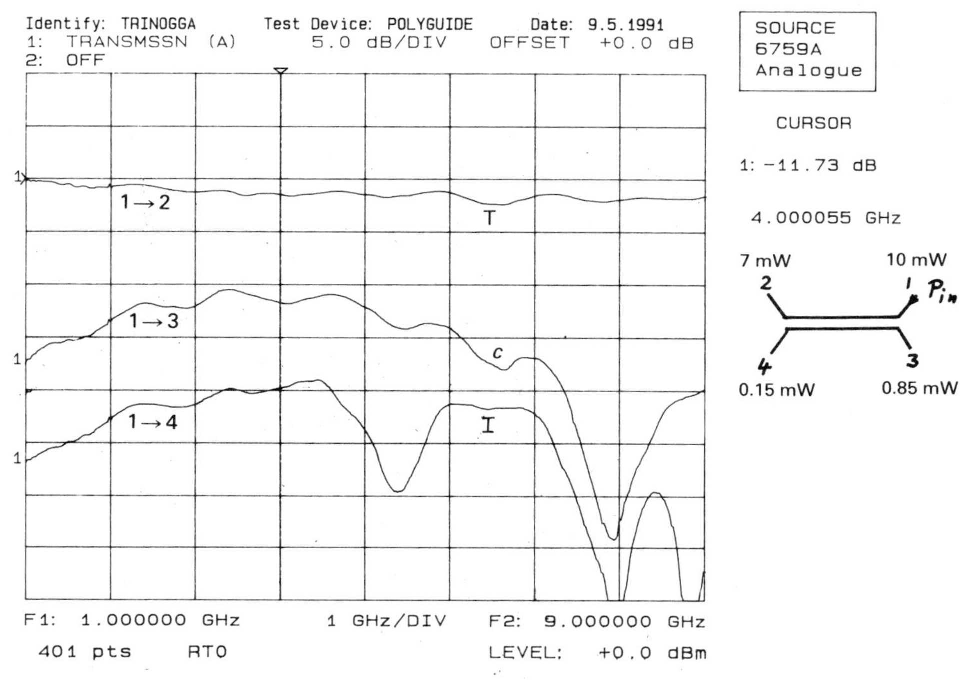

Fig. 9.5 — (b) Parallel line coupler response, $\varepsilon_r = 2.32$, $h = 0.7874$ mm, $s = 0.8$ mm, $w =$ 1.7mm, $l = 22$ mm.

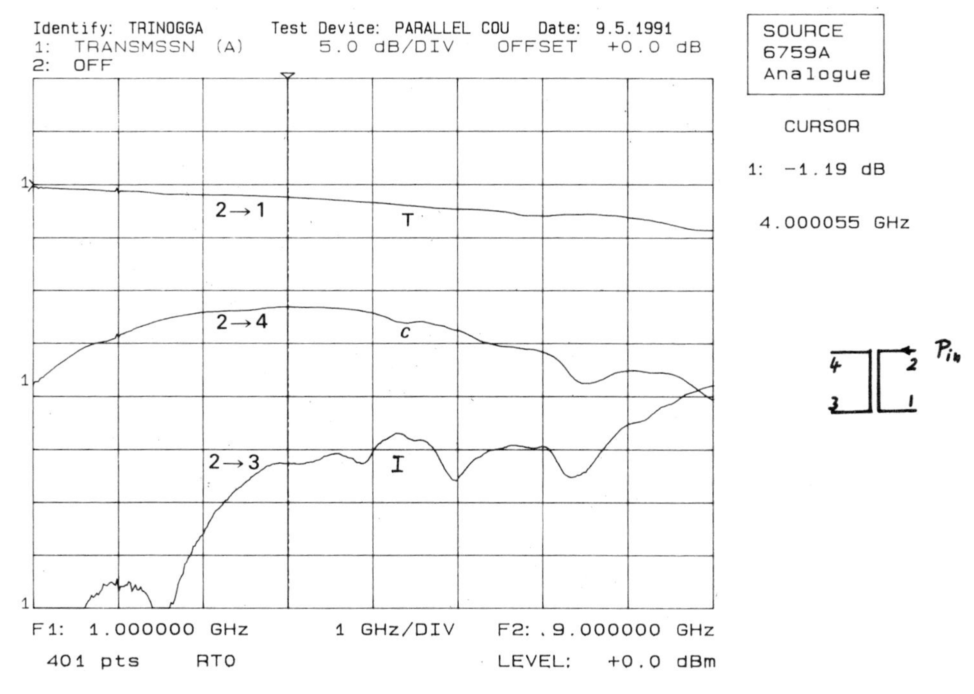

Fig. 9.5 — (c) Parallel line coupler response, $\varepsilon_r = 4.8$, $h = 1.58$ mm, $s = 0.3$ mm, $w = 2.7$ mm, $l = 7.5$ mm.

```
RUN
                     DESIGN OF A PARALLEL LINE COUPLER



   COUPLER SPECIFICATION
   ---------------------

Enter the characteristic impedance (Ohm)?50
Enter the coupling coefficient  = > 10  (dB) ?10
Enter the mid-band frequency (GHz)?5


  SUBSTRATE SPECIFICATION
  -----------------------


The substrate permittivity  Er ?4.8
The substrate height h[mm]?1.57
```

Calculated results

even mode impedance ZOE [Ohm]	= 69.371
odd mode impedance ZOO [Ohm]	= 36.038
space to height ratio S/H	= 0.134
width to height ratio W/H	= 1.521
width w1 in coupled region [mm]	= 2.389
space s in coupled region [mm]	= 0.210
quarter wavelength LGM/4 [mm]	= 8.247
50 Ohm line width w2 [mm]	= 2.811

see Fig. 9.3

press RETURN key for drawing?

```
ANOTHER SCALE (Y/N)?N
>*LOAD"SCRNDMP"
>CALL&900
```

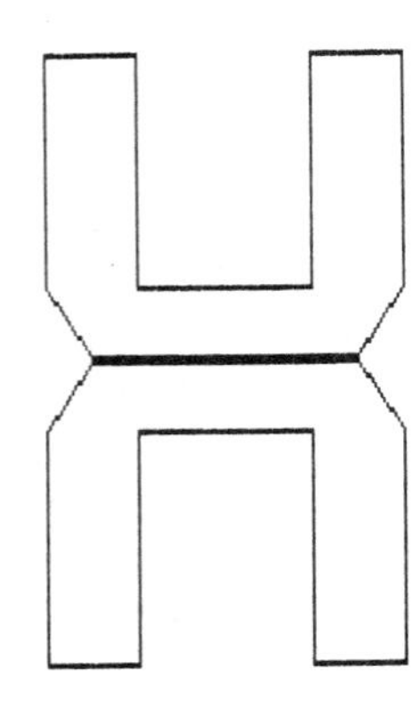

```
L.
   10 REM This programme is called COUPDES
   20 REM This programme calculates Zoe,Zoo,s/h,w/h,l,s,lambda,w for a given Zo,
C,fo,Er,h
   30 MODE 3
   40 CLS:@%=131850
   50 PRINT"                    DESIGN OF A PARALLEL LINE COUPLER"
   60 PRINT''
   70 PRINT''
   80 PRINT"    COUPLER SPECIFICATION"
   90 PRINT"    ---------------------"
  100PRINT'
  110 INPUT"Enter the characteristic impedance (Ohm)";ZO
  120 INPUT"Enter the coupling coefficient  = > 10  (dB) ";C
  130 IF C>0 THEN C=0-C ELSE 140
  140 INPUT"Enter the mid-band frequency (GHz)";F
  150 PRINT''
  160 PRINT"  SUBSTRATE SPECIFICATION"
  170 PRINT"  -----------------------"
  180 PRINT''
  190 INPUT"The substrate permittivity  Er ";ER
  200 INPUT"The substrate height h[mm]";H
  210 PROCimp
  220 ZO=ZOE/2
  230 PROCcalwh
  240 WHSE=WHS
  250 ZO=ZOO/2
  260 PROCcalwh
  270 WHSO=WHS
  280 PROCcalsh
  290 TEMP1=SH*H
  300 PROCleng
  310 ZO=50
  320 PROCcalwh
  330 WX=WHS*H
  340 CLS:PRINT'
  350 PRINT"             Calculated results     "
  370 PRINT"________________________________________"
  380 PRINT"even mode impedance  ZOE  [Ohm]  = ";ZOE
  390 PRINT"________________________________________"
  400 PRINT"odd mode impedance   ZOO  [Ohm]  = ";ZOO
  410 PRINT"________________________________________"
  420 PRINT"space to height ratio  S/H       = ";SH
  430 PRINT"________________________________________"
  440 PRINT"width to height ratio  W/H       = ";WH
  450 PRINT"________________________________________"
  460 PRINT"width w1 in coupled region [mm]  = ";WH*H
  470 PRINT"________________________________________"
  480 PRINT"space s in coupled region [mm]   = ";SH*H
  490 PRINT"________________________________________"
  500 PRINT"quarter wavelength LGM/4 [mm]    = ";LX
  510 PRINT"________________________________________"
  520 PRINT"50 Ohm  line width w2 [mm]       = ";WX
  530 PRINT"________________________________________"
  540 PRINT"              see Fig. 9.3              "
  550 PRINT"________________________________________"
  560 INPUT"      press RETURN key for drawing";A$
  570 IF A$="O" THEN 580 ELSE 580
  580 W=WH*H:S=SH*H:W2=WX:L=LX
  590 MODE 128
  600 XF=1280/225.5:YF=1024/163.5:L=LX*XF:W=W*YF:S=S*YF:W2=WX*XF
  610 INPUT"SCALE FACTOR";SF:PROCscale:CLS
```

```
  620 MOVE 640,512+S/2
  630 DRAW 640+L/2,512+S/2:MOVE 640+L/2-W2/2,512+S/2+W:DRAW 640-L/2+W2/2,512+S/2
+W:MOVE 640-L/2,512+S/2:DRAW 640,512+S/2:MOVE 640+L/2-W2/2,512+S/2+W
  640 DRAW 640+L/2-W2/2,512+S/2+W+L:MOVE 640+L/2-W2/2,512+S/2+W:DRAW 640+L/2-W2/
2,512+S/2+W+L:DRAW 640+L/2+W2/2,512+S/2+W+L:DRAW 640+L/2+W2/2,512+S/2+W:DRAW 640
+L/2,512+S/2
  650 MOVE 640-L/2+W2/2,512+S/2+W:DRAW 640-L/2+W2/2,512+S/2+W+L:DRAW 640-L/2-W2/
2,512+S/2+W+L:DRAW 640-L/2-W2/2,512+S/2+W:DRAW 640-L/2,512+S/2
  660 MOVE 640,512-S/2:DRAW 640+L/2,512-S/2:MOVE 640+L/2-W2/2,512-S/2-W:DRAW 640
-L/2+W2/2,512-S/2-W:MOVE 640-L/2,512-S/2:DRAW 640,512-S/2
  670 MOVE 640+L/2-W2/2,512-S/2-W
  680 DRAW 640+L/2-W2/2,512-S/2-W-L:DRAW 640+L/2+W2/2,512-S/2-W-L:DRAW 640+L/2+W
2/2,512-S/2-W:DRAW 640+L/2,512-S/2
  690 MOVE 640-L/2+W2/2,512-S/2-W:DRAW 640-L/2+W2/2,512-S/2-W-L:DRAW 640-L/2-W2/
2,512-S/2-W-L:DRAW 640-L/2-W2/2,512-S/2-W:DRAW 640-L/2,512-S/2
  700 INPUT"ANOTHER SCALE (Y/N)";Z$:IF Z$="N" THEN 710 ELSE 580
  710 END
  720 DEF PROCimp
  730 ZOE=ZO*SQR((1+10^(C/20))/(1-10^(C/20)))
  740 ZOO=ZO*SQR((1-10^(C/20))/(1+10^(C/20)))
  750 ENDPROC
  760 DEF PROCcalwh
  770 IF ZO>(44-2*ER) THEN 780 ELSE 810
  780 H1=ZO*SQR(2*(ER+1))/119.9+.5*((ER-1)/(ER+1))*(LN(PI/2)+(1/ER)*LN(4/PI))
  790 WHS=((EXP(H1))/8-1/(4*EXP(H1)))^-1
  800 GOTO 830
  810 D1=59.95*PI*PI/(ZO*SQR(ER))
  820 WHS=(2/PI)*((D1-1)-LN(2*D1-1))+((ER-1)/(PI*ER))*(LN(D1-1)+.293-.517/ER)
  830 ENDPROC
  840 DEF PROCcalsh
  850 MN1=(PI*WHSE/2):NM1=(PI*WHSO/2)
  860 MN=(EXP(MN1)+EXP(-MN1))/2:NM=(EXP(NM1)+EXP(-NM1))/2
  870 TEMP=(MN+NM-2)/(NM-MN):SH=(2/PI)*LN(TEMP+SQR(TEMP^2-1)):IF SH<.005 THEN 88
0 ELSE 890
  880 ENDPROC
  890 g=(EXP(PI*SH/2)+EXP(-PI*SH/2))/2
  900 d=(g-1+((g+1)*MN))/2
  910 WH=(LN(d+SQR(d^2-1))-(PI*SH/2))/PI
  920 TEMP=(2*d-g-1)/(g-1):z=(2/PI)*LN(TEMP+SQR(TEMP^2-1)):TEMP=1+2*(WH/SH)
  930 TEMP2=LN(TEMP+SQR(TEMP^2-1)):IF (ER<6) THEN 940 ELSE 960
  940 WHSO2=z+(4*TEMP2/(PI*(1+(ER/2))))
  950 GOTO 970
  960 WHSO2=z+(TEMP2/PI)
  970 IF (WHSO2-WHSO)>.1 OR (WHSO-WHSO2)>.1  THEN 980 ELSE 1020
  980 IF (WHSO2<WHSO) THEN 990 ELSE 1010
  990 SH=SH-.01
 1000 GOTO 1020
 1010 SH=SH+.01
 1020 IF (WHSO2-WHSO)>.1 OR (WHSO-WHSO2)>.1 THEN 890 ELSE 1030
 1030 IF (WHSO2-WHSO)>.01 OR (WHSO-WHSO2)>.01 THEN 1040 ELSE 1080
 1040 IF (WHSO2<WHSO) THEN 1050 ELSE 1070
 1050 SH=SH-.001
 1060 GOTO 1080
 1070 SH=SH+.001
 1080 IF (WHSO2-WHSO)>.01 OR (WHSO-WHSO2)>.01 THEN 890 ELSE 1090
 1090 ENDPROC
 1100 DEF PROCleng:LOCAL ER,C
 1110 EEFF=1:ER=1:E0=8.85E-12:C=3E8
 1120 g=(EXP(PI*SH/2)+EXP(-PI*SH/2))/2
 1130 d=(EXP(PI*WH+PI*SH/2)+EXP(-(PI*WH+PI*SH/2)))/2
 1140 x=(2*d-g+1)/(g+1)
 1150 PROCikosh:WHSE=2*TEMP1/PI:x=(2*d-g-1)/(g-1):PROCikosh
```

```
1160 x=(1+WH/SH)
1170 PROCikosh2
1180 WHSO=2*TEMP1/PI+(4/(PI*(1+ER/2)))*TEMP2
1190 IF (WH<3.3) GOTO 1290
1200 TEMP3=LN(EXP(1)*PI*PI/16)/(2*PI)*(ER-1)/(ER^2)
1210 TEMP1=WHSO/2+LN(4)/PI+TEMP3
1220 TEMP4=(ER+1)/(2*PI*ER)
1230 TEMP2=TEMP4*(LN(PI*EXP(1)/2)+LN(WHSO/2+.94))
1240 ZOSO=(119.9*PI/(2*SQR(ER)))*(1/(TEMP1+TEMP2))
1250 TEMP1=WHSE/2+LN(.4)/PI+TEMP3
1260 TEMP2=TEMP4*(LN(PI*EXP(1)/2)+LN(WHSE/2+.94))
1270 ZOSE=(119.9*PI/(2*SQR(ER)))*(1/(TEMP1+TEMP2))
1280 GOTO 1340
1290 TEMP1=LN(4*(1/WHSO)+SQR(16*(1/WHSO)*(1/WHSO)+2))
1300 TEMP2=(.5*(ER-1)/(ER+1))*(LN(PI/2)+LN(4/PI)/ER)
1310 ZOSO=(119.9/SQR(2*(ER+1)))*(TEMP1-TEMP2)
1320 TEMP1=LN(4*(1/WHSE)+SQR(16*(1/WHSE)*(1/WHSE)+2))
1330  ZOSE=(119.9/SQR(2*(ER+1)))*(TEMP1-TEMP2)
1340 ZO1=SQR(ZOSE*ZOSO)*2
1350 CP=EO*ER*WH
1360 CF=SQR(EEFF)/(C*ZO1*2)-CP/2
1370 TEMP1=EXP(-.1*EXP(2.33-2.53*WH))
1380 TAH=(EXP(8*SH)-EXP(-8*SH))/(EXP(8*SH)+EXP(-8*SH))
1390 CF1=CF*SQR(ER/EEFF)/(1+(TEMP1*(1/SH)*TAH))
1400 K=SH/(SH+2*WH)
1410 K1=SQR(1-K^2)
1420 IF (K^2<.5) GOTO 1450 ELSE 1430
1430 TEMP1=PI/(LN(2*(1+SQR(K))/(1-SQR(K))))
1440 GOTO 1460
1450 TEMP1=LN(2*(1+SQR(K1))/(1-SQR(K1)))/PI
1460 CGA=EO*TEMP1
1470 TAH1=(EXP(PI*SH/4)-EXP(-PI*SH/4))/(EXP(PI*SH/4)+EXP(-PI*SH/4))
1480 TEMP1=LN(1/TAH1)
1490 TEMP2=.02*SQR(ER)/SH+1-1/(ER^2)
1500 CGD=EO*ER*TEMP1/PI+.65*CF*TEMP2
1510 CE=CP+CF+CF1
1520 CO=CP+CF+CGA+CGD
1530 ZO1E=1/(C*CE)
1540 ZO1O=1/(C*CO)
1550 LGE=300*ZOE/(F*ZO1E)
1560 LGO=300*ZOO/(F*ZO1O)
1570 LGM=(LGE+LGO)/2
1580 LX=LGM/4
1590 ENDPROC
1600 DEF PROCikosh
1610 TEMP1=LN(x+SQR(x*x-1))
1620 ENDPROC
1630 DEF PROCikosh2:TEMP2=LN(x+SQR(x*x-1)):ENDPROC
1640 DEF PROCscale
1650 S=(S*SF):W=(W*SF)-(.5*YF):L=L*SF:W2=W2*SF:ENDPROC
```

Example

Design a 10 dB parallel line coupler for a mid-band frequency of 4 GHz. The glass–epoxy substrate from which the coupler is to be constructed has a permittivity of 4.8 and a thickness of 1.57 mm. Assume $Z_o = 50\,\Omega$.

Solution

Load the program COUPDES and run. Enter the above data. A printout of the entered data and calculated results is shown below together with a printout of a coupler layout.

The measured results of the two parallel line couplers designed earlier on are shown in Fig. 9.5(b). The first coupler was fabricated on a proper microwave substrate, Polyguide. At the design frequency of 4 GHz the coupler has a transmission loss of about 1.5 dB. The coupling at the same frequency is 11.5 dB. The return loss for transmission is around 20 dB which equals a VSWR of around 1.2 dB. This result can be improved upon through better close tolerance fabrication. Above and below the design centre frequency coupling decreases symmetrically. The 3 dB coupling bandwidth may be read off as approximately 4 GHz.

The coupler fabricated on an ordinary glass epoxy printed circuit board shown in Fig. 9.5(c) displays a surprisingly good response. Against this we must remember the cheapness of the substrate and that we used it for the purpose of trial measurements, i.e. to get a feel for the suitability of the coupler design. If judged suitable, better microwave substrates can be used. The performance in this case is similar to that of the Polyguide coupler. Isolation over the 3 dB bandwidth is better than 24 dB.

The Polyguide coupler output powers were measured with a power meter giving the readings shown in the inset of Fig. 9.5(b). The example is again given for illustrative purposes and a 3 W powerhead was used to measure the powers. To compare the results obtained with the network analyser with those of power meter measurements we have to perform the following calculations:

$T\,[\text{dB}] = 10 \log T = 10 \log (P_2/P_1) = 10 \log (7/10) = -1.55$ dB
$C\,[\text{dB}] = 10 \log (0.85/10) = -\ 10.7$ dB
$I\,[\text{dB}] = 10 \log (0.15/10) = -18.24$ dB
This shows good correlation of the coupler parameters.

BRANCH LINE COUPLER

Tight coupling values, typically less than 3 dB, are difficult to achieve with the parallel line coupler owing to the extremely narrow gap requirement. This has led to the development of the branch line coupler, also known as hybrid coupler. It has the added advantage of maintaining d.c. continuity between all ports as well as handling high power.

The schematic of a branch line coupler is shown in Fig. 9.6. Because of the shunting arms (branches) it is more convenient to represent the coupler in terms of admittances. The mean length of each arm is $\lambda/4$ or $\lambda_g/4$ when physically constructed. Branch line couplers may be laid out in diverse ways and two possibilities are shown

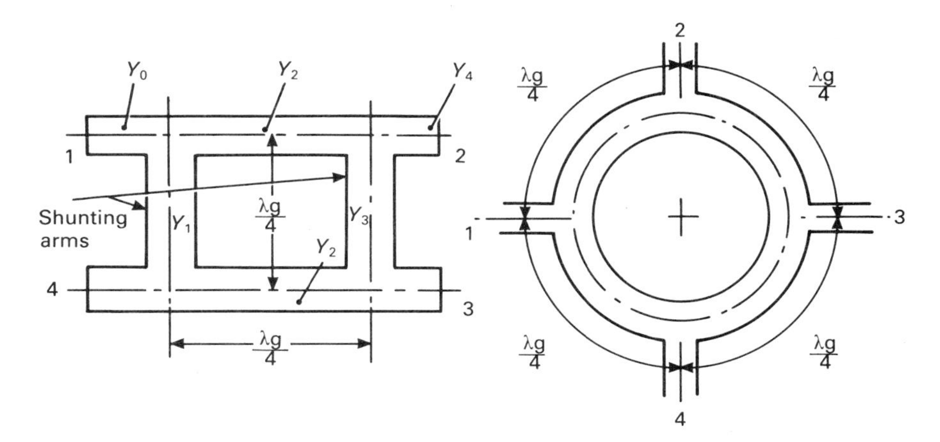

Fig. 9.6 — Branch line coupler, schematic.

in Fig. 9.7. Chamfering of sharp corners will improve coupler performance. Usually several prototypes are being made in order to obtain the best design. As can be seen, a hybrid coupler has two sets of arms of different width and hence impedance. The basic practical problem is thus to identify the most appropriate reference points for the quarter wavelength arms.

The operation of the coupler may be best described with reference to Fig. 9.7(a). Like the case of a ring coupler which will be described in the next section, power fed into port 1 will split and travel both clockwise and anticlockwise around the branches.Hence, the power available at any given port will depend on the phase relationships of these two waves travelling in opposite directions at that port. Frequently one is interested in a coupler with equal power split at port 2 and 3 and with all ports terminated in an impedance or resistance equal to the microstrip impedance which is usually 50 Ω. This case is now considered in more detail with Fig. 9.7(b).

If a hybrid coupler is to be designed for a resistive load R_L at all ports, then two of its arms must have an impedance which is different from Z_o as shown in Fig. 9.7(b). Two of the arms have then for example an impedance of $Z_a = Z_c = Z_o/(2)^{0.5}$ and the two other arms an impedance $Z_b = Z_d = Z_o$. For a 50 Ω system this means $Z_a = Z_c = 35.35\ \Omega$ and $Z_b = Z_d = 50\ \Omega$. When power is now applied to port 1, a standing wave is set up on the branch ring, whose total circumference is λ. This is by virtue of the mismatch by Z_a and Z_c. The λ/4 section between ports 1 and 2 with impedance Z_a acts as an impedance inverter (see Chapter 4). The voltage at port 2 will be $V_2 = V_1/(2)^{0.5}$. Between ports 2 and 3 where $Z_b = Z_o$ there will be no standing wave and as a result there will be an output voltage $V_3 = V_1/(2)^{0.5}$ at port 3. The phase difference between V_1 and V_2 is 90°. The power at each port is then

$$(V_2/(2)^{0.5})^2 = (V_3/(2)^{0.5})^2 = P_1/2$$

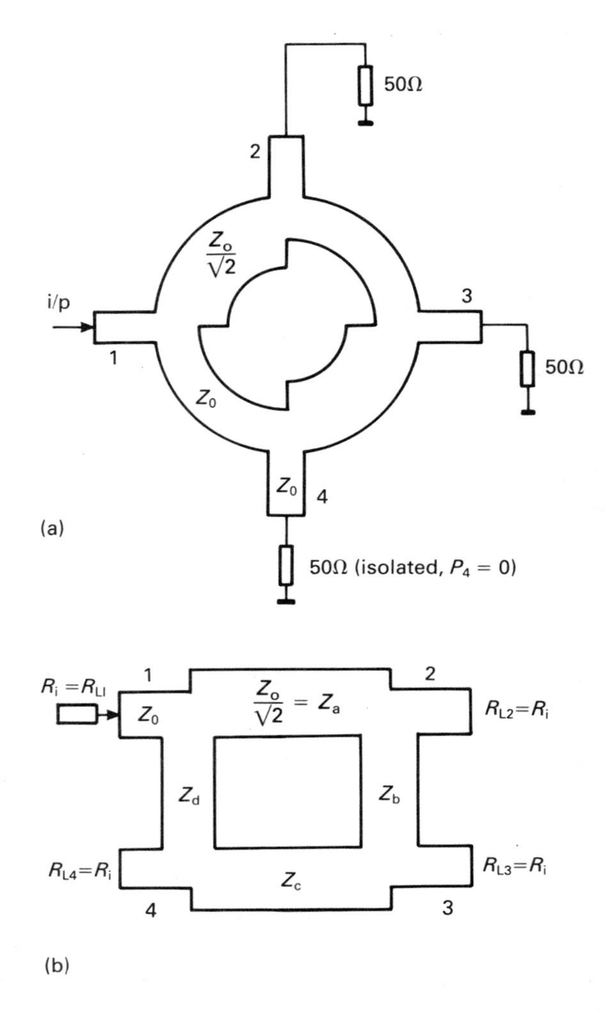

Fig. 9.7 — Actual layout of branch line coupler.

The power at port 4 is zero since the waves from port 1 travelling in opposite directions around the branch ring arrive out of phase (180°) at port 4 and thus cancel. Port 4 is said to be isolated or decoupled with respect to port 1. The same reasoning can be applied to the case where power is applied to ports 2, 3 or 4. There exist no standing waves in the $\lambda/4$ section Z_b. Looking into port 2 towards port 3, the $\lambda/4$ section is terminated by the load R_{L2} in parallel with the coupler arm impedance Z_c. Since $V_4 = 0$, there exists a short circuit for microwave frequencies at port 4. A short circuited $\lambda/4$ line, however, means that the impedance as seen from port 3 into this line is infinity (see Chapter 4). The $\lambda/4$ section of impedance Z_b is thus a properly terminated transmission line section, which supports the previous statement that there exists no standing wave between port 2 and 3.

Apart from its use for power splitting and combining, a branchline coupler can be used for impedance matching between unequal impedances.

BRANCH LINE COUPLER DESIGN

Consider the schematic of the branch line coupler as shown in Fig. 9.6. Admittances are obviously the most appropriate quantities to work with. Y_1 and Y_3 are the shunt arm admittances and Y_2 are the series arm admittances. These are all normalized with respect to the input admittance Y_o. If the coupler is perfectly matched at all four ports then it can be shown [4] that

$$Y_1 = Y_3\ Y_4 \tag{9.22}$$

where Y_4 is the output admittance. Additionally, if one asumes perfect directivity, then

$$Y_2^2 = Y_4 + Y_1\ Y_3 \tag{9.23}$$

As a result, all input power P_1 will split between output ports 2 and 3. Port 4 is isolated with respect to port 1. The power split between ports 2 and 3 may be defined as

$$k = P_2/P_3 \tag{9.24}$$

With this one obtains the following coupler admittances:

$$Y_1 = 1/(k)^{0.5} \tag{9.25}$$

$$Y_2 = \left[\frac{(k+1)\ Y_4}{k}\right]^{0.5} \tag{9.26}$$

$$Y_3 = Y_4/(k)^{0.5} \tag{9.27}$$

From this and the knowledge of the substrate parameters the microstrip dimensions of the coupler can be calculated. The following are two examples.

Example 1
Design a branchline coupler with a power coupling ratio of 3 dB. Assume the input and output impedances of the coupler to be 50 Ω.

Solution
From Fig. 9.6 we obtain the normalized admittances

$$Y_o = 1/Z_o = 1/50 = 0.02\ S$$

$$Y_4 = 0.02/0.02 = 1$$

For a coupling ratio of −3 dB we have

$$P_{31}\ [\mathrm{dB}] = 10\ \log\ P_3/P_1$$

or

$$(P_3/P_1) = \mathrm{alog}\ (-0.3) = 0.5$$

or

$$P_3 = 0.5\ P_1$$

Since $P_1 = P_2 + P_3$ we obtain $P_2 = P_3 = 0.5\ P_1$ or $k = 1$.
From eqns (9.25), (9.26) and (9.27)

$$y_1 = 1,\ y_2 = (2)^{0.5},\ y_3 = 1$$

After denormalization we obtain for the coupler shunt and series arms

$$Y_1 = y_1 \times 1/50 = 0.02\ \mathrm{S}\ \text{or}\ Z_1 = 50\ \Omega$$

$$Y_2 = y_2 \times 1/50 = 0.02828\ \mathrm{S}\ \text{or}\ Z_2 = 35.35\ \Omega$$

$$Y_3 = y_3 \times 1/50 = 0.02 \text{ S or } Z_3 = 50\ \Omega$$

With the help of the microstrip program POLSTRI the physical dimensions can be calculated for a given substrate.

Example 2

Design a branchline coupler with a power coupling coefficient of -10 dB, assuming an input and output impedance of 50 Ω.

Solution

$$P_{31}\ [\text{dB}] = 10 \log (P_3/P_1) = -10$$

$$(P_3/P_1) = \text{alog}\ (-0.1) = 0.8$$

$$P_3 = 0.8\ P_1 \text{ and } P_2 = 0.2\ P_1$$

$$k = P_2/P_3 = 0.25$$

From eqns (9.25), (9.26) and (9.27) we obtain

$$Y_1 = Y_3 = 1/(\text{k})^{0.5} = 1/(0.25)^{0.5} = 2\ S$$

$$Y_2 = 2.236\ S$$

After denormalization we have

$$Y_1 = Y_3 = y_1/50 = 0.04\ S \text{ or } Z_1 = Z_3 = 25\ \Omega$$

$$Y_2 = 0.0447\ S \text{ or } Z_2 = 22.36\ \Omega$$

$Z_o = Z_4 = 50\ \Omega$ as specified in the problem.

The flowchart for a directional coupler design is shown in Fig. 9.8(a) and examples of actual printouts with a typical measured response are given in Fig. 9.8 (b to d).

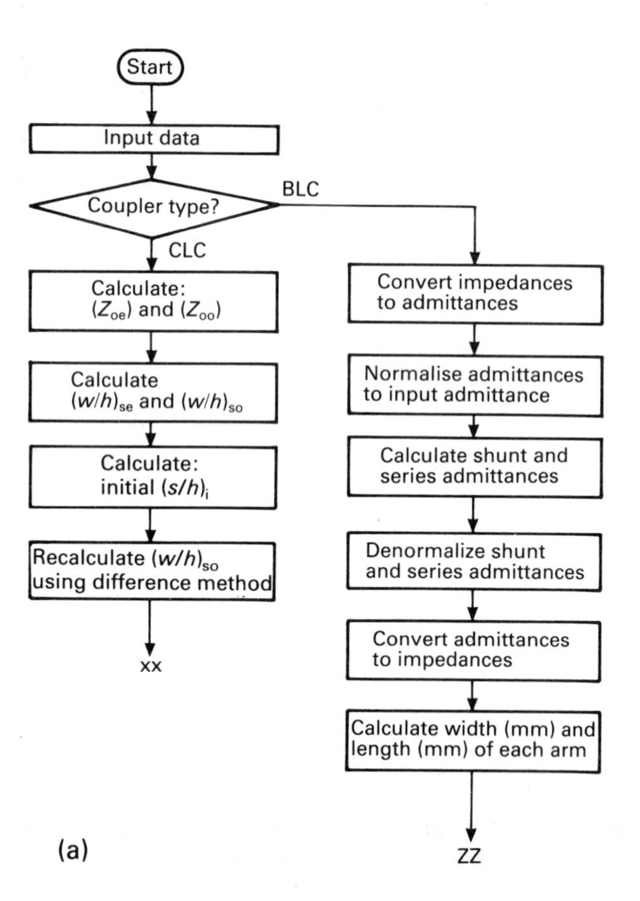

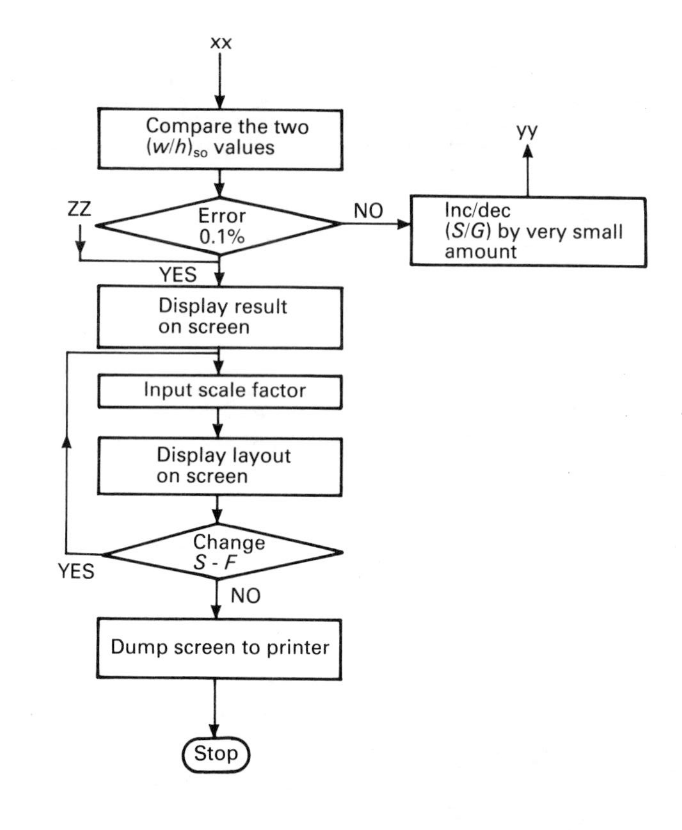

Fig. 9.8 — (a) Flowchart of a directional coupler, outline.

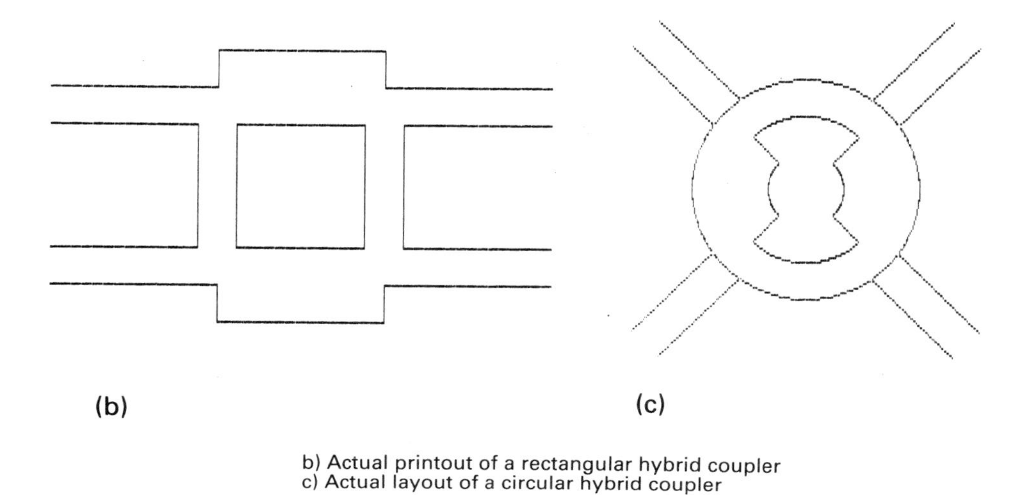

b) Actual printout of a rectangular hybrid coupler
c) Actual layout of a circular hybrid coupler

Fig. 9.8 — (b and c) DRAWHYB and DRAWRIN.

RATRACE

A special version of the branch line coupler is the hybrid magic ring, also known as 'ratrace'. It has a circumference of 1.5 λ and may be constructed in a variety of techniques using either coaxial lines, strip lines, microstrip or waveguide. It is the possibility of planar design which makes this coupler interesting and suitable for integration into microcircuits.

The principle of operation in either construction is the same. If power is applied to port 1, Fig. 9.9(a), it will split and travel around the ring in either direction. Whether or not there is an output at any of the other ports depends on the phase relationship of the waves travelling in opposite directions. Thus, at port 4 the waves are in phase and will reinforce each other giving an output. The same applies to port 2. The wave travelling from port 1 counter clockwise does 1.25 λ and that travelling clockwise from the same source travels 0.25 λ, resulting in an reinforcement and thus an output at port 2. For the same reason the output at port 3 is zero. In summary, power applied to port 1 results in an equal power split at ports 2 and 4 and no power at port 3. This means that ports 1 and 3 are mutually decoupled. From Fig. 9.9(a) it is also evident that the signals at port 2 and 4 are 180° out of phase which each other. Note that the λ/4 sections of the ring have all the same impedance.

The performance of a 1.3 GHz ratrace constructed on a glass epoxy substrate of permittivity 4.8 is shown in Fig. 9.10. This graph shows clearly the power split of about 3 dB at the design centre frequency. For 3 dB coupling over a wider frequency range more complex designs are required. The graph shows also clearly the isolation experienced at port 3 (Fig. 9.9(a)) with respect to port 1. The return loss of all three output ports is fairly uniform. Better results can be obtained with closer manufacturing tolerances and a better microwave substrate.

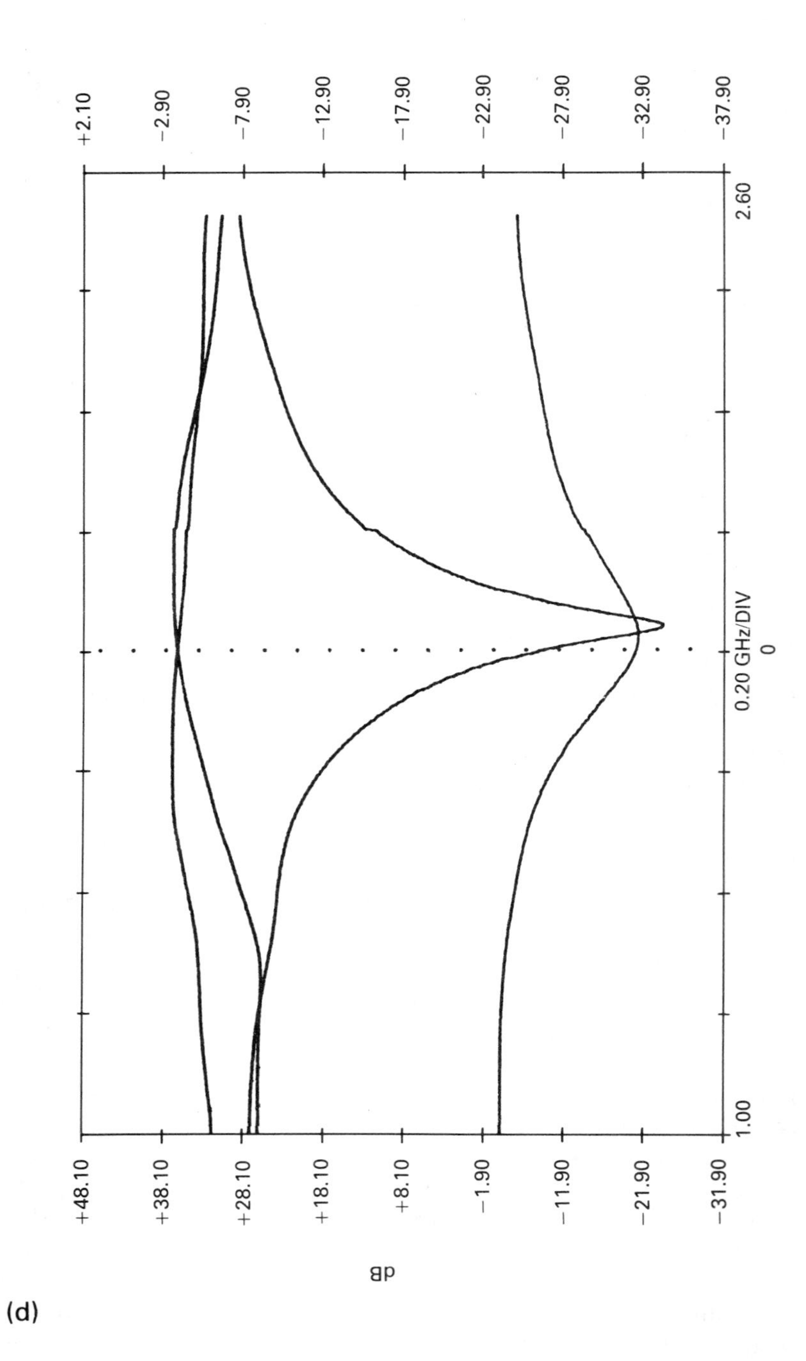

Fig. 9.8 — (d) Measured 3dB hybrid coupler response.

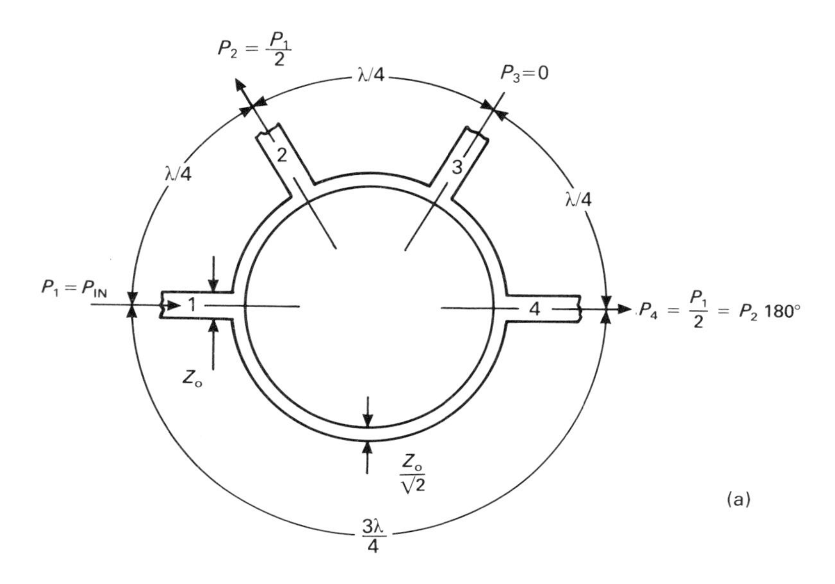

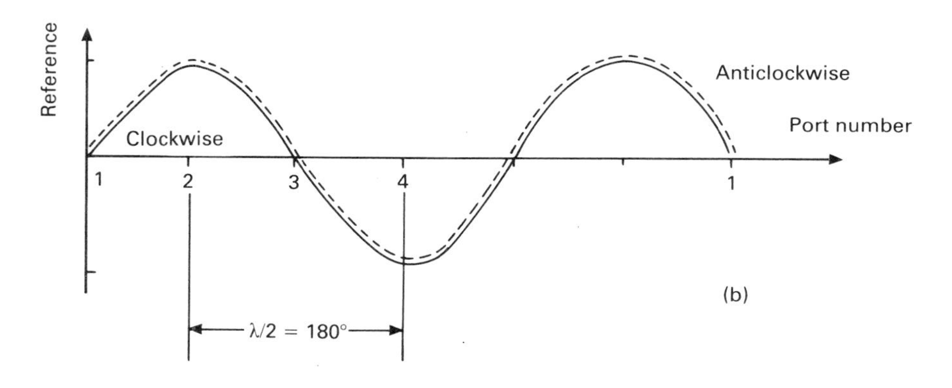

Fig. 9.9 — (a) Schematic diagram of the hybrid ring (ratrace). (b) Drawing explaining power and phase relationship between ports.

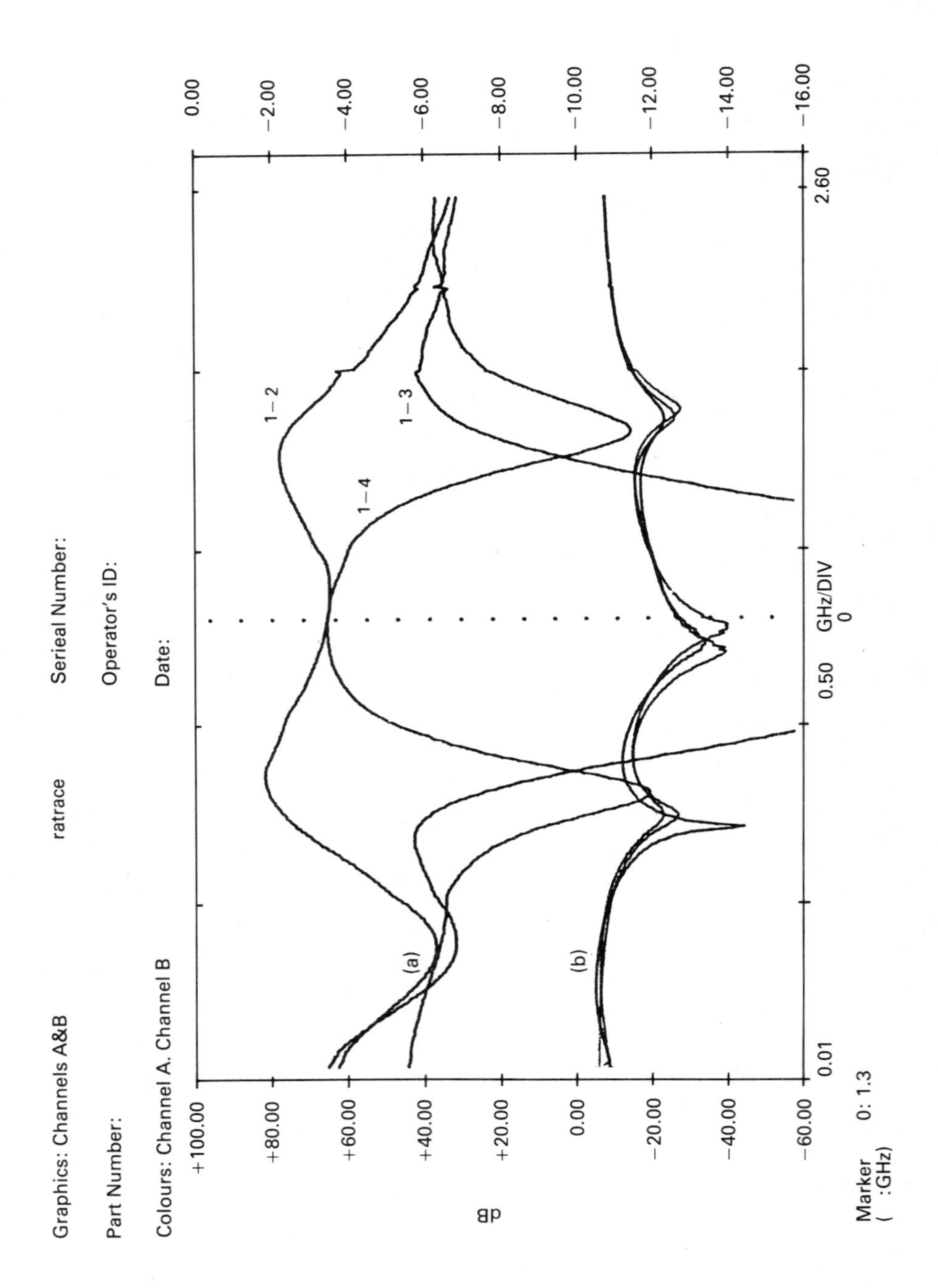

Fig. 9.10 — Response of a ratrace (a) transmission (b) return loss.

```
This program will draw a HYBRID COUPLER on the screen

and to do this, the following inputs are required :-

   a) Length L1 in mm.

   b) Length L2 in mm.

   c) Width Z0 in mm.

   d) Width Z1 in mm.

L1 , L2 , Z0 and Z1 are shown in Fig.9.7

Input L1 (mm)?30


Input L2 (mm)?30


Input WIDTH Z0 (mm)?6


Input WIDTH Z1 (mm)?12


Would you like this to be drawn to scale or not ?

If the drawing is to be drawn to scale then enter '1'

otherwise input the magnification factor.
?2
```

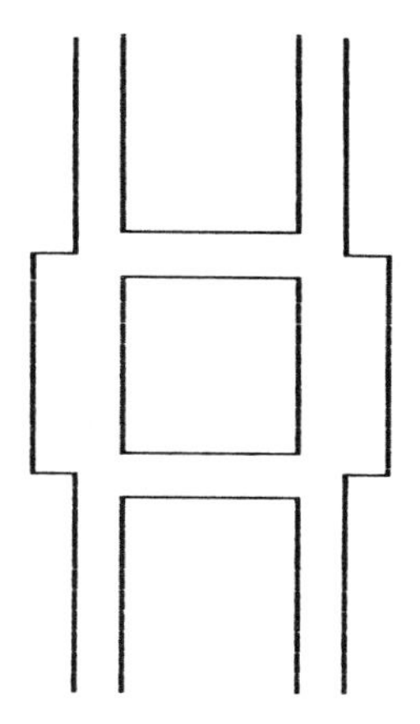

```
L.
  10 REM This programme is called DRAWHYB
  20  REM  *£  Program to draw a HYBRID COUPLER on the screen £*
  30 MODE0
  40 CLS
  50 PRINT ""''''''''""
  60 PRINT "    This program will draw a HYBRID COUPLER on the screen"
  70PRINT
  80PRINT "    and to do this, the following inputs are required :-"
  90PRINT
 100PRINT
 110PRINT "        a) Length L1 in mm."
 120PRINT
 130PRINT "        b) Length L2 in mm."
 140PRINT
 150PRINT "        c) Width Z0 in mm."
 160PRINT
 170PRINT "        d) Width Z1 in mm."
 180PRINT
 190 PRINT"  L1 , L2 , Z0 and Z1 are shown in Fig.9.7"
 200PRINT
 210PRINT "  Input L1 (mm)";
 220INPUT L1
 230CLS
 240PRINT ""''''''''""
 250PRINT "  Input L2 (mm)";
 260INPUT L2
 270CLS
 280PRINT ""''''''''""
 290PRINT "  Input WIDTH Z0 (mm)";
 300INPUT ZOTHICK
 310CLS
 320PRINT ""''''''''""
 330PRINT "  Input WIDTH Z1 (mm)";
 340INPUT ZITHICK
 350 CLS
 360 PRINT ""''''''''""
 370 PRINT "  Would you like this to be drawn to scale or not ? "
 380 PRINT
 390 PRINT "  If the drawing is to be drawn to scale then enter '1' "
 400 PRINT
 410 PRINT "  otherwise input the magnification factor. "
 420 INPUT MAG
 430 CLS
 440 SPX=225
 450 SPY=162
 460 XFACTOR=MAG*1280/225
 470 YFACTOR=MAG*1024/162
 480 THIGHT=(L2+2*ZITHICK-ZOTHICK)*YFACTOR
 490 YOFFSET=(1024-THIGHT)/2
 500 ZIEXTRA=ZITHICK-ZOTHICK
 510 PLOT 4,0,YOFFSET+ZIEXTRA*YFACTOR
 520 PLOT 1,L1*XFACTOR,0
 530 PLOT 1,0,-ZIEXTRA*YFACTOR
 540 PLOT 1,L1*XFACTOR,0
 550 PLOT 1,0,ZIEXTRA*YFACTOR
 560 PLOT 1,L1*XFACTOR,0
 570 PLOT 4,0,YOFFSET+(ZITHICK+L2)*YFACTOR
 580 PLOT 1,L1*XFACTOR,0
 590 PLOT 1,0,ZIEXTRA*YFACTOR
 600 PLOT 1,L1*XFACTOR,0
 610 PLOT 1,0,-ZIEXTRA*YFACTOR
 620 PLOT 1,L1*XFACTOR,0
 630 PLOT 4,0,YOFFSET+ZITHICK*YFACTOR
 640 PLOT 1,(L1-ZOTHICK/2)*XFACTOR,0
 650 PLOT 1,0,(L2-ZOTHICK)*YFACTOR
 660 PLOT 1,-L1*XFACTOR,0
 670 PLOT 4,3*L1*XFACTOR,YOFFSET+ZITHICK*YFACTOR
 680 PLOT 1,(ZOTHICK/2-L1)*XFACTOR,0
 690 PLOT 1,0,(L2-ZOTHICK)*YFACTOR
 700 PLOT 1,(L1-ZOTHICK/2)*XFACTOR,0
 710 PLOT 4,(L1+ZOTHICK/2)*XFACTOR,YOFFSET+ZITHICK*YFACTOR
 720 PLOT 1,(L1-ZOTHICK)*XFACTOR,0
 730 PLOT 1,0,(L2-ZOTHICK)*YFACTOR
 740 PLOT 1,(ZOTHICK-L1)*XFACTOR,0
 750 PLOT 1,0,(ZOTHICK-L2)*YFACTOR
 760 END
```

```
RUN
This program will draw a RING SHAPED COUPLER on the screen
and to do this, the following inputs are required :-

   a)  Diameter of inner ring in mm.
   b)  ZOTHICK in mm.
   c)  Z1THICK in mm.

Input Diameter of inner ring in mm. ?30

Input ZOTHICK in mm.?6

Input Z1THICK in mm.?3
```

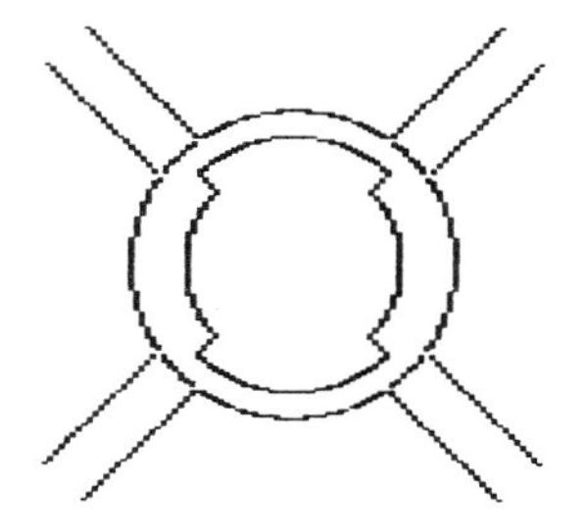

```
L.
   10 REM This programme is called DRAWRIN
   20 REM ** Program to draw a Ring coupler on screen and/or printer **
   30 MODE0
   40 CLS
   50 PRINT "  This program will draw a RING SHAPED COUPLER on the screen "
   60 PRINT
   70 PRINT "  and to do this, the following inputs are required :-"
   80 PRINT
   90 PRINT
  100 PRINT "     a)  Diameter of inner ring in mm."
  110 PRINT
  120 PRINT "     b)  ZOTHICK in mm."
  130 PRINT
  140 PRINT "     c)  Z1THICK in mm."
  150 PRINT ""'""
  160 PRINT "  Input Diameter of inner ring in mm. ";
  170 INPUT DIAMETER
  180 CLS
  190 PRINT ""'""
  200 PRINT "  Input ZOTHICK in mm.";
  210 INPUT ZOTHICK
  220 CLS
  230 PRINT ""'""
  240 PRINT "  Input Z1THICK in mm.";
  250 INPUT ZITHICK
  260 CLS
  270 REM
  280 REM ** Now draw the coupler **
  290 REM
  300 YFACTOR=1024/162
  310 XFACTOR=1280/225
  320 REM
  330 REM ** Draw inner circle **
  340 R=DIAMETER/2
  350 D=ZITHICK-ZOTHICK+R
  360 MOVE (112.5+DIAMETER/2)*XFACTOR,81*YFACTOR
  370
  380     FOR X=PI/500 TO 2*PI STEP PI*5/1000
  390       IF X>PI/4 AND X<3*PI/4 THEN R=D ELSE R=R
  400       IF X>5*PI/4 AND X<7*PI/4 THEN R=D ELSE R=R
  410      DRAW (112.5+COS(X)*R)*XFACTOR,(81+SIN(X)*R)*YFACTOR
  420      R=DIAMETER/2
  430     NEXT X
  440 REM
  450 REM ** Now draw the outer circle **
  460 OUTERDIA=DIAMETER+ZITHICK*2
  470 MOVE (112.5+OUTERDIA/2)*XFACTOR,81*YFACTOR
  480     FOR X=PI/500 TO 2*PI STEP PI*5/1000
  490      DRAW (112.5+COS(X)*(OUTERDIA/2))*XFACTOR,(81+SIN(X)*(OUTERDIA/2))*YFAC
TOR
  500     NEXT X
  510 REM ** Now draw parrallel lines **
  520 REM * Calc thickness of Zo WRT 2PI (or 360 deg) *
  530 ZOANGLE=(ZOTHICK/(PI*OUTERDIA))*2*PI
  540 ZOANGLE1=(ZOTHICK/(PI*2*OUTERDIA))*2*PI
  550 FOR W=0 TO 1
  560   FOR Q=0 TO 3
  570      A1=(PI/4)-(ZOANGLE/2)+Q*PI/2+W*ZOANGLE
  580      A2=(PI/4)-(ZOANGLE1/2)+Q*PI/2+W*ZOANGLE1
  590      X1=(112.5+COS(A1)*(OUTERDIA/2))*XFACTOR
  600      Y1=(81+SIN(A1)*(OUTERDIA/2))*YFACTOR
  610      X2=(112.5+COS(A2)*(OUTERDIA))*XFACTOR
  620      Y2=(81+SIN(A2)*(OUTERDIA))*YFACTOR
  630      PLOT 4,X1,Y1
  640      PLOT 6,X2,Y2
  650   NEXT Q
  660 NEXT W
  670 END
>
```

BALANCED COUPLER CONFIGURATIONS

The coupler is frequently used in balanced amplifier configurations. This reduces the VSWR of the circuit and allows for amplifier cascading to obtain higher gains. A coupler most suitable for such applications is the Lange coupler which will be discussed shortly. A single stage balanced amplifier configuration is shown in Fig. 9.11. The microwave signal is fed to the input coupler and half of it appears at either

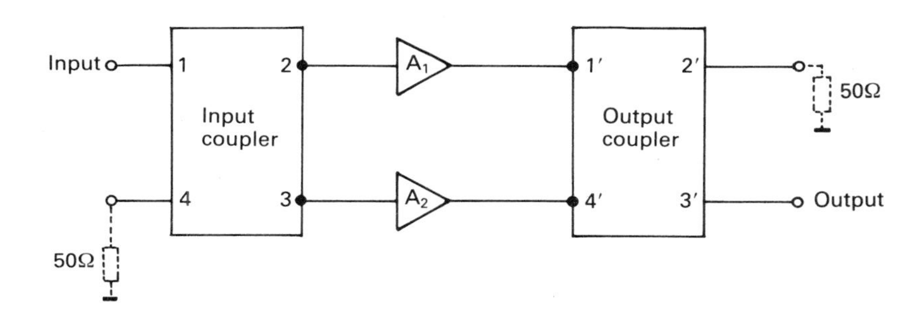

Fig. 9.11 — Single stage balanced amplifier configuration.

input of amplifier A_1 and A_2. After amplification another coupler of the same type is used to combine the signal which is then available at the coupler output for further processing. More detailed information is now given in the following.

The symbol of a general coupler is shown in Fig. 9.12 and from this it is evident

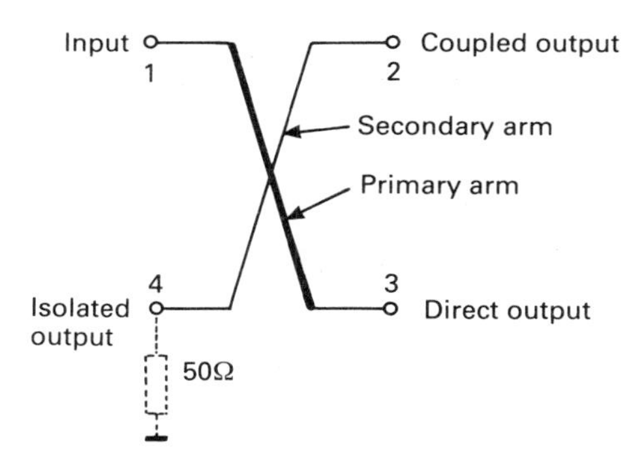

Fig. 9.12 — Symbol of a coupler with isolated output terminated in $R = Z_o = 50\ \Omega$.

that the coupler is a 4-port. The magnitude and phase conditions existing on a Lange coupler are now briefly described with respect to Fig. 9.13. Assume an incident

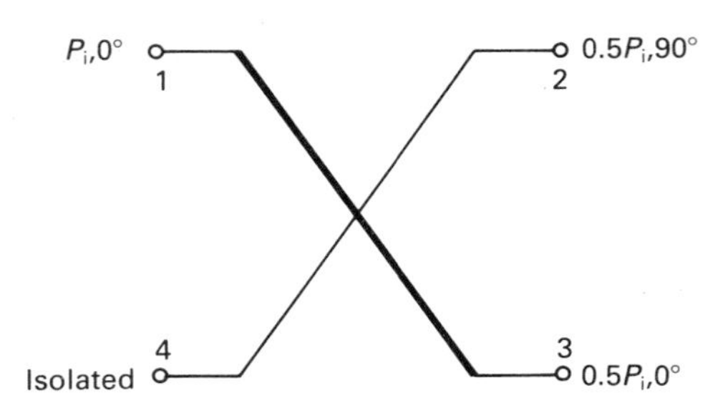

Fig. 9.13 — Magnitude and phase relationship, power splitting.

power P_i at port 1 which for convenience of explanation has a phase angle of 0°. The signal appears at port 2 and 3 as half of the incident power and with a phase shift of 90°, thus the other names of 3 dB or 90° coupler. The powers are $P_2 = 0.5\ P_1$, 90° and $P_3 = 0.5\ P_1$, 0°. Hence we note that in a Lange coupler there is no phase shift in the primary arm and a 90° shift in the coupled or secondary arm. Since the device is symmetrical the same applies to the reciprocal case, i.e. where the coupled arm becomes the primary arm and vice versa.

The phase shift between ports 2 and 3 is then 90°. Port 4 is isolated with respect to the input, i.e. no power flows into this port. In a circuit application as that in Fig. 9.11 the coupler can also be used to perform the opposite function to power splitting, namely power combining. This is now explained with Fig. 9.14, where a dash has

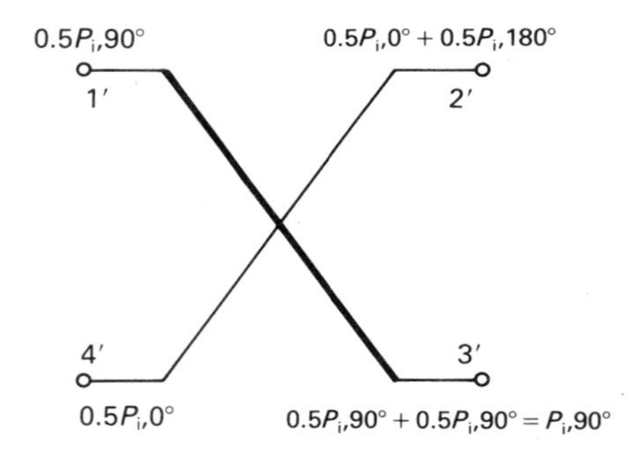

Fig. 9.14 — Magnitude and phase relationship, power combining.

been used with the port number to indicate that we deal with the output coupler and hence power combination. Now imagine ports 2 and 3 of Fig. 9.13 connected to ports 1′ and 4′ of Fig. 9.14. The power incident on port 1′ is 0.5 P_i, 90° and the power

```
L.
   10 REM This programme is called DRAWRIN
   20 REM ** Program to draw a Ring coupler on screen and/or printer **
   30 MODE0
   40 CLS
   50 PRINT "  This program will draw a RING SHAPED COUPLER on the screen "
   60 PRINT
   70 PRINT "  and to do this, the following inputs are required :-"
   80 PRINT
   90 PRINT
  100 PRINT "     a)  Diameter of inner ring in mm."
  110 PRINT
  120 PRINT "     b)  ZOTHICK in mm."
  130 PRINT
  140 PRINT "     c)  Z1THICK in mm."
  150 PRINT ""'""
  160 PRINT "  Input Diameter of inner ring in mm. ";
  170 INPUT DIAMETER
  180 CLS
  190 PRINT ""'""
  200 PRINT "  Input ZOTHICK in mm.";
  210 INPUT ZOTHICK
  220 CLS
  230 PRINT ""'""
  240 PRINT "  Input Z1THICK in mm.";
  250 INPUT ZITHICK
  260 CLS
  270 REM
  280 REM ** Now draw the coupler **
  290 REM
  300 YFACTOR=1024/162
  310 XFACTOR=1280/225
  320 REM
  330 REM ** Draw inner circle **
  340 R=DIAMETER/2
  350 D=ZITHICK-ZOTHICK+R
  360 MOVE (112.5+DIAMETER/2)*XFACTOR,81*YFACTOR
  370
  380    FOR X=PI/500 TO 2*PI STEP PI*5/1000
  390      IF X>PI/4 AND X<3*PI/4 THEN R=D ELSE R=R
  400      IF X>5*PI/4 AND X<7*PI/4 THEN R=D ELSE R=R
  410     DRAW (112.5+COS(X)*R)*XFACTOR,(81+SIN(X)*R)*YFACTOR
  420     R=DIAMETER/2
  430    NEXT X
  440 REM
  450 REM ** Now draw the outer circle **
  460 OUTERDIA=DIAMETER+ZITHICK*2
  470 MOVE (112.5+OUTERDIA/2)*XFACTOR,81*YFACTOR
  480    FOR X=PI/500 TO 2*PI STEP PI*5/1000
  490     DRAW (112.5+COS(X)*(OUTERDIA/2))*XFACTOR,(81+SIN(X)*(OUTERDIA/2))*YFAC
TOR
  500    NEXT X
  510 REM ** Now draw parrallel lines **
  520 REM * Calc thickness of Zo WRT 2PI (or 360 deg) *
  530 ZOANGLE=(ZOTHICK/(PI*OUTERDIA))*2*PI
  540 ZOANGLE1=(ZOTHICK/(PI*2*OUTERDIA))*2*PI
  550 FOR W=0 TO 1
  560   FOR Q=0 TO 3
  570     A1=(PI/4)-(ZOANGLE/2)+Q*PI/2+W*ZOANGLE
  580     A2=(PI/4)-(ZOANGLE1/2)+Q*PI/2+W*ZOANGLE1
  590     X1=(112.5+COS(A1)*(OUTERDIA/2))*XFACTOR
  600     Y1=(81+SIN(A1)*(OUTERDIA/2))*YFACTOR
  610     X2=(112.5+COS(A2)*(OUTERDIA))*XFACTOR
  620     Y2=(81+SIN(A2)*(OUTERDIA))*YFACTOR
  630     PLOT 4,X1,Y1
  640     PLOT 6,X2,Y2
  650   NEXT Q
  660 NEXT W
  670 END
>
```

called an 'interdigitated stripline quadrature hybrid'. The large benefit to be derived from this coupler, also known as Lange coupler, is that it can work over an octave of bandwidth or more.

In actual fact, Lange constructed his coupler, as shown in Fig. 9.16, in microstrip,

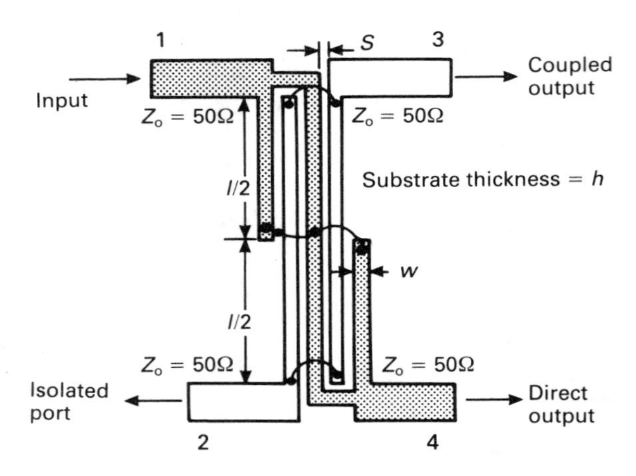

Fig. 9.16 — Interdigitated (Lange) coupler.

having four ports. Like any ideal hybrid, power incident on port 1 couples equally to ports 3 and 4 with no power appearing at port 2. Owing to the coupling length, the phase difference between ports 3 and 4 is 90°, hence also the name of quadrature coupler. Interdigitated couplers consist of three or more lines (Fig. 9.17), with alternate fingers joined together. This results in the tight coupling with this particular type of coupler. In an n-finger coupler there are always n-1 bridging wire connections and in Fig. 9.17 it is shown where these wires need to be connected. A two-finger coupler is a special case in that it is also known as a parallel line coupler. It can be constructed without a bridging wire, although the port numbers differ from those of the multi-finger couplers, i.e. three fingers or more.

Compared with the branch line coupler the Lange coupler has not only a wider bandwidth, but requires much less substrate area for a given specification. In order to obtain good performance, the bondwires, i.e. the wires which join alternate fingers, should be as short as possible. Freqently two or more bonds are made in parallel in order to minimize bondwire inductance. At S-band Lange reported a directivity of 27 dB, a return loss larger than 25 dB, an insertion loss less than 0.13 dB and an imbalance between the output ports of less than 0.25 dB over a 40% bandwidth. Furthermore, the phase variation between outputs 3 and 4 was within 2° of 90°.

As can be seen from Fig. 9.16, the feedlines of the coupler are of different impedance to that of the coupling fingers. Also, the finger width w and the finger spacing s is constant throughout the entire coupling structure. The coupling length l is a quarter of a wavelength of the centre frequency, although one may also use the lower or upper operating frequency to bring out or modify the coupler response in a

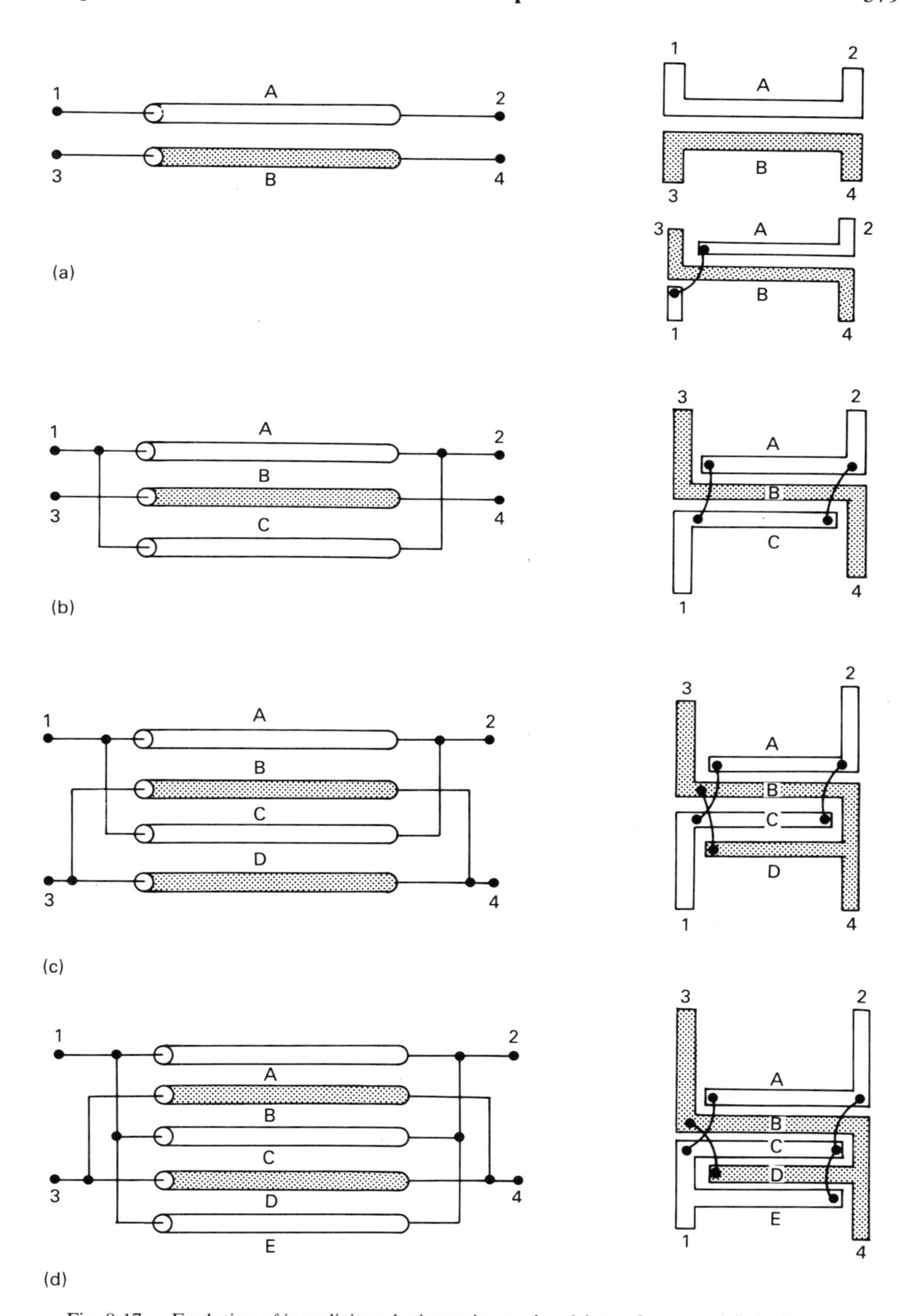

Fig. 9.17 — Evolution of interdigitated microstrip coupler, (a) two fingers to (e) six fingers.

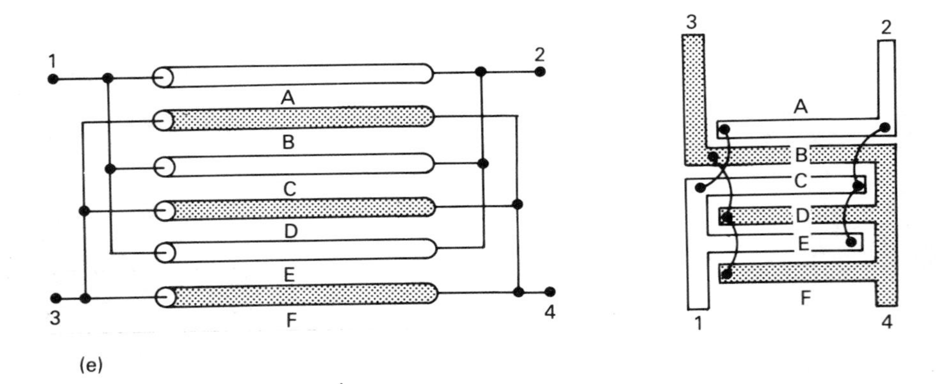

Fig. 9.17 — Evolution of interdigitated microstrip coupler, (a) two fingers to (e) six fingers.

particular way. The odd and even mode field distribution of a four-finger interdigitated coupler is shown in Fig. 9.18. It is thus evident that coupler analysis will involve even and odd-mode impedances.

The practical work by Lange prompted research by others [2,3,6,7,8] which resulted in a numerical analysis, even wider bandwidth, or coupler variations in that the fingers were all of the same length. Based on Lange's experimental work Ou [7] derived the voltage–current relationships for an interdigitated directionnal coupler based on Fig. 9.19, resulting in the following matrix:

$$\begin{bmatrix} I_A \\ I_B \\ I_C \\ I_D \end{bmatrix} = \begin{bmatrix} 0 & 0 & jN & jM \\ 0 & 0 & jM & jN \\ jN & jM & 0 & 0 \\ jM & jN & 0 & 0 \end{bmatrix} \begin{bmatrix} V_A \\ V_B \\ V_C \\ V_D \end{bmatrix} \quad (9.28)$$

If all four ports of the coupler are perfectly matched in the characteristic admittance Y_o or characteristic impedance Z_o, then the following condition holds:

$$Y_o^2 = M^2 - N^2 \quad (9.29)$$

where

$$M = \frac{k\,Y_{11}}{2} + \left(\frac{k}{2} - 1\right)\frac{Y_{12}^2}{Y_{11}} \quad (9.30)$$

and

$$N = (k-1)\,Y_{12} \quad (9.31)$$

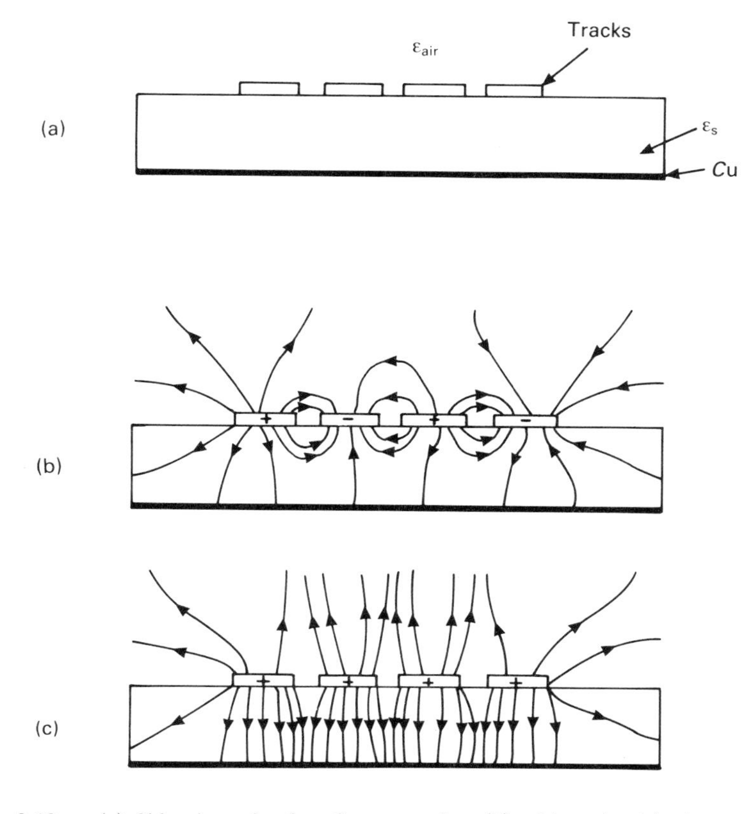

Fig. 9.18 — (a) Side view of a four-finger coupler, (b) odd mode, (c) even mode field distribution of four-finger coupler.

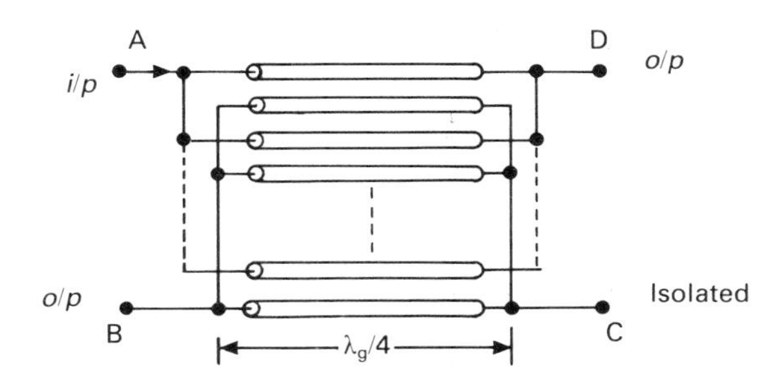

Fig. 9.19 — Interdigitated coupler.

For an input at A the following terms are obtained for coupling C, isolation I and transmission T:

$$C = P_B/P_A = N^2/M^2 \tag{9.32}$$

$$I = P_C/P_A = 0 \tag{9.33}$$

$$T = P_D/P_A = (M^2 - N^2)/M^2 \tag{9.34}$$

Equation (9.33) refers thus to perfect isolation, i.e. there is no output at port C for an input at port A. The number of coupler fingers is denoted by the symbol k. Substituting eqns (9.30) and (9.31) into eqns (9.29) and (9.32) yields the following design equations from (9.29):

$$Y_o^2 = M^2 - N^2 = \left[\frac{k\ Y_{11}}{2} + \left(\frac{k}{2} - 1\right)\frac{Y_{12}^2}{Y_{11}}\right]^2 - \left[(k-1)\ Y_{12}\right]^2 \tag{9.35}$$

and

$$C = \frac{P_B}{P_A} = \left[\frac{2(k-1)\ Y_{11}\ Y_{12}}{k\ Y_{11}^2 + (k-2)\ Y_{12}^2}\right]^2 \tag{9.36}$$

Based on the theory of parallel coupled lines (Chapter 2), the admittances Y_{11} and Y_{12} can be expressed in terms of the odd and even modes via eqns (9.37) and (9.38), namely:

$$Y_{11} = 1/2\ (Y_{oo} + Y_{oe}) \tag{9.37}$$

$$Y_{12} = -\ 1/2\ (Y_{oo} - Y_{oe}) \tag{9.38}$$

Substitution of eqns (9.10) and (9.11) into eqns (9.8) and (9.9) gives:

$$Y_o^2 = \frac{[(k-1)\ Y_{oo}^2 + Y_{oo}\ Y_{oe}][(k-1)\ Y_{oe}^2 + Y_{oo}\ Y_{oe}]}{(Y_{oo} + Y_{oe})^2} \tag{9.39}$$

and

$$C = \frac{P_B}{P_A} = \left[\frac{(k-1)\ Y_{oo}^2 - (k-1)\ Y_{oe}^2}{(k-1)\ Y_{oo}^2 + 2\ Y_{oo}\ Y_{oe} + (k-1)\ Y_{oe}^2}\right]^2 \tag{9.40}$$

A check on this equation may be made by letting $k=2$, i.e. assuming two parallel coupled lines. Hence substituting $k=2$ into eqns (9.39) and (9.40) gives

$$Y_o^2 = Y_{oo} Y_{oe} \tag{9.41}$$

and

$$C = \frac{P_B}{P_A} = \left(\frac{Y_{oo} - Y_{oe}}{Y_{oo} + Y_{oe}}\right)^2 \tag{9.42}$$

Equations (9.41) and (9.42) are the same equations as those stated in Chapter 2, namely eqns (2.21) and (2.22).

Ou then lists a table from which two lines are extracted and added to, as shown in Table 9.1.

Table 9.1

No. of lines, k	C (dB)	Y_{oo} (S)	Z_{oo} (Ω)	Y_{oe} (S)	Z_{oe} (Ω)	Y_o (S)	Z_o (Ω)
2	3	0.0483	20.70	0.00828	120.77	0.02	50
4	6	0.0147	68.02	0.00702	142.45	0.02	50

Because of fabrication difficulties and small tolerance requirements couplers with more than six fingers are rarely used. In any case, good results are obtained with six or fewer fingers.

If the coupler ports are terminated in $Z_o = 50\,\Omega$ then $Y_o = 0.02$S. The same applies to Y_{oo}, Z_{oo} *and* Y_{oe}, Z_{oe}. To check the first line in Table 9.1 we use Ou's analysis, eqn. (9.42)

$$C = \left(\frac{Y_{oo} - Y_{oe}}{Y_{oo} + Y_{oe}}\right)^2 = \left(\frac{0.0483 - 0.00828}{0.0482 + 0.00828}\right)^2 = 0.50029$$

$$C[\text{dB}] = 10 \log C = 10 \log 0.50029 = -3 \text{ dB}$$

This confirms that coupling is 3 dB. Similarly, we can confirm that the analysis for a four-finger coupler results in a coupling of 6 dB by using eqn. (9.40). Naturally, Z_{oo} and Z_{oe} are not normally available and must be obtained from the shape ratios w/h and s/h of the finger structure. This is shown in the following.

Osmani [8] extended the analysis work by Ou and that of Lange and derived a set of simple synthesis equations which offer themselves readily for computation or CAD. His coupler synthesis includes the underlying theory for a pair of parallel coupled microstrips as published by Akhtarzad [1]. This was dealt with in Chapter 2. Osmanis's synthesis equations are stated as follows:

$$Z_{oo} = Z_o \sqrt{\frac{1-C}{1+C}} \frac{(k-1)(1+q)}{(C+q)+(k-1)(1-C)} \tag{9.43}$$

$$Z_{oe} = Z_{oo} \frac{C+q}{(k-1)(1-C)} \tag{9.44}$$

where

$$q = \sqrt{C^2 + (1-C^2)(k-1)^2} \tag{9.45}$$

Here k are the even number of fingers, i.e. $k = 2,4,6$, etc., C is the coupling factor (voltage ratio) and Z_{oo}, Z_{oe} the odd and even mode impedances. The following is an example illustrating the design or synthesis of an interdigitated coupler.

First of all the program MENU is loaded which then asks for the desired coupler and substrate data to be inputted. From this the physical coupler parameters are then being calculated. Finally, the program asks for the scale factor of the coupler to be displayed on the VDU. The program scales the drawing automatically according to the operating frequency and permittivity used for the coupler construction. With the aid of a dump program the coupler can be printed on a standard printer such as, for example, an EPSON LX800. The measured performance of a coupler designed in this way is shown in Fig. 9.20.

WIRELINE COUPLERS

Couplers may also be made from a pair of twin conductor lines. These are commercially available under the trade name of Sage Wireline and can be usefully employed at frequencies up to about 3 GHz. Sage Wireline can be used to fabricate quadrature hybrid couplers and directional couplers. Furthermore, the wireline can for example be employed in mixers, power monitors, power dividers/combiners and interstage coupling of amplifiers.

The cross-sectional area of a wireline is shown in Fig. 9.21. The two centre wires are running in parallel and are separated only by the insulation of one of the wires. The two wires are thus fairly closely coupled. The terminal diagram of a coupler made from Sage Wireline is shown in Fig. 9.22 and a more physical presentation in Fig. 9.23. With regard to Fig. 9.22, the first wire, which is also known as primary arm, is marked AB and the second wire, also known as secondary arm, is marked CD. There is no d.c. path between AB and CD.

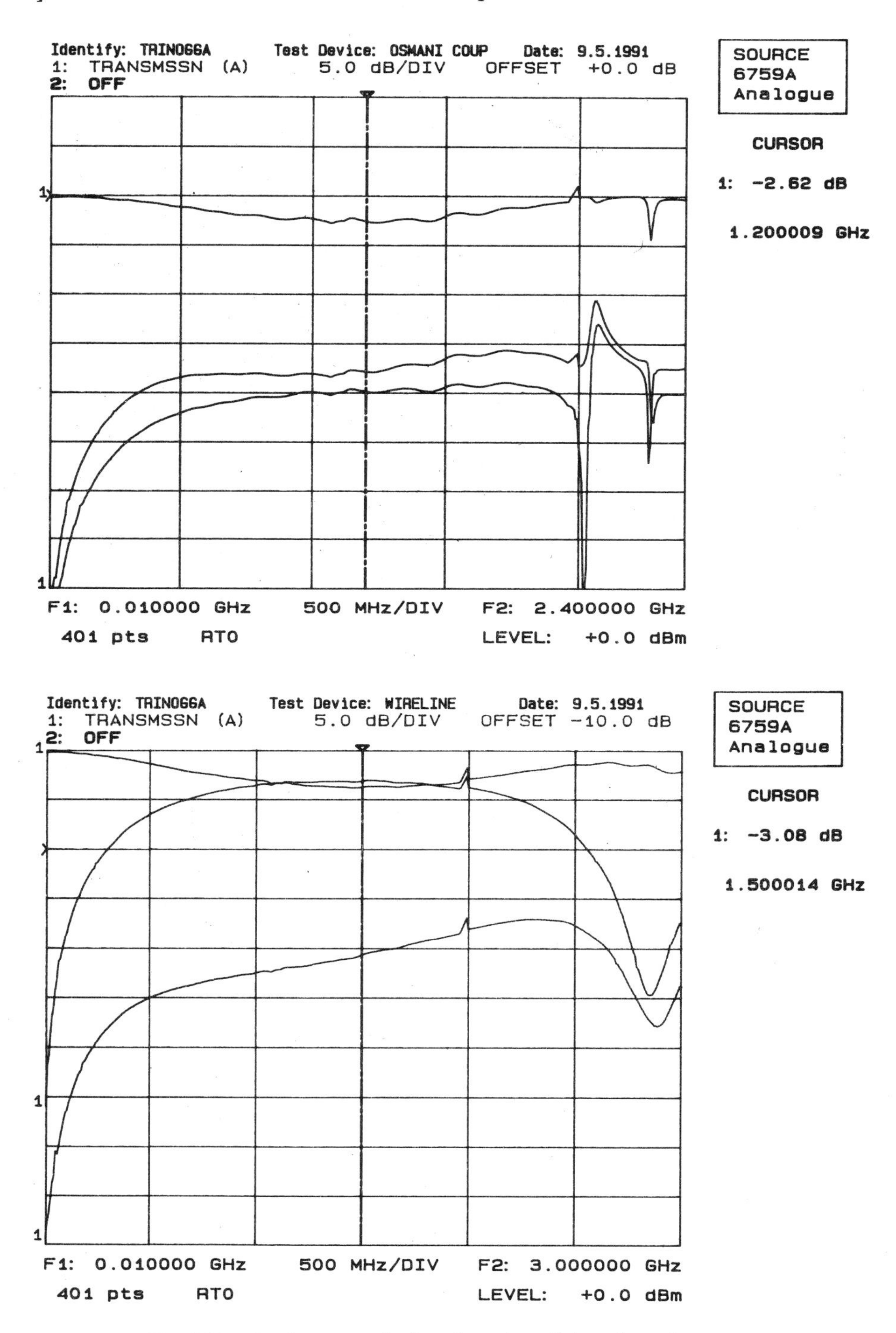

Fig. 9.20 — Performance of a four-finger interdigitated coupler.

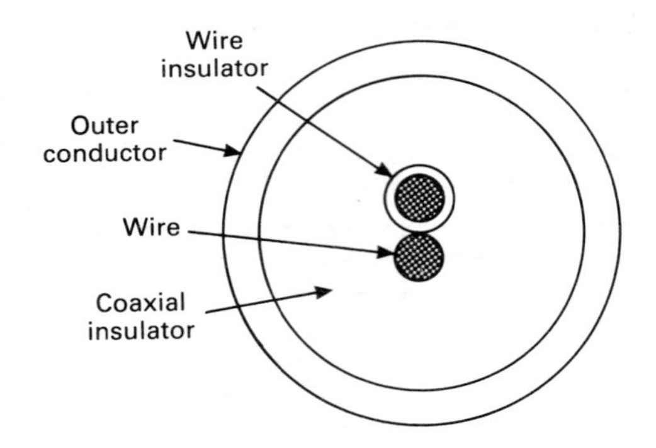

Fig. 9.21 — Cross-sectional area of a wireline.

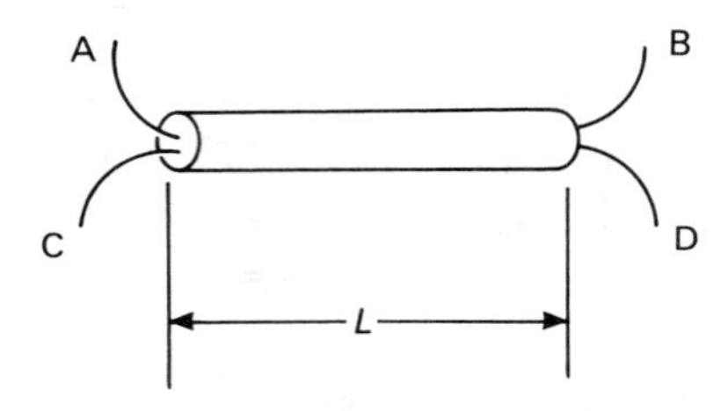

Fig. 9.22 — Terminal diagram of a wireline.

```
RUN
     LANGE COUPLER calculation program
     =================================
 USING EQUATIONS OUTLINED BY OSMANI AND AKHTARZAD THIS PROGRAM WILL
 CALCULATE BOTH THE STRIP SEPARATION AND THE WIDTH OF THE STRIP FOR
 FAST,ACCURATE MICROWAVE COUPLER DESIGN OF LANGE COUPLERS.
PRESS ANY KEY TO CONTINUE
```

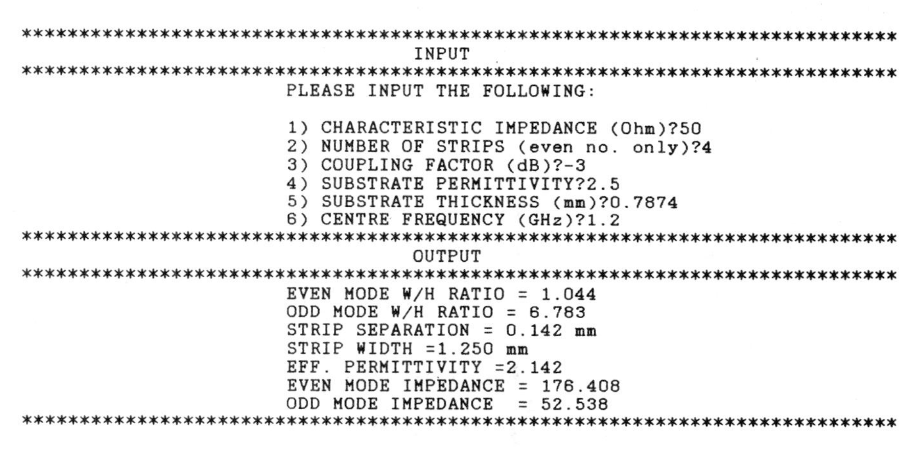

```
****************************************************************************
                                  INPUT
****************************************************************************
                       PLEASE INPUT THE FOLLOWING:

                       1) CHARACTERISTIC IMPEDANCE (Ohm)?50
                       2) NUMBER OF STRIPS (even no. only)?4
                       3) COUPLING FACTOR (dB)?-3
                       4) SUBSTRATE PERMITTIVITY?2.5
                       5) SUBSTRATE THICKNESS (mm)?0.7874
                       6) CENTRE FREQUENCY (GHz)?1.2
****************************************************************************
                                  OUTPUT
****************************************************************************
                       EVEN MODE W/H RATIO = 1.044
                       ODD MODE W/H RATIO = 6.783
                       STRIP SEPARATION = 0.142 mm
                       STRIP WIDTH =1.250 mm
                       EFF. PERMITTIVITY =2.142
                       EVEN MODE IMPEDANCE = 176.408
                       ODD MODE IMPEDANCE  = 52.538
****************************************************************************

                         PRESS ANY KEY TO CONTINUE
```

```
WOULD YOU LIKE TO:-
1) DRAW THE SPECIFIED COUPLER
2) RUN THE PROGRAM AGAIN
3) RETURN TO MENU
PLEASE ENTER YOUR CHOICE 1,2 OR 31
```

```
ENTER YOUR SCALE FACTOR ?3
THIS COUPLER HAS A SCALE FACTOR OF 3.000

LIKE TO TRY ANOTHER SCALE FACTOR(Y/N)?

>*LOAD"SCRNDMP"
>CALL&900
```

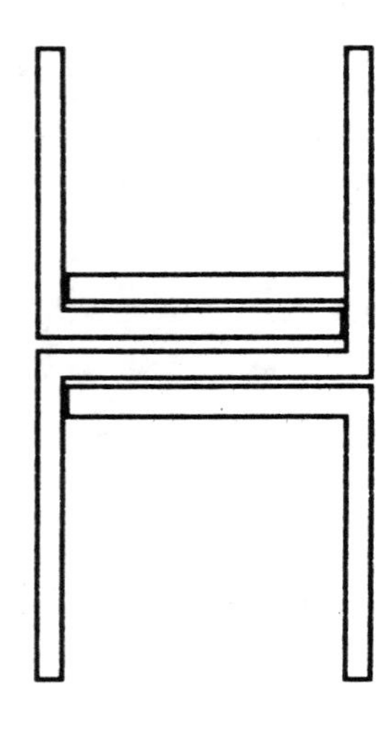

```
>LOAD"MENU"
>L.
   10 REM This is the MENU programme
   20 MODE 3
   30 PRINT"              MICROWAVE COUPLER SYNTHESIS"
   40 PRINT"              ---------------------------"
   50 PRINT
   60 PRINT"             Do you require :"
   70 PRINT
   80 PRINT"             1) PARALLEL LINE coupler synthesis ? "
   90 PRINT
  100 PRINT"             2) LANGE coupler synthesis (Osmani approach) ? "
  110 PRINT
  120 INPUT"             Please enter MENU number 1 OR 2 " ; IN
  130 IF IN<1 OR IN>2 THEN 120
  140 IF IN=1 THEN CHAIN"WORK2"
  150 IF IN=2 THEN CHAIN "LANGER3"
  160 END

>LOAD"DUMPER"
>L.
   10 REM This program is called DUMPER
   20 REM This is part of program MENU
   30 REM Screen dump
   40 :
   50 REM DUMPER
   60 REM To cause any graphics screen to be dumped when @ key is pressed.
   70 *FX14,2
   80 ?&220=00
   90 ?&221=&09
  100 *LOAD"CODE1"
  110 *LOAD"CODE2"
  120 CHAIN"MENU"
  130 END
>LOAD"WORK2"

>L.
   10 REM This sub-routine is called WORK2
   20 REM This is part of program MENU
   30 CLS
   40 MODE3
   50 REM *******************************************************************
   60 REM ** PROGRAM TO CALCULATE THE SEPARATION BETWEEN STRIPS OR STRIP    **
   70 REM ** WIDTH OF AN INTERDIGITATED COUPLER                             **
   80 REM *******************************************************************
   90 REM *****************
  100 REM ** INSTRUCTIONS *
  110 REM *****************
  120 CLS
  130 PRINT TAB(16,5)"   PARALLEL COUPLER design program"
  140 PRINT TAB(16,6)"   =============================="
  150 PRINT TAB(5,9)" Based on the work of Lange,Osmani and Akhtarzad,this"
  160 PRINT TAB(5,11)"programme will calculate the strip separation and the "
  170 PRINT TAB(5,13)"width of microstrip for coupler synthesis "
  175 PRINT " "
  180 PRINT TAB(20,17)"PRESS ANY KEY TO CONTINUE"
  190 LET A$=GET$
  200 IF A$=" " THEN 210
  210 CLS
  220 REM ** DATA INPUT **
  230 CLS
  240 PRINT"                                   INPUT"
```

```
 250 PRINT"***************************************************************
***********"
 260 PRINT
 270 PRINT"                     Please INPUT the following  :"
 280 PRINT
 290 PRINT
 300 INPUT"                     1) CHARACTERISTIC IMPEDANCE [OHM]";ZO
 310 PRINT
 320 INPUT"                     2) CENTRE FREQUENCY (GHz)";FREQ
 330 PRINT
 340 INPUT"                     3) COUPLING FACTOR >=10 dB";CF
 350 IF CF>0 THEN CF=0-CF
 360 PRINT
 370 INPUT"                     4) SUBSTRATE PERMITTIVITY";ER
 380 PRINT
 390 INPUT"                     5) SUBSTRATE THICKNESS (mm)";H
 400 PRINT
 410
 420 PRINT "**************************************************************
************"
 430 PRINT
 440 PRINT"                      CALCULATING... PLEASE WAIT."
 450 REM ** DEFINE PROCEDURES **
 460 PROCimp
 470 ZO=ZOE/2
 480 PROCwh
 490 WHSE=WHS
 500 ZO=ZOO/2
 510 PROCwh
 520 WHSO=WHS
 530 PROCwave
 540 ZO=50
 550 PROCsh
 560 PROCwh
 570 WX=WHS*H
 580 REM ***  DATA  OUTPUT          ***
 590 CLS
 600 PRINT "                                        OUTPUT"
 610 PRINT"***************************************************************
***********"
 620 PRINT"                        SPACE TO HEIGHT RATIO     = ";SH;" mm"
 630 PRINT"---------------------------------------------------------------
-----------"
 640 PRINT"                        WIDTH TO HEIGHT RATIO     = ";WH;" mm"
 650 PRINT"---------------------------------------------------------------
-----------"
 660 PRINT"                        EFFECT. PERMITTIVITY      = ";EFF
 670 PRINT"---------------------------------------------------------------
-----------"
 680 PRINT"                        EVEN MODE IMPEDANCE       = ";ZOE;" Ohm"
 690 PRINT"---------------------------------------------------------------
-----------"
 700 PRINT"                        ODD MODE IMPEDANCE        = ";ZOO;" Ohm"
 710 PRINT"---------------------------------------------------------------
-----------"
 720 PRINT"                        LENGTH OF COUPLED REGION = ";WH*H;" mm"
 730 PRINT"---------------------------------------------------------------
-----------"
 740 PRINT"                        SPACE OF COUPLED REGION  = ";SH*H;" mm"
 750 PRINT"---------------------------------------------------------------
-----------"
 760 PRINT"                        QUARTER WAVELENGTH        = ";QW;" mm"
```

```
  770 PRINT"--------------------------------------------------------------------
-----------"
  780 PRINT"                              50 Ohm LINE WIDTH          = ";WX;" mm"
  790 PRINT"********************************************************************
***********"
  800 PRINT
  810 PRINT"                                PRESS ANY KEY TO CONTINUE"
  820 LET A$=GET$
  830 IF A$=" " THEN 840
  840 CLS
  850 PRINT TAB(12,5);"WOULD YOU LIKE TO:-"
  860 PRINT TAB(22,8);"1) DRAW THE SPECIFIED COUPLER"
  870 PRINT TAB(22,11)"2) RUN THE PROGRAM AGAIN"
  880 PRINT TAB(22,14)"3) RETURN TO MENU"
  890 INPUT TAB(20,17)"PLEASE ENTER YOUR CHOICE 1,2 OR 3"CH
  900 IF CH=1 THEN 930
  910 IF CH=2 THEN 40
  920 IF CH=3 THEN CHAIN "MENU"
  930 REM
  940 REM ** DRAWING THE COUPLER **
  950 MODE 4
  960 CLS
  970 INPUT"ENTER YOUR SCALE FACTOR ";SCALE
  980 CLS
  990 GOSUB 1050
 1000 PRINT "THIS COUPLER HAS A SCALE FACTOR OF ";SCALE
 1010 INPUT "LIKE TO TRY ANOTHER SCALE FACTOR ? (Y/N)",SC$
 1020 IF SC$="YES" OR SC$="Y" THEN LET SCALE=1:GOTO 930
 1030 IF SC$="NO" OR SC$="N" THEN MODE3:GOTO 840
 1040 END
 1050 XF=1000/177
 1060 YF=1024/162
 1070 W=WH*H*SCALE
 1080 S=SH*H*SCALE
 1090 L=QW*SCALE
 1100 X=80:Y=80
 1110 MOVE XF*(X+W/2),YF*Y
 1120 DRAW XF*(X+W/2),YF*(Y+L/2+W/2)
 1130 DRAW XF*(X+W/2),YF*(Y-L/2+W/2)
 1140 DRAW XF*(X+W-8*W-W/4),YF*(Y-L/2+W/2)
 1150 DRAW XF*(X+W-8*W-W/4),YF*(Y-L/2+W/2-W)
 1160 DRAW XF*(X+W/2+W),YF*(Y-L/2+W/2-W)
 1170 DRAW XF*(X+W/2+W),YF*(Y+L/2+W/2+W)
 1180 DRAW XF*(X+W-8*W-W/4),YF*(Y+L/2+W/2+W)
 1190 DRAW XF*(X+W-8*W-W/4),YF*(Y+L/2+W/2)
 1200 DRAW XF*(X+W/2),YF*(Y+L/2+W/2)
 1210 MOVE XF*(X+W/2+W+S),YF*Y
 1220 DRAW XF*(X+W/2+W+S),YF*(Y+L/2+W+W/2)
 1230 DRAW XF*(X+W/2+W+S),YF*(Y-L/2-W/2)
 1240 DRAW XF*(X+W/2+2*W+8*W),YF*(Y-L/2-W/2)
 1250 DRAW XF*(X+W/2+2*W+8*W),YF*(Y-L/2-W/2+W)
 1260 DRAW XF*(X+W/2+2*W+S),YF*(Y-L/2-W/2+W)
 1270 DRAW XF*(X+W/2+2*W+S),YF*(Y+L/2-W/2+W)
 1280 DRAW XF*(X+W/2+2*W+8*W),YF*(Y+L/2-W/2+W)
 1290 DRAW XF*(X+W/2+2*W+8*W),YF*(Y+L/2-W/2+2*W)
 1300 DRAW XF*(X+W/2+S+W),YF*(Y+L/2-W/2+2*W)
 1310 RETURN
 1320 END
 1330 DEF PROCimp
 1340 ZOE=ZO*SQR((1+10^(CF/20))/(1-10^(CF/20)))
 1350 ZOO=ZO*SQR((1-10^(CF/20))/(1+10^(CF/20)))
 1360 ENDPROC
```

```
1370 DEF PROCwh
1380 IF ZO>=(44-2*ER) THEN 1390 ELSE 1430
1390 C=ZO*SQR(2*(ER+1))/119.9+.5*((ER-1)/(ER+1))*(LN(PI/2)+(1/ER)*LN(4/PI))
1400 WHS=((EXP(C))/8-1/(4*EXP(C)))^-1
1410 EFF=((ER+1)/2)*(1-1/(2*C)*((ER-1)/(ER+1))*(LN(PI/2)+((1/ER)*LN(4/PI))))^-2
1420 GOTO 1460
1430 D=(59.95*(PI*PI))/(ZO*SQR(ER))
1440 WHS=(2/PI)*((D-1)-LN(2*D-1))+((ER-1)/(PI*ER))*(LN(D-1)+0.293-0.517/ER)
1450 EFF=(ER/(0.96)+ER*(0.109-0.004*ER)*(LOG(10+ZO)-1))
1460 ENDPROC
1470 DEF PROCsh
1480 MN1=(PI*WHSE/2):NM1=(PI*WHSO/2)
1490 MN=(EXP(MN1)+EXP(-MN1))/2:NM=(EXP(NM1)+EXP(-NM1))/2
1500 TEMP=(MN+NM-2)/(NM-MN)
1510 SH=(2/PI)*LN(TEMP+SQR(TEMP^2-1)):IF SH<0.005 THEN 1520 ELSE 1530
1520 ENDPROC
1530 FOR LOOP=1 TO 100
1540 IF LOOP=100 THEN 1760
1550 g=(EXP(PI*SH/2)+EXP(-PI*SH/2))/2
1560 d=(g-1+((g+1)*MN))/2
1570 WH=(LN(d+SQR(d^2-1))-(PI*SH/2))/PI
1580 TEMP=(2*d-g-1)/(g-1)
1590 Z=(2/PI)*LN(TEMP+SQR(TEMP^2-1))
1600 TEMP=1+2*(WH/SH)
1610 TEMP2=LN(TEMP+SQR(TEMP^2-1)):IF (ER<6) THEN 1620 ELSE 1640
1620 WHSO2=Z+(4*TEMP2/(PI*(1+(ER/2))))
1630 GOTO 1650
1640 WHSO2=Z+(TEMP2/PI)
1650 IF(WHSO2-WHSO)>.1 OR (WHSO-WHSO2)>.1 THEN 1660 ELSE 1700
1660 IF (WHSO2<WHSO) THEN 1670 ELSE 1690
1670 SH=SH-.01
1680 GOTO 1700
1690 SH=SH+.01
1700 IF (WHSO2-WHSO)>.01 OR (WHSO-WHSO2)>.01 THEN 1710 ELSE 1750
1710 IF (WHSO2<WHSO) THEN 1720 ELSE 1740
1720 SH=SH-.001
1730 GOTO 1750
1740 SH=SH+.001
1750 NEXT LOOP
1760 ENDPROC
1770 DEF PROCwave
1780 LEF=FREQ*SQR(EFF)
1790 QW=81/LEF
1800 ENDPROC

>LOAD"LANGER3"
>L.
   10 REM This sub-routine is called LANGER3
   20 REM This is part of program MENU
   30 CLS
   40 MODE 3
   50 REM ********************************************************************
   60 REM ** PROGRAM TO CALCULATE THE SEPARATION BETWEEN STRIPS OR STRIP    **
   70 REM ** WIDTH RELATED TO LANGE MICROWAVE COUPLERS.
   80 REM ********************************************************************
   90 REM
  100 REM
  110 REM ******************
  120 REM ** INSTRUCTIONS **
  130 REM ******************
  140 CLS
  150 PRINT TAB(16,5)"      LANGE COUPLER calculation program"
```

```
  160 PRINT TAB(16,6)"      ================================="
  170 PRINT TAB(5,9)" USING EQUATIONS OUTLINED BY OSMANI AND AKHTARZAD THIS PROG
RAM WILL"
  180 PRINT TAB(5,11)" CALCULATE BOTH THE STRIP SEPARATION AND THE WIDTH OF THE
STRIP FOR"
  190 PRINT TAB(5,13)" FAST,ACCURATE MICROWAVE COUPLER DESIGN OF LANGE COUPLERS.
"
  200 PRINT TAB(20,17)"PRESS ANY KEY TO CONTINUE"
  210 LET A$=GET$
  220 IF A$=" " THEN GOTO 230
  230 CLS
  240 REM ****************
  250 REM ** DATA INPUT **
  260 REM ****************
  270 CLS
  280 PRINT
  290 PRINT
  300 PRINT"********************************************************************
***********"
  310 PRINT"                                   INPUT"
  320 PRINT"********************************************************************
***********"
  330 PRINT"                        PLEASE INPUT THE FOLLOWING:"
  340 PRINT
  350 INPUT"                        1) CHARACTERISTIC IMPEDANCE (Ohm)";ZO
  360 INPUT"                        2) NUMBER OF STRIPS (even no. only)";K
  370 INPUT"                        3) COUPLING FACTOR (dB)";CF
  380 IF CF>0 THEN CF=0-CF
  390 INPUT"                        4) SUBSTRATE PERMITTIVITY";ER
  400 INPUT"                        5) SUBSTRATE THICKNESS (mm)";H
  410 INPUT"                        6) CENTRE FREQUENCY (GHz)";FREQ
  420 PRINT "*******************************************************************
************"
  430 REM ********************************
  440 REM ***    CALCULATIONS          ***
  450 REM ********************************
  460 REM
  470 REM
  480 REM **************************************
  490 REM * CALC. ODD AND EVEN MODE IMPEDANCES *
  500 REM **************************************
  510 C1=10^(CF/20)
  520 q=((C1^2)+((1-(C1^2))*((K-1)^2)))^.5
  530 ZOO=ZO*(((1-C1)/(1+C1))^.5)*(K-1)*(1+q)/((C1+q)+((K-1)*(1-C1)))
  540 ZOE=ZOO*(C1+q)/((K-1)*(1-C1))
  550 REM *************************************
  560 REM ** CALCULATION OF STRIP SEPARATION **
  570 REM *************************************
  580 P=EXP((ZOO/84.8)*((ER+1)^0.5))-1
  590 OS=(8/P)*((((P/11)*(7+(4/ER)))+((1+(1/ER))/0.81))^0.5)
  600 R=EXP((ZOE/84.8)*((ER+1)^0.5))-1
  610 ES=(8/R)*((((R/11)*(7+(4/ER)))+((1+(1/ER))/0.81))^0.5)
  620 ODD=OS*(PI/2)
  630 EVEN=ES*(PI/2)
  640 T=(EXP(EVEN)+EXP-(EVEN)+EXP(ODD)+EXP-(ODD)-4)/(EXP(ODD)+EXP-(ODD)-EXP(EVEN
)-EXP-(EVEN))
  650 X=((2*T)+SQR((4*(T^2))+4))/2
  660 Y=((2*T)-SQR((4*(T^2))+4))/2
  670 IF X>0 THEN CC=X
  680 IF Y>0 THEN CC=Y
  690 SH=(H*2/PI)*LN(CC)
  700 REM ********************************
```

```
 710 REM ** CALCULATION OF STRIP WIDTH **
 720 REM ********************************
 730 REM *****************
 740 REM ** CALCULATIONS **
 750 REM ** CONSTANTS *****
 760 REM *****************
 770 A=ZO*SQR(2*(ER+1))/119.9
 780 B=0.5*((ER-1)/(ER+1))*(LN(PI/2)+(1/ER)*LN(4/PI))
 790 C=A+B
 800 D=59.95*(PI*PI)/ZO*SQR(ER)
 810 REM ********************
 820 REM ** NARROW STRIPS  **
 830 REM ********************
 840 IF ZO<=44-(2*ER) THEN 880 ELSE 850
 850 WH=H*((EXP(C))/8-1/(4*EXP(C)))^-1
 860 EFF=((ER+1)/2)*(1-(1/(2*C))*((ER-1)/(ER+1))*(LN(PI/2)+((1/ER)*LN(4/PI))))^
-2
 870 GOTO 930
 880 REM *****************
 890 REM ** WIDE STRIPS **
 900 REM *****************
 910 WH=H*((2/PI)*((D-1)-(LN(2*D)-1))+(((ER-1)/(PI*ER))*((LN(D-1))+0.293-(0.517
/ER))))
 920 EFF=(ER/(0.96)+ER*(0.109-0.004*ER)*(LOG(10+ZO)-1))
 930 REM ****************************
 940 REM ***  DATA  OUTPUT          ***
 950 REM ****************************
 960 IF SH>.35 THEN SH=SH-.3 ELSE SH=SH
 970 IF WH>1.1 THEN WH=WH-1 ELSE WH=WH
 980 PRINT "                              OUTPUT"
 990 PRINT"*******************************************************************
***********"
1000 PRINT"                        EVEN MODE W/H RATIO = ";ES
1010 PRINT"                        ODD MODE W/H RATIO = ";OS
1020 PRINT"                        STRIP SEPARATION = ";SH;" mm"
1030 PRINT"                        STRIP WIDTH =";WH;" mm"
1040 PRINT"                        EFF. PERMITTIVITY =";EFF
1050 PRINT"                        EVEN MODE IMPEDANCE = ";ZOE
1060 PRINT"                        ODD MODE IMPEDANCE  = ";ZOO
1070 PRINT"*******************************************************************
***********"
1080 PRINT
1090 PRINT"                          PRESS ANY KEY TO CONTINUE"
1100 LET A$=GET$
1110 IF A$=" " THEN 1130
1120 CLS
1130 PRINT TAB(12,5);"WOULD YOU LIKE TO:-"
1140 PRINT TAB(22,8);"1) DRAW THE SPECIFIED COUPLER"
1150 PRINT TAB(22,11)"2) RUN THE PROGRAM AGAIN"
1160 PRINT TAB(22,14)"3) RETURN TO MENU"
1170 INPUT TAB(20,17)"PLEASE ENTER YOUR CHOICE 1,2 OR 3"CH
1180 IF CH=1 THEN 1210
1190 IF CH=2 THEN 40
1200 IF CH=3 THEN CHAIN"MENU"
1210 REM
1220 REM ************************
1230 REM ** DRAWING THE COUPLER **
1240 REM ************************
1250 MODE 4
1260 CLS
1270 INPUT"ENTER YOUR SCALE FACTOR ";SCALE
1280 C=2.99793*10^8
```

```
 1290 LX=C/(FREQ*10^6*4)
 1300 XF=1000/177
 1310 YF=1024/162
 1320 W=WH*SCALE
 1330 S=SH*SCALE
 1340 L=LX*SCALE
 1350 X=100:Y=80
 1360 REM ********************************************
 1370 REM ** ALL DIMENSIONS IN mm. FREQUECIES IN GHZ **
 1380 REM ********************************************
 1390 REM ***************************
 1400 REM ** PLOTTING DIRECT ROUTE **
 1410 REM ***************************
 1420 MOVE XF*(X+W/2),YF*Y
 1430 DRAW XF*(X+W/2),YF*(Y+L/8+W/2)
 1440 DRAW XF*(X+W/2),YF*(Y-L/8+W/2)
 1450 DRAW XF*(X+W/2+(L/4)+W/4),YF*(Y-L/8+W/2)
 1460 DRAW XF*(X+W/2+(L/4)+W/4),YF*(Y-L/8+W/2-W)
 1470 DRAW XF*(X-W+W/4),YF*(Y-L/8+W/2-W)
 1480 DRAW XF*(X-W+W/4),YF*Y
 1490 DRAW XF*(X-W+W/4),YF*(Y+L/8-W/2)
 1500 DRAW XF*(X-(L/4)-W),YF*(Y+L/8-W/2)
 1510 DRAW XF*(X-(L/4)-W),YF*(Y+L/8-W/2+W)
 1520 DRAW XF*(X+W/2),YF*(Y+L/8-W/2+W)
 1530 REM ********************************
 1540 REM **  PLOTTING ISOLATED FINGER  **
 1550 REM ********************************
 1560 MOVE XF*(X-W-S),YF*Y
 1570 DRAW XF*(X-W-S),YF*(Y+L/8-W/2-S-W/8)
 1580 DRAW XF*(X-W-S),YF*(Y-L/8-W/2)
 1590 DRAW XF*(X-W-(L/4)),YF*(Y-L/8-W/2)
 1600 DRAW XF*(X-W-(L/4)),YF*(Y-L/8-W/2+W)
 1610 DRAW XF*(X-2*W-S-W/4),YF*(Y-L/8-W/2+W)
 1620 DRAW XF*(X-2*W-S-W/4),YF*Y
 1630 DRAW XF*(X-2*W-S-W/4),YF*(Y+L/8-W/2-S-W/8)
 1640 DRAW XF*(X-W-S),YF*(Y+L/8-W/2-S-W/8)
 1650 REM ******************************
 1660 REM ** PLOTTING COUPLED FINGER  **
 1670 REM ******************************
 1680 MOVE XF*(X+W/2+S+W/3.5),YF*Y
 1690 DRAW XF*(X+W/2+S+W/3.5),YF*(Y-L/8+W/2+S+W/8)
 1700 DRAW XF*(X+W/2+S+W/3.5),YF*(Y+L/8+W/2)
 1710 DRAW XF*(X+(L/4)+W/2+W/4),YF*(Y+L/8+W/2)
 1720 DRAW XF*(X+(L/4)+W/2+W/4),YF*(Y+L/8+W/2-W)
 1730 DRAW XF*(X+2*W+S),YF*(Y+L/8+W/2-W)
 1740 DRAW XF*(X+2*W+S),YF*Y
 1750 DRAW XF*(X+2*W+S),YF*(Y-L/8+W/2+S+W/8)
 1760 DRAW XF*(X-W/2+S+W+W/3.5),YF*(Y-L/8+W/2+S+W/8)
 1770 IF K=4 OR K=5 OR K=6 OR K=7 THEN GOSUB 2050
 1780 IF K=5 OR K=6 OR K=7 THEN GOSUB 2100
 1790 IF K=6 OR K=7 THEN GOSUB 2150
 1800 IF K=7 THEN GOSUB 2200
 1810 PRINT"THIS COUPLER HAS A SCALE FACTOR OF ";SCALE
 1820 INPUT"LIKE TO TRY ANOTHER SCALE FACTOR(Y/N)",SC$
 1830 SCALE=0
 1840 IF SC$="YES" OR SC$="Y":INPUT"SPECIFY YOUR NEW SCALE FACTOR ";SCALE:CLS:GO
TO 1290
 1850 IF SC$="N" OR SC$="NO" THEN CLS
 1860 MODE3
 1870 PRINT"''''''''"
 1880 PRINT"   DO YOU WISH TO:-"
 1890 PRINT
```

```
1900 PRINT"                        1) RUN THE PROGRAM AGAIN"
1910 PRINT
1920 PRINT"                        2) RETURN TO MENU"
1930 PRINT
1940 PRINT"                        3) QUIT THE PROGRAM"
1950 PRINT
1960 INPUT"                        ENTER YOUR CHOICE 1,2 OR 3";CH
1970 IF CH<1 OR CH>3 THEN 1870
1980 IF CH=1 THEN 30
1990 IF CH=2 THEN CHAIN"MENU"
2000 IF CH=3 THEN END
2010 END
2020 REM ******************************
2030 REM *** PLOTTING EXTRA FINGERS ***
2040 REM ******************************
2050 MOVE XF*(X+(2*W)+(2*S)+W/4),YF*(Y-L/8+W/2)
2060 DRAW XF*(X+(2*W)+(2*S)+W/4),YF*(Y+L/8-W/2-S-W/8)
2070 DRAW XF*(X+(2*W)+(2*S)+W/2+W),YF*(Y+L/8-W/2-S-W/8)
2080 DRAW XF*(X+(2*W)+(2*S)+W/2+W),YF*(Y-L/8+W/2)
2090 RETURN
2100 MOVE XF*(X-(2*W)-(2*S)-W/2),YF*(Y+L/8-W/2)
2110 DRAW XF*(X-(2*W)-(2*S)-W/2),YF*(Y-L/8+W/2+S+W/8)
2120 DRAW XF*(X-(2*W)-(2*S)-W/2-W-W/4),YF*(Y-L/8+W/2+S+W/8)
2130 DRAW XF*(X-(2*W)-(2*S)-W/2-W-W/4),YF*(Y+L/8-W/2)
2140 RETURN
2150 MOVE XF*(X+(4*W)+(3*S)-W/4),YF*(Y+L/8-W/2)
2160 DRAW XF*(X+(4*W)+(3*S)-W/4),YF*(Y-L/8+W/2+S+W/8)
2170 DRAW XF*(X+(4*W)+(3*S)+W),YF*(Y-L/8+W/2+S+W/8)
2180 DRAWXF*(X+(4*W)+(3*S)+W),YF*(Y+L/8-W/2)
2190 RETURN
2200 MOVE XF*(X-(4*W)-(3*S)),YF*(Y-L/8+W/2)
2210 DRAW XF*(X-(4*W)-(3*S)),YF*(Y+L/8-W/2-S-W/8)
2220 DRAW XF*(X-(4*W)-(3*S)-W/4-W),YF*(Y+L/8-W/2-S-W/8)
2230 DRAW XF*(X-(4*W)-(3*S)-W/4-W),YF*(Y-L/8+W/2)
2240 RETURN
```

If we apply now a signal to terminal A, then a predetermined fraction of the applied signal power appears at terminal B and a fraction at terminal C with very little signal power reaching D. This terminal D is usually terminated with the characteristic impedance of the line, e.g. 50 Ω. From the foregoing it is evident that

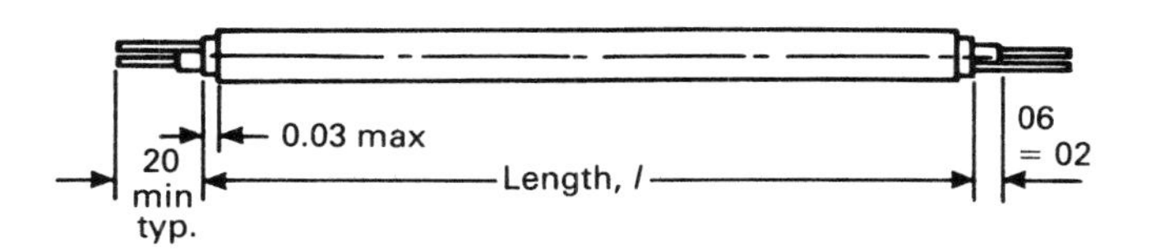

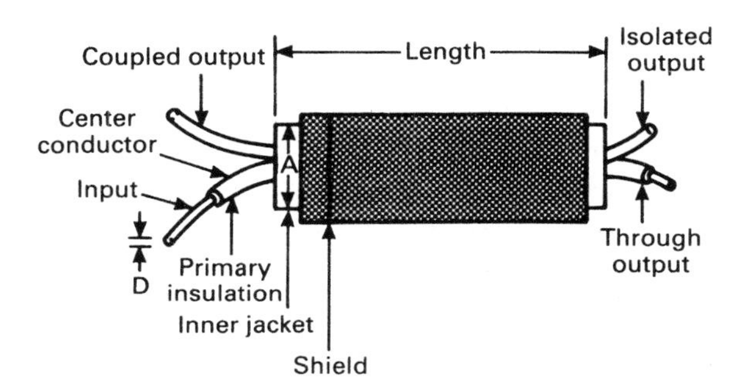

Fig. 9.23 — Physical aspects of a wireline.

there can be equal power split between B and C, or in other words between the outputs of the primary (direct output P_t) and secondary (coupled output P_c) arms of the coupler. The phase relationship of the signals between B and C is 90°, hence the name 'quadrature coupler' for this type of coupler. If we include the notation of coupling, then the coupler is properly specified by or termed as a '3 dB quadrature coupler'.

Like any coupler, a coupler made from a wireline may be characterized in terms of transmission, coupling, isolation and directivity.

Transmission is defined as the ratio of

$$T = P_A/P_B = 10 \log (P_A/P_B) \text{ [dB]} \tag{9.46}$$

Similarly, coupling is defined as

$$C = P_A/P_C = 10 \log (P_A/P_C) \text{ [dB]} \tag{9.47}$$

Having defined the power relationships between A, B and C, consideration must be given to the power at terminal D. This can be done in two ways in terms of either isolation or directivity. Isolation is given by

$$I = P_A/P_D = 10 \log (P_A/P_D) \text{ [dB]} \tag{9.48}$$

and the directivity by

$$D = P_C/P_D = 10 \log (P_C/P_D) \text{ [dB]} \tag{9.49}$$

The typical performance of a 3 dB coupler made from wireline is shown in Fig. 9.24

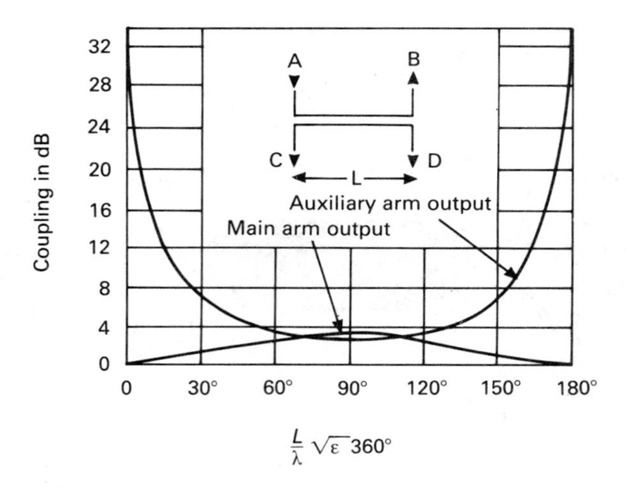

Fig. 9.24 — $C = f(L)$ for a 3dB coupled line.

with the coupler centre frequency located at 90°. From this we see that the primary arm output or main arm coupling to B is slightly larger than 3 dB whilst the secondary arm output at C is lightly less than 3 dB. As is evident, this approximate 3 dB coupling exists over the range from about 60° to 120°.

The manufacturer of the wireline provides design equations for couplers and the following is an example of how these may be used. Assume we want to design a 3 dB coupler for a 1.5 GHz balanced amplifier subsystem with a bandwidth of say 200 MHz. The quarter wavelength frequency F_q can then be obtained from

$$F_q = \frac{F_{min} + F_{max}}{2} \tag{9.50}$$

Hence

$$F_q = \frac{1400 + 1600}{2} = 1500 \text{ MHz}$$

In order to avoid large coupler unbalances the inequality

$$\frac{F_{max} - F_{min}}{F_q} < 1 \tag{9.51}$$

should be met. Substituting the upper and lower frequency limits, we obtain $0.13 < 1$. The coupling length L can be calculated from the following formula

$$L\ [\text{cm}] = \frac{4700}{F_q[\text{MHz}]} \tag{9.52}$$

with an error of about $\pm 10\%$. For a 1.5 GHz coupler the length becomes 3.13 cm. Table 9.2 gives equations for computing the relative power output of a coupler as a function of frequency F.

Table 9.2 — Equations for calculating P_t and P_c as a function of frequency (Sage Wireline)

Parameter	Symbol	Equation	
Quarter wavelength frequency	F_q	$\dfrac{90\,F}{K_2}$	(9.53)
Relative coupled power to input power at F_q	P_q	$\dfrac{1}{K_1\left(\dfrac{1}{P_c}-1\right)+1}$	(9.54)
Operating frequency	F	$\dfrac{K_2\,F_q}{90}$	(9.55)
Relative coupled power to input power at F	P_c	$\dfrac{K_1}{K_1-1+\dfrac{1}{P_q}}$	(9.56)
Relative directly transmitted power output	P_t	$1-P_c$	(9.57)

where

$$P[\text{dB}] = -10 \log P \text{ or } P = \text{alog}\,(-P\,[\text{dB}]/10) \tag{9.58}$$

$$K_1 = \sin^2\left(\frac{90\,F}{F_q}\right) \tag{9.59}$$

$$K_2 = \sin^{-1}\sqrt{\frac{P_c}{P_c-1} * \frac{P_q-1}{P_q}} \tag{9.60}$$

Having calculated the line length for this 3 dB coupler as 3.13 cm we can now check the coupler outputs for equal power split. From eqns (9.53) to (9.60)

$P_c = \text{alog}\,(-0.3) = 0.5$
$P_q = \text{alog}\,(-0.3) = 0.5$

$$K_2 = \sin^{-1} \sqrt{\frac{0.5}{0.5-1} \times \frac{0.5-1}{0.5}} = 90$$

$$F_q = \frac{90\ F}{K_2} = \frac{90 \times 1500}{90} = 1500\ \text{MHz}$$

$$L = \frac{4700}{1500} = 3.13\ \text{cm}$$

$$K_1 = \sin^2 \frac{90 \times 1500}{1500} = 1$$

$$P_c = \frac{1}{1 - 1 + (1/0.5)} = 0.5$$

or

$$P_c\ [\text{dB}] = -\ 10 \log 0.5 = 3$$

$$P_t = 1 - P_c = 1 - 0.5 = 0.5$$

or

$$P_t\ [\text{dB}] = -\ 10 \log 0.5 = 3$$

This means that at the coupler center frequency of 1.5 GHz we have equal powersplit as shown in Fig. 9.25 where an input signal power of 10 mW was assumed for the purpose of illustration.

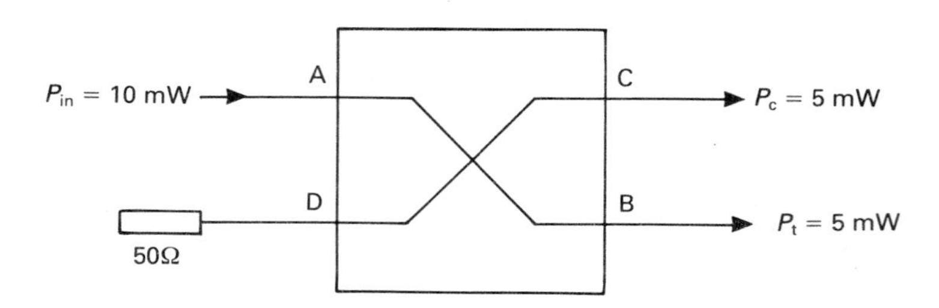

Fig. 9.25 — 3 dB coupler.

If we now change the frequency only of the signal applied to the coupler, then this will result in a change of P_t and P_c. The change becomes more pronounced as we depart from the centre frequency. Again, using eqns (9.53) to (9.60) we arrive at results tabulated in Table 9.3.

Table 9.3 — Calculated coupler parameters P_c and P_t

F (MHz)	K_1	P_c	P_c (dB)	P_t	P_t (dB)
300	0.10	0.09	10.31	0.90	0.42
500	0.26	0.21	6.76	0.78	1.02
1000	0.77	0.43	3.57	0.56	2.50
1500	1.00	0.50	3.00	0.50	3.00
2300	0.37	0.27	5.64	0.72	1.38
2500	0.16	0.14	8.40	0.85	0.67

The coupler discussed above was built and a frequency response taken on the network analyser as shown in Fig. 9.26. The calculated results as shown in Table 9.3 are added and show good agreement. In order that the coupler gives best performance it is essential to trim the coupling length correctly and to ground the outer conductor properly. A thin copper strip is most suitable for this. A plan view of a typical mounting for a Sage Wireline coupler is shown in Fig. 9.27. A slot is usually cut into the substrate to accept the coupled line.

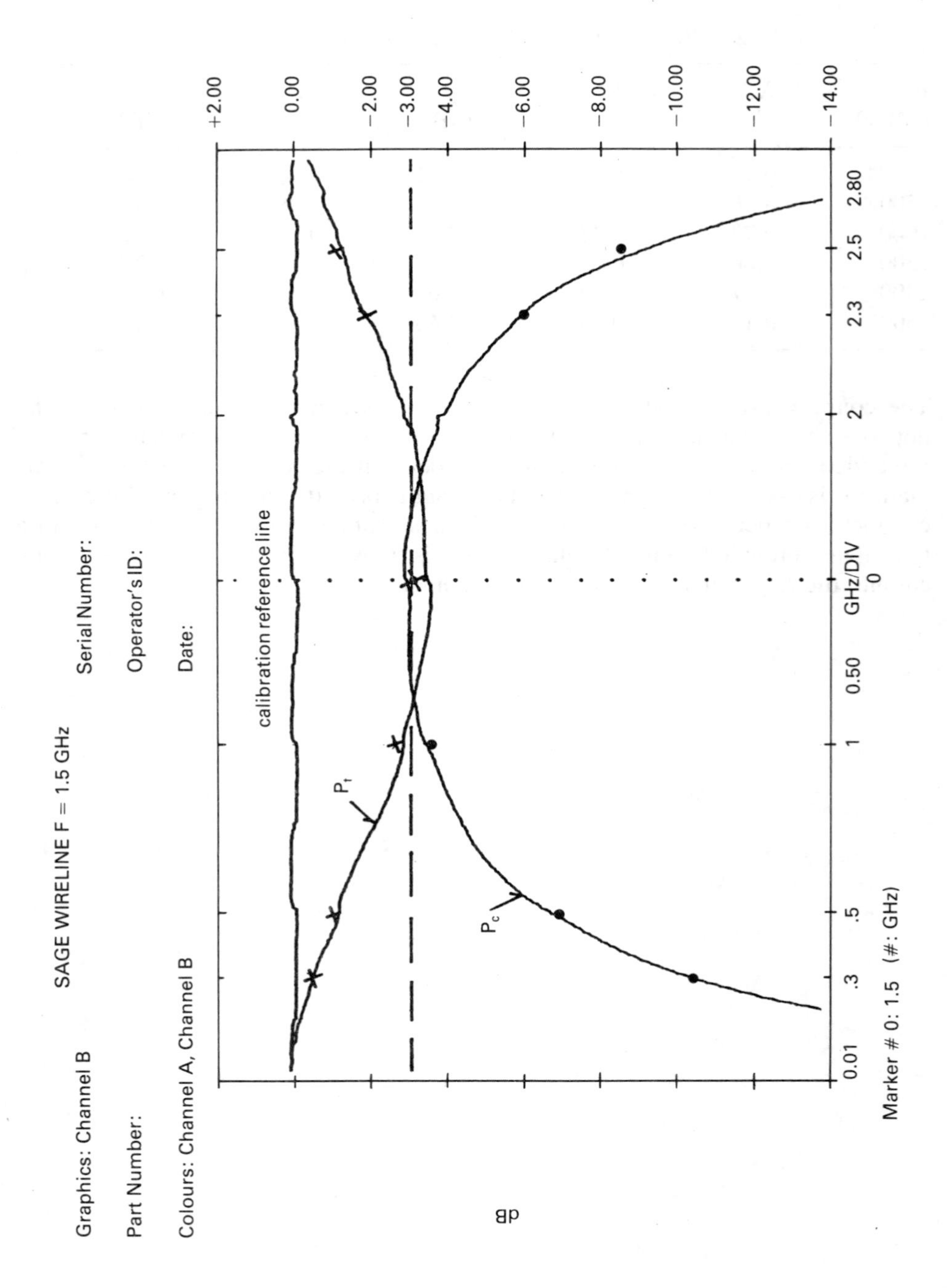

Fig. 9.26 — Comparison of measured and calculated results of a wireline coupler.

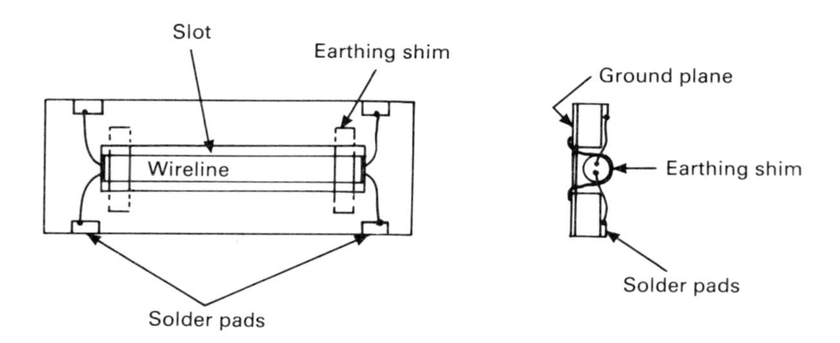

Fig. 9.27 — Mounting a wireline.

REFERENCES

[1] Akhtarzad, S. *et al.*, The design of coupled microstrip lines, *IEEE Trans.* **MTT23**, No. 6, June 1975, p. 486.
[2] Kemp, G. *et al.*, Ultra-wideband quadrature coupler, *Electronics Letters*, **19**, No. 6, 17th March 1983, p. 197.
[3] Muraguchi, M., Optimum design of 3-dB branch line couplers using microstrip lines, *IEEE Trans.* **MTT 31**, No. 8, p. 674.
[4] Fusco, V. F., *Microwave circuits: Analysis and computer aided design*, Prentice-Hall International, 1987.
[5] Lange J., Interdigitated stripline quadrature hybrid, *IEEE Trans.* **MTT**, Dec. 1969, pp. 150.
[6] Tajima, Y., Multiconductor couplers, *IEEE Trans.* **MTT 26**, No. 10, October 1978, p. 795.
[7] Ou, W. P., Design equations for an interdigitated directional coupler, *IEEE Trans*, **MTT**, Feb. 1975, p. 253.
[8] Osmani R. M., Synthesis of Lange couplers, *IEEE Trans.* **MTT 29**, No. 2, Feb. 1981, p. 168.

Index